Green Energy and Technology

More information about this series at http://www.springer.com/series/8059

Md. Rabiul Islam · Youguang Guo · Jianguo Zhu

Power Converters for Medium Voltage Networks

Springer

Md. Rabiul Islam
Youguang Guo
Jianguo Zhu
School of Electrical, Mechanical and Mechatronic Systems
Faculty of Engineering and Information Technology
University of Technology Sydney (UTS)
Ultimo
NSW
Australia

ISSN 1865-3529 ISSN 1865-3537 (electronic)
ISBN 978-3-662-44528-0 ISBN 978-3-662-44529-7 (eBook)
DOI 10.1007/978-3-662-44529-7

Library of Congress Control Number: 2014947125

Springer Heidelberg New York Dordrecht London

Printed on acid-free paper

Springer is part of Springer Science+Business Media (www.springer.com)

Preface

Medium and large-scale renewable power plants require large areas of land, and are usually installed in offshore or remote areas, far from cities, in order to capture large amount of energy. For power transmission and grid integration, the low voltage, e.g., 690 V for a typical wind turbine generator, at which the electricity is generated from renewable energy sources is transformed to the medium-voltage (e.g., 11–33-kV) commonly by using the step-up transformers. In many cases, the power transformers operated at the frequency of 50/60 Hz are heavy, large, and inefficient. These technological drawbacks would have significant impacts on offshore and remote renewable power plants where the costs of installation and regular maintenance are extremely high.

With the fast development of power electronics, it is becoming a reality to replace the combination of low voltage inverter and step-up-transformer by a medium-voltage converter for direct grid connection to reduce the system volume and weight, as well as the cost. Traditionally, two- and three-level converters are commonly used for high power applications. When applied to medium-voltage systems, very expensive semiconductor switching devices of high-voltage ratings are required. To reduce the cost, multilevel converters made of cascaded switching devices of relatively low-voltage ratings are developed. In comparison with the conventional two- and three-level converters, the multilevel converters present lower switching losses, lower voltage stress on switching devices, and higher quality output power, and thus suit better the medium-voltage applications. Although several multilevel converter topologies have been developed in the last few decades, most of them are not suitable for medium-voltage applications as their number of auxiliary components scales quadratically with the number of levels. Because of some special features (e.g., the number of components scales linearly with the number of levels, and individual modules are identical and completely modular in construction hence enabling high-level number attainability) the modular multilevel cascaded (MMC) converter topology can be considered as a possible candidate for medium voltage applications. However, the MMC converter requires multiple isolated DC sources that must be balanced, and as a result its application is not straightforward, especially in wind power generation systems. Moreover, the multilevel converter requires a number of switching and control PWM signals, which cannot be generated by a digital signal processor (DSP) because the currently available DSP can only provide about six pairs of PWM channels. In this instance, a field programmable gate array (FPGA) becomes the natural choice. Most of the available design techniques require special software such as HDL coder, System generator, PSIM and ModelSim, which increases the developmental time and cost.

Chapter 1 of this book discusses various aspects, such as historical growth of two dominating renewable, i.e., wind and solar, energy sources, the existing technologies, technical challenges, and possible solutions for large-scale power generation from these energy sources.

An extensive literature survey has been conducted in Chap. 2 focusing on various aspects of medium-voltage converter development for step-up-transformer-less direct grid integration of photovoltaic (PV) power plants. The main objective is to show how power electronic converter topologies, power electronic devices, and control complexities have affected the development of medium-voltage converters, and how to make an excellent choice of the suitable converter topologies for step-up-transformer-less grid integration through medium-voltage converters, which is really a critical problem and highly affects the converter performance and cost.

The main aim of Chap. 3 is to find out a suitable converter topology, which can interconnect the renewable generation units directly to the medium-voltage grid with the commercially available matured semiconductor devices. Different multilevel converter topologies, such as the neutral point clamped (NPC), flying capacitor (FC), and MMC converters, have been considered and compared for the design of an 11-kV converter system. The comparison is made in terms of the number of semiconductors, semiconductor cost and commercial availability, total harmonic distortions (THDs), filter size and control complexity of the converters. The performance is analyzed and compared in the MATLAB/Simulink environment.

To couple a renewable energy source to the MMC converter, a high-frequency magnetic-link with multiple secondary windings can be an excellent option and is explored in Chap. 4. The high-frequency magnetic-link is used to generate the isolated balanced multiple DC supplies for all of the H-bridge inverter cells of the MMC converter from a single low voltage power source. Compared with the conventional power frequency transformers operated at 50 or 60 Hz, the high-frequency magnetic-links (in the range of a few kilohertz to megahertz) have much smaller and lighter magnetic cores and windings, and thus much lower costs.

The capability of parallel processing of the FPGA affords the opportunity to the switching controller to update all gate signals simultaneously. Various design techniques and software environments are available for the modeling of switching control schemes with the FPGA technology. Most of the techniques require special software, which increases the development time and cost. In Chap. 5, the most common software such as the MATLAB/Simulink and Xilinx ISE-based alternative design technique is proposed, which may reduce the developmental time and cost of the switching controller.

To verify the feasibility of the high-frequency MMC converter, a scaled down 1-kV laboratory prototype test platform with a 5-level MMC converter is developed. Furthermore, the prototype design and implementation, test platform, and experimental results are analyzed and discussed in Chap. 6. The component selection, converter fabrication, and experimentation techniques are equally applicable to any other converter applications.

In Chap. 7, an 11-kV system and a 33-kV system are designed and analyzed taking into account the specified system performance, control complexity, and cost and market availability of the power semiconductors. It is found that, the 19-level and 43-level converters are the optimal choice for the 11-kV and 33-kV systems, respectively. Besides the design and analysis of medium voltage converters, the traditional low voltage converters with power frequency step-up-transformers are also discussed.

Chapter 8 concludes the book. Future directions have been recommended for further research and development. It is expected that the technology presented in this book has a great potential to be implemented in future wind farms, PV power plants, and smart micro-grid applications.

Contents

Figures

Tables

Symbols

A	Cross-sectional area (m^2)
V	Air velocity (m/s)
ρ	Mass density (kg/m^3)
P	Power (W)
ω_r	Rotor speed
ω_1	Synchronous speed
P_p	Number of poles of power winding
f_p	Frequency of power winding current (Hz)
P_c	Number of poles of control winding
f_c	Frequency of the control winding current (Hz)
n_r	Rotational speed (RPM)
G	Generator
C	Capacitor
D	Diode
S	Switching device
L	Inductor
$\mathrm{R_{scell}}$	Intrinsic series resistance of a solar cell (Ω)
$\mathrm{R_{shcell}}$	Equivalent shunt resistance of a solar cell (Ω)
I_L	Light-generated current (A)
I_{cell}	Cell output current (A)
I_{PV}	PV array output current (A)
n_s	Number of cells connected in series
n_p	Number of cells connected in parallel
V_{PV}	PV array output voltage (V)
R_s	PV array series resistance (Ω)
R_{sh}	PV array shunt resistance (Ω)
I_o	Cell saturation current (A)
q	Electronic charge (1.602×10^{-19} coulomb for an electron)
A_i	Ideality factor (1.92)
K	Boltz-man's constant (1.38×10^{-23} J/K)

T	Temperature (°C)
ΔP	Change of PV array output power (W)
ΔV	Change of PV array output voltage (V)
V_{rms}	PV inverter output AC voltage (V)
R	Reservation factor
V_A	Minimum PV array voltage (V)
V_{dc}	DC-link voltage of converter (V)
V_{an}, V_{bn}, V_{cn}	Phase voltages of converter
n	Neutral point
a, b, c	Terminals of three-phase system
$V_{com@100FIT}$	Device commutation voltage for a device reliability of 100 FIT (V)
m	Converter number of levels
m_f	Frequency modulation index
f_m	Frequency of reference signals (Hz)
f_{ca}	Frequency of carrier signals (Hz)
A_m	Peak to peak amplitude of reference signal
A_c	Peak to peak amplitude of carrier signal with phase-shifting
m_{ap}	Amplitude modulation index with phase-shifting
B_n	Number of H-bridge inverter cell on a particular phase leg
θ_{ps}	Carrier phase-shifting (Degree)
m_{al}	Amplitude modulation index with level-shifting
A_{cl}	Peak-to-peak amplitude of carrier signal with level-shifting
$V_{ll(rms)}$	Grid line to line voltage (V)
$V_{dc(min)}$	Minimum voltage of each DC-link capacitor (V)
$V_{dc(nom)}$	Nominal voltage of each DC-link capacitor (V)
$I_{p(rms)}$	Converter phase current (A)
S_c	Apparent output power of the converter (VA)
n_c	Number of series-connected flying capacitor cells
ΔV_c	Capacitor voltage ripple (V)
I_{dc}	Converter DC output current (A)
I_d	Normalized index value
y	Given value
y_{min}	Minimum value
y_{max}	Maximum value
P_{core}	Core loss (W)
f	Frequency (Hz)
B	Flux density (T)
k', y', z'	Coefficients of Steinmetz's equation
V_{max}	Maximum excitation voltage (V)
φ_{max}	Maximum flux (Wb)

T	Time period (s)
$v(t)$	Time varying voltage (V)
φ	Flux (Wb)
$\varphi(t)$	Time varying flux (Wb)
N	Number of turns
B_{max}	Maximum flux density (T)
N_p	Number of turns in the primary winding
N_s	Number of turns in the secondary winding
μ	Permeability (H/m)
ρ	Resistivity (Ωm)
R_{dc}	DC resistance (Ω)
P_{skin}	Skin effect loss (W/m)
d	Diameter of the conductor (mm)
δ	Skin depth (mm)
$P_{\mathrm{p\_int}}$	Power loss due to internal proximity effect (W/m)
σ	Conductivity (Siemens per meter)
U	Number of layers
n_s	Total number of strands in a bundle
l_s	Total length of a single strand (m)
r_b	Radius of the bundle (mm)
H_e	External magnetic field strength (A/mm)
P_{p_ext}	Power loss due to external proximity effect (W/m)
K_r	AC/DC resistance ratio
R_{ac}	AC resistance (Ω)
i_p, i_s	Current of primary and secondary coils (A)
J	Current density (A/mm^2)
a_p	Cross-sectional area of a strand in primary winding (mm^2)
d_p	Diameter of a strand in primary winding (mm)
a_s	Cross-sectional area of a strand in secondary winding (mm^2)
d_s	Diameter of a strand in secondary winding (mm)
R_p, R_s	Resistances of the primary and secondary windings (Ω)
C_{po}, C_{so}	Self capacitances of primary and secondary windings (F)
C_{pso}	Mutual capacitance between primary and secondary windings (F)
N_p, N_s	Number of turns in primary and secondary windings
R_m, L_m	Core loss resistance and magnetization inductance
C_p, C_s, C_{ps}	Equivalent capacitances referred to the primary side (F)
L_{ex}	External inductance (H)
C_{eqp}, C_{eqs}	Equivalent capacitances seen from primary and secondary sides (F)

L_{dl}	Differential leakage inductance (H)
L_{dlp}, L_{dls}	Differential leakage inductances of primary and secondary windings
φ_{lp}, φ_{ls}	Leakage fluxes of primary and secondary windings (Wb)
A_w	Area required by the primary and secondary windings (mm^2)
N_A, N_B, N_C, N_D, N_E, N_F	Number of turns in secondary windings
v_A, v_B, v_C, v_D, v_E, v_F	Secondary coils induced voltages (V)
TR_A, TR_B, TR_C, TR_D, TR_E, TR_F	Voltage transformation ratios
$i(t)$	Excitation current (A)
l_e	Mean length of the core (m)
N_1, N_2	Number of turns of excitation and pick-up coils
A_e	Cross-sectional area of the core (cm^2)
V_L	Pick-up coil voltage (V)
k_1	Clock divider value in carrier generator unit
pi	π
$\Delta\theta$	Angular resolution (Degree)
L_n	Number of entries for a cycle
A_{out}	Amplitude of the output signal
Ψ	Phase angle (Degree)
k_2	Clock divider value for reference generator unit
D_{off}	Diode off time (s)
$V_{r(P\text{-}P)}$	Peak-to-peak rms output voltage (V)
f_s	Frequency of the source voltage (Hz)
γ_r	Ripple factor
T_J, T_C, T_A	Temperatures of junction, case, and ambient (°C)
θ_{JA}, θ_{JC}, θ_{CS}, θ_{SA}	Thermal resistances of total, transistor, insulator, and heatsink (°C/W)
P_D	Power dissipation (W)
R_T, C_T	Timing resistor and capacitor (Ω, F)
D_{max}	Desired maximum duty cycle
$D_{\%}$	Percentage of duty cycle
T_{high}, T_{high}	Switch on and off time (s)
f_H	Frequency of the medium frequency inverter (Hz)
P_{loss_inv}	Power losses in the inverter section (W)
P_{c_inv}, P_{sw_inv}	Conduction and Switching losses (W)
$P_{sw\_inv\_2}$, $P_{sw\_inv\_M}$	Switching loss of 2-level and multilevel converters (W)
f_{c_2}, f_{c_M}	Carrier frequencies of 2-level and multilevel converters (Hz)
P_{c_sw}, P_{c_D}	Conduction losses of IGBT and diode (W)
m_a	Amplitude modulation index
p_f	Power factor

V_t, V_f	Voltage drop at zero current of IGBT and diode (V)
R_{CE}, R_{AK}	Forward resistances of IGBT and Diode (Ω)
η	Efficiency
P_{out}	Output power (W)
P_{in}	Input power (W)
$\eta_{\%}$	Percentage of efficiency
P_{c_9}	Conduction loss of 9-level converter (W)
$V_{CE(sat)}$	Collector-emitter saturation voltage (V)

Acronyms

A	Ampere
ABB	ASEA Brown Boveri
AC	Alternating current
ALOs	Arithmetic and logic operations
AMSC	American Superconductor Corporation
ASIC	Application-specific integrated circuit
AUD	Australian dollar
BDFM	Brushless doubly-fed machine
BSCCO	Bismuth strontium calcium copper oxide
DC	Direct current
DDS	Direct digital synthesis
DFIG	Doubly-fed induction generator
DMM	Digital Multimeter
DSP	Digital signal processor
DTC	Direct torque control
DVUF	Device voltage utilization factor
EMC	Electromagnetic compatibility
EMI	Electromagnetic interference
FC	Flying capacitor
FPGA	Field programmable gate array
g/cm^3	Gram per cubic centimeter
GW	Gigawatt
HDL	Hardware description language
HFT	High frequency transformer
HT	Height
HTS	High temperature superconductor
HVAC	High voltage alternating current
HVDC	High voltage direct current
Hz	Hertz
I/O	Input and (or) output
IBM	International Business Machines

IC	Integrated circuit
ID	Inner diameter
IGBTs	Insulated gate bipolar transistors
I-V	Current versus voltage
k	Kilo
kB	Kilobyte
kg	kilogram
kHz	Kilohertz
km	Kilometer
km^2	Square kilometer
kV	kilovolt
kVA	Kilo volt ampere
kVAR	Kilo volt ampere reactive
kW	Kilo watt
kWh	Kilowatt hour
LC	Inductor-capacitor
LED	Light emitting diode
LEs	Logic elements
LUT	Look-up table
LVRT	Low voltage ride through
m^2	Square meter
m^3	Cubic meter
MHz	Megahertz
MMC	Modular multilevel cascaded
MOSFET	Metal oxide semiconductor field effect transistor
MPP	Maximum power point
MPPT	Maximum power point tracker
MRI	Magnetic resonance imaging
MVA	Mega volt ampere
MW	Megawatt
NCD	Native circuit description
nF/kWp	Nino farad per kilo watt peak
NGD	Native generic database
NPC	Neutral point clamped
NRE	Nonrecurring engineering
OD	Outer diameter
PM	Permanent magnet
PMG	Permanent magnet generator
PMSG	Permanent magnet synchronous generator
PV	Photovoltaic
P-V	Power versus voltage
PWM	Pulse width modulation
rms	Root mean square
RTL	Register transfer level
SCIG	Squirrel-cage induction generator

SDFM	Single doubly-fed machine
SDPWM	Sixty degree pulse width modulation
SliM	Spoked lightweight machine
SPI	Serial peripheral interface
SPWM	Sine pulse width modulation
SVM	Space vector modulation
THDs	Total harmonic distortions
THPWM	Third harmonic injected pulse width modulation
TRPWM	Trapezoidal pulse width modulation
US	United States
USD	United States dollar
V	Volt
VHDL	Very high speed integrated hardware description language
VSC	Voltage source converter
VSIs	Voltage source inverters
W	Watt
W/kg	Watt per kilogram
WRSG	Wound rotor synchronous generator
μF/kWp	Micro farad per kilowatt peak

Authors Biography

Md. Rabiul Islam (M'14, IEEE) received the B.Sc. and M.Sc. in Electrical and Electronic Engineering (EEE) from Rajshahi University of Engineering and Technology (RUET), Rajshahi, Bangladesh, in 2003 and 2009, respectively; and the Ph.D. in Electrical and Electronics from University of Technology Sydney (UTS), Sydney, Australia, in 2014.

From 2005 to 2008, he lectured with the Department of EEE, RUET, where he became an Assistant Professor in June 2008. He is currently a Research Associate with the School of Electrical, Mechanical and Mechatronic Systems, UTS. He has authored and coauthored more than 40 technical papers and 2 book chapters. His research interests are in the fields of power electronic converters, renewable energy technologies, and smart grid.

Dr. Islam is a member of the Institution of Engineers, Bangladesh and the Australian Institute of Energy, Australia. He received the University Gold Medal and Joynal Memorial Award from RUET for his outstanding academic performance while pursuing the B.Sc. engineering degree. He also received the Best Paper Award at IEEE PECon-2012. He acts as a reviewer for several prestigious international journals.

Youguang Guo (S'02-M'05-SM'06, IEEE) received the B.E. from Huazhong University of Science and Technology (HUST), Wuhan, China, in 1985, the M.E. from Zhejiang University, Zhejiang, China, in 1988, and the Ph.D. from University of Technology Sydney (UTS), Sydney, Australia in 2004, all in Electrical Engineering.

From 1988 to 1998, he lectured in the Department of Electric Power Engineering, HUST. From March 1998 to July 2008, he was a Visiting Research Fellow, Ph.D. candidate, Postdoctoral Fellow, and Research Fellow in the Center for Electrical Machines and Power Electronics, Faculty of Engineering, UTS. He is currently an Associate Professor at the School of Electrical, Mechanical and Mechatronic Systems, UTS. His research fields include measurement and modeling of magnetic properties of magnetic materials, numerical analysis of electromagnetic field, electrical machine design and optimization, power electronic drives and control.

Jianguo Zhu (S'93-M'96-SM'03, IEEE) received the B.E. from the Jiangsu Institute of Technology, Zhenjiang, China, in 1982, the M.E. from Shanghai University of Technology, Shanghai, China, in 1987, and the Ph.D. from University of Technology Sydney (UTS), Sydney, Australia, in 1995.

He is currently a Professor of Electrical Engineering and the Head of the School of Electrical, Mechanical and Mechatronic Systems, UTS. His research interests include electromagnetics, magnetic properties of materials, electrical machines and drives, power electronics, renewable energy systems, and smart micro-grids.

Keywords

Wind farms · PV power plants · Power converter topologies · Multilevel converters · Modulation techniques · Medium-voltage smart micro-grids · Step-up-transformers · Amorphous alloys · Nanocrystalline · High-frequency magnetic-links · Multiple isolated and balanced DC supplies · Digital switching controller · Field programmable gate array (FPGA) · Design and implementation

Chapter 1
Introduction

Abstract This chapter presents a succinct preface focusing on various aspects, including the historical growth of two dominating renewable energy sources, i.e., wind and solar sources, the existing technologies for large-scale generation of these renewable powers, technical challenges, and possible solutions. Reduction of volume and weight of power conversion system would solve effectively some technical problems of the existing technologies for large-scale renewable power generation systems, which is the foremost intention of this book.

Keywords Energy and environment challenges · Renewable energy · Offshore wind farms · Photovoltaic (PV) power plants · Remote area power supply · Grid connection of renewable power plants

1.1 Most Dominating Renewable Power Sources

The energy and environment represent two major areas of current global crisis, and it is more and more widely recognized that renewable energy sources, especially wind and solar energy, can offer effective solutions to these enormous challenges [1] as the wind and solar power development is experiencing dramatic growth [2–6]. Over 318 GW of wind power generation and over 130 GW solar photovoltaic (PV) power generation have been installed by 2013. Accordingly, the cumulative growth of wind and solar PV installation has directly pushed the wind turbine and solar PV converter technologies into a more competitive area [7–12]. In this propitious climate, it is therefore essential for scientists and engineering researchers to find the most effective converter technologies for integration of the wind and solar power generation systems into the main power grids.

M.R. Islam et al., *Power Converters for Medium Voltage Networks*,
Green Energy and Technology, DOI 10.1007/978-3-662-44529-7_1

1.1.1 Historical Growth of Wind Power Generation Capacity

Wind speed varies continuously with time and height because of the changes in the thermal conditions of air masses. The motion of air masses is not only a global phenomenon but also a regional and local phenomenon. The annual peak hours are normally around 2,500–3,000 h at good sites. A wind turbine generator converts wind energy to electricity energy. If A is the cross-sectional area through which the air of velocity V flows, and σ_d is the air density, the theoretical power P available in a wind stream can be calculated from [13, 14]

$$P = \frac{1}{2}\sigma_d A V^3. \tag{1.1}$$

Usually, offshore winds tend to flow at higher speeds than onshore winds. This allows the turbine to produce more electricity as the possible energy produced from the wind is proportional to the cube of the wind speed. Also unlike onshore wind, offshore breezes can be strong in the afternoon, matching the time when load demands are at peak level. Moreover, wind farms cover large areas of land. The land area covered by a 3.6-MW turbine can be almost 0.37 km^2, such that 54 turbines would cover about a land area of 20 km^2. Table 1.1 summarizes the land covered by some offshore wind farms [15]. For example, the Anholt offshore wind power plant is a Danish offshore wind power plant with nameplate capacity of 400 MW officially inaugurated in September 2013, which covers a land area of 88 km^2. It is the third largest offshore wind farm in the world and the largest in Denmark. Siemens supplied, installed, and commissioned 111 wind turbines, each with a capacity of 3.6 MW and a rotor diameter of 120 m. Figure 1.1 shows a photograph of the Anholt offshore wind power plant. Since offshore wind farms can save land rental expense which is equivalent to 10–18 % of the total operating and maintenance costs of a wind farm, offshore-based wind farms have attracted great attention in the last few years. Considering the wind speed and capacitance effect of the transmission-line cables, offshore wind turbines are usually installed about

Table 1.1 Land area covered by offshore wind farms

Wind farms	Power capacity (MW)	Distance from shore (km)	Number of turbines	Area covered (km^2)
Anholt	400	21	111	88.00
Barrow	108	7	30	10.00
Gunfleet Sands	172	7	48	17.50
Horns Rev	160	15	80	20.00
Horns Rev 2	200	27	91	35.00
Lynn and Inner	194	9	54	20.00
Ormonde	150	10	30	8.70
Princess Amalia	120	23	60	14.00

Fig. 1.1 A photograph of Anholt offshore wind power plant was officially inaugurated in September 2013 [16]

7–27 km from the shore. Figure 1.2 shows an ideal model of smart power systems with offshore wind farms.

According to the statistical data, the cumulative installed capacity of offshore wind farms in 2008, 2010, and 2012 is 1.50, 3.08, and 5.41 GW, respectively. The global annually installed capacity of offshore wind farms is summarized in Table 1.2. The capacity therefore almost doubled in every 2 years [18], whereas the cumulative installed capacity of total (onshore and offshore) wind power generation in 2008, 2010, and 2012 was substantially less than this at 120.26, 197.68, and 282.43 GW, respectively. The global annually installed capacity of farms is summarized in Table 1.3. It is expected that the global offshore installed capacity will increase to approximately 20 GW by 2015 and rise sharply to 104 GW by 2025. According to the Global Wind Energy Council and Green Peace International estimations, it is possible to mitigate 20 % of global electricity demand with wind power [19].

1.1.2 Technical Challenges and Possible Solutions

To integrate scattered wind turbine generators into a medium-voltage grid (e.g., 11–33 kV), a power frequency (50 or 60 Hz) transformer is commonly used to

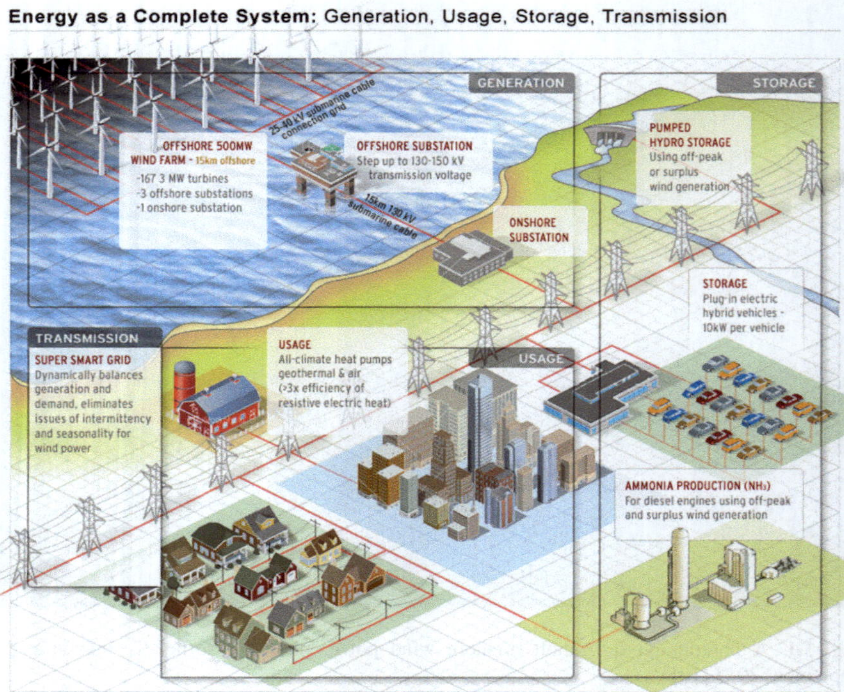

Fig. 1.2 An ideal model of renewable power source-based smart power systems [17]

Table 1.2 Global offshore wind farms installed capacity

Years	Annual addition (MW)	Cumulative (MW)
2005	90	723
2007	318	1,134
2009	577	2,084
2011	1,037	4,120
2013	1,647	7,062

Table 1.3 Global wind farms (onshore and offshore) installed capacity

Years	Annual addition (MW)	Cumulative (MW)
2005	11,471	59,091
2007	19,951	93,889
2009	38,351	158,975
2011	40,125	238,126
2013	34,923	318,117

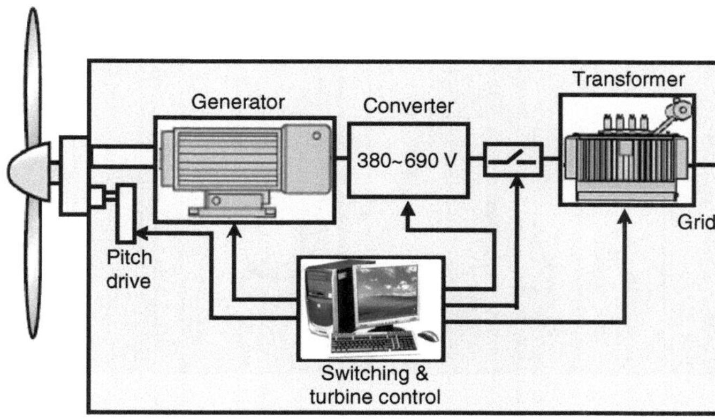

Fig. 1.3 Wind turbine with step-up transformer

step up the voltage for long-distance transmission. Figure 1.3 shows the traditional wind turbine with step-up transformer. In an offshore wind turbine power generation system, this transformer is usually installed at a height of about 80 m inside the nacelle together with other equipment, such as the generator and power converter. The tower provides support to the rotating parts and nacelle (the stationary parts). The nacelle weight of a 5-MW turbine is about 300 t, while the rotor represents only about 120 t. Therefore, the tower diameter and strength depend mostly on the weight of the nacelle and expected wind loads. The tower cost accounts for 26 %, the largest component, of the total turbine component cost [20]. The tower is normally held by a heavy foundation to ensure that it withstands the overturning moment created from the turbine. The foundation size of a 2.3-MW wind turbine is 314 m^2, and the approximate weight is 2,000 t or more [21]. In an offshore area, the cost of installation is extremely high. On average, approximately 20 % of the capital costs are associated with the installation [22, 23]. During the past decade, Van Oord has been involved in many offshore wind farm projects in North West Europe, providing construction services such as foundation and turbine installation, cable installation, and scour protection. Van Oord has an innovative and advanced transport and installation vessel named “Aeolus.” The vessel is 139 m long and 38 m wide with a draft of 5.7 m. The vessel is equipped with a crane capable of hoisting 900 t at 30 m radius and has accommodation for 74 persons. Aeolus is used for the installation of foundations and turbines. Figure 1.4 shows a photograph of wind turbine installation process. Therefore, a reduction in mechanical loading represents an enormous saving of tower construction and turbine installation costs. Moreover, regular monitoring and maintenance are also critical in offshore applications. For example, an offshore wind farm requires about 20 % higher operating and maintenance cost compared with an onshore farm. Figure 1.5 shows photograph of offshore wind power plant under maintenance. As the power rating and the distance of the wind park from the shore increase, the engineering

Fig. 1.4 A photograph of installation process of offshore wind power plant with the vessel Aeolus [24]

Fig. 1.5 A photograph of offshore wind farm under maintenance [25]

challenges associated with the installation and maintenance of the power generation systems also grow. Recently, many researchers/scientists have begun to focus on a number of areas including proposing new converters to eliminate the step-up transformers, applying medium-frequency transformers to reduce the weight and volume of step-up transformers, and superconducting wind generators to reduce the weight and volume of wind turbine generators. These steps are aimed at reducing the weight and volume of the wind turbine generator system. The weight and volume reduction of wind turbine power generation system still needs further investigation because the wind power has become one of the main renewable energy sources for future electricity supply within smart microgrids.

1.1.3 Historical Growth of Solar PV Power Generation Capacity

Solar energy is the second main renewable energy source for future electricity supply. Solar PV generates electricity in well over 100 countries and continues to be the fastest growing renewable resource in the world. By the end of 2011, a total of 69.68 GW PV power capacity had been installed, sufficient to generate around 80 billion kWh/year and enough to cover the annual power supply needs for more than 20 million households as reported by European Photovoltaic Industry Association (EPIA). Solar PV installations substantially increased over the last 5 years. The annual installation of new PV power capacity rose from 16.80 GW in 2010 to 29.66 GW in 2011, and in 2009, it was only 7.43 GW. The globally installed PV power capacity is summarized in Table 1.4 [26]. According to statistical data, the cumulative installed capacity of solar PV generation in 2007, 2009, and 2011 was 9.44, 23.21, and 69.68 GW, respectively. This means that the capacity of solar PV generation effectively tripled every 2 years.

Since 2007, medium- and large-scale PV power plants have attracted great interest and PV power plants of more than 10 MW in capacity have now become a reality. More than 200 PV power plants have already been installed in the world, each of them generating an output of more than 10 MW. Of these plants, 34 are located in Spain and 26 in Germany. The number of PV power plants will continue to rise. The literature indicates that more than 250 PV power plants will be installed within the next few years [27]. Future PV power plants will have higher power capacity. Indeed, some are to have a capacity in excess of 250 MW. These multi-megawatt PV power plants require large areas of land. Owing to this, they

Table 1.4 Global solar PV installations [26]

Years	Annual addition (MW)	Cumulative (MW)
2001	328	1,753
2003	578	2,798
2005	1,429	5,340
2007	2,528	9,443
2009	7,438	23,210
2011	29,665	69,684
2012	32,472	102,156

Fig. 1.6 A photograph of Beneixama 20 MW PV power plant [27]

are usually installed in remote areas, far from cities. The 20 MW PV power plant in Beneixama, Spain, used about 200 SINVERT 100 M inverters and installed approximately 100,000 PV modules in a land area of 500,000 m^2 [27]. Therefore, for power transmission, a medium-voltage grid is usually used. Figure 1.6 shows the photograph of Beneixama 20 MW PV power plant.

1.1.4 Technical Challenges of Solar PV Power Generation Systems

Although different power electronic converters have been developed in the last few decades using conventional topologies for solar PV systems, it is hard to connect the traditional converters to the grids directly as the distortion in generated output voltages is high and a single switch cannot stand at grid voltage level. With the rapid growth of grid-connected PV generations, the total harmonic distortions (THDs) generated from PV inverters is becoming a major concern [28]. In this regard, conventional systems utilizing the power frequency step-up transformer, filter, and booster not only increase the size, weight, and loss but also increase the cost and complexity of the system installation and operation. Swiss solar company Tritec built a PV power plant with 5.2 MW of power for the commercial company Migros. The installation feeds directly from the roofs of the Migros-Verteilbetriebs

Fig. 1.7 A photograph of step-up transformer installation of Tritec 5.2 MW solar PV power plant [30]

Neuendorf AG into the medium-voltage grid. Tritec installed SolarMax transformer compact stations from the Swiss manufacturer Sputnik Engineering on three roof areas of the factory. To reduce transmission losses, Tritec feeds solar power directly from the roof into a medium-voltage grid through step-up transformer. Figure 1.7 shows photograph of step-up transformer installation of Tritec 5.2 MW solar PV power plant.

These heavy and large-size power frequency step-up transformers significantly increase the weight and volume of the renewable power generation systems. Because of the heavy weight and large size of the power frequency transformer, the wind turbine generator and PV inverter system can be expensive and complex in terms of installation and maintenance especially in offshore and remote area applications. For example, the weight and volume of a 0.69/33 kV, 2.6 MVA transformer are typically in the range of 6–8 t and 5–9 m^3, respectively, and the weight and volume of a 0.4/33 kV, 1 MVA power frequency transformer are typically in the range of 3–5 t and 4–5 m^3, respectively [29]. A liquid-filled 2-MVA step-up transformer uses about 900 kg of liquid as the coolant and insulator, which requires regular monitoring and replacement. These levels are critical in offshore and remote area applications where the costs of installation and regular maintenance are extremely high.

1.1.5 *Possible Solution to Technical Challenges for Large-Scale Renewable Generation Systems*

To ensure that the goals of smart grids are met, the current industrial trend is to move away from these heavy and large-size passive components to compact and lightweight systems that use more and more semiconductor devices controlled by advanced digital controls. In comparison with the conventional two-level converters, multilevel converters present lower switching losses, lower voltage stress on switching devices, and better power quality. They also enable the development of medium-voltage converters using matured and cheap power semiconductor devices. The development of multilevel medium-voltage converters enables the connection of renewable energy systems directly to the grid without using large, heavy, and costly power transformers. Although several multilevel converter topologies have been used in low-voltage applications, most of them are not suitable for medium-voltage applications. Because of some special features, such as the number of components scaling linearly with the number of levels and high-level number attainability by the use of identical modules, the modular multilevel-cascaded (MMC) converter topology can be considered as a possible candidate for medium-voltage applications [31] to connect renewable power generation units to the medium-voltage grid directly. Since the component number and control complexity increase linearly with the number of levels, the optimal selection of the number of converter levels is important in order to achieve the best performance/cost ratio for the medium-voltage converter systems. For example, the 19-level and 43-level converters are found to be optimal for 11- and 33-kV systems, respectively [32]. However, the MMC converter requires multiple isolated DC sources that must be balanced, and therefore, its application is not straightforward, especially in wind power generation systems [33, 34]. Moreover, the multilevel converter requires a number of switching and control pulse width modulation (PWM) signals, which cannot be generated by the available digital signal processor (DSP) because the available DSP can only at present provide about 6 pairs of PWM channels. Furthermore, the DSP runs a sequential program in its microprocessor, i.e., all the gate pulses cannot be updated with the same clock pulse [35]. Therefore, the DSP-based system requires an additional control strategy to compensate the time delay [36].

1.2 Major Objectives of the Book

The major objectives of this book are as follows:

1. to find a suitable power converter for the compact and lightweight direct grid integration of renewable power generation systems with matured and low-price semiconductor devices, a modern digital controller, and advanced magnetic materials with high saturation flux density and low core losses;

2. to investigate the different multilevel converter topologies, taking into account the availability of power semiconductors, output power quality, control complexity, and the cost of semiconductors so as to find a suitable multilevel converter topology for medium-voltage applications;
3. to research the different number of MMC converter levels so as to introduce a way to design a medium-voltage MMC converter with an optimal number of converter levels;
4. to design and analyze a fully digital switching controller for a multilevel converter within the most common software environment so as to reduce the developmental time and cost of the switching controller;
5. to design and analyze a high-frequency magnetic link with multiple secondary windings so as to verify the new concept of generating multiple isolated and balanced DC supplies for all of the H-bridge inverter cells of the MMC converter; and
6. to design and construct a scaled down test prototype of a high-frequency magnetic-link MMC converter to validate the new concept of step-up-transformer-less direct grid integration of renewable generation systems with medium-voltage converters.

1.3 Contribution of the Book

The major contributions of this book include the following:

1. an in-depth investigation into the commercially available medium- and large-scale renewable power generation technologies and the research and development of converter topologies leading to the conclusion that medium-voltage converter may be the best possible option to reduce the weight and volume of renewable power generation systems;
2. the comparison of different available multilevel converter topologies with medium-voltage applications leading to the conclusion that the MMC converter is the best possible candidate for medium-voltage applications;
3. the detailed study of a design process for the MMC converter which focuses on the optimal selection of the number of converter levels;
   (*To the best knowledge of the authors, they are the first who have presented a technique to select the optimal number of converter levels for medium-voltage applications*.)
4. the detailed study of a new design process for the switching controller to reduce developmental time and cost;
   (*An alternative technique to design the switching controller to save development time and cost are presented.*)
5. the detailed study of high-frequency magnetic link with multiple secondary windings to generate multiple isolated and balanced DC supplies for all the H-bridge inverter cells of the MMC converter; and

(To the best knowledge of the authors, they are the first who have presented high-frequency magnetic link to generate multiple isolated and balanced DC supplies for the MMC converter.)

6. the development and testing within a laboratory of a scaled down high-frequency-link MMC converter which converters 210 V DC into 3-phase 1 kV rms 50 Hz AC. *(To the best knowledge of the authors, they are the first who have experimentally verified a high-frequency magnetic-link MMC medium-voltage converter for direct grid integration of renewable generation systems.)*

1.4 Organization of the Book

Chapter 2 presents a review of the traditional system, which requires a step-up transformer to interconnect the renewable generation systems to the grid, and the other recently proposed converter topologies for transformer-less grid interconnection in order to show a complete picture of the importance of size and weight reduction, and performance improvement of power converter systems. In order to develop a common converter for both wind and PV power generation systems, after a thorough investigation of all possible converter topologies covering almost all types of existing power converters, it is concluded that the multilevel converter topology with high-frequency magnetic link would be the most feasible option for developing medium-voltage converters for direct grid connection of wind and PV power systems.

Chapter 3 is dedicated to the method to find a suitable multilevel converter topology, which can connect the renewable generation units directly to the medium-voltage grid with mature semiconductor devices. Different multilevel converter topologies, such as the neutral point clamped (NPC), flying capacitor (FC), and MMC converters, have been considered for the design of an 11-kV converter system. The comparison is made in terms of the number of semiconductors, semiconductor cost and availability, THDs, filter size, and control complexity of the converters. The performance is analyzed and compared through numerical analysis in the MATLAB/Simulink environment. To generate the switching pulses, a level-shifted carrier-based switching scheme is used for NPC topologies and a phase-shifted carrier-based switching scheme is used for the FC and MMC converter topologies with a carrier frequency of 1–2 kHz and a modulation index of 0.8–0.98.

To couple the renewable energy source to the MMC converter, a common high-frequency magnetic link with multiple secondary windings is considered and electromagnetic performance is reported in Chap. 4. The medium-frequency link is used to generate the isolated balanced multiple DC supplies for the MMC converter from a single low-voltage power source. Compared with the conventional transformers operated at the power frequency (50 or 60 Hz), the high-frequency (in the range of a few kHz to MHz) transformers have much smaller and lighter magnetic cores and windings and thus much lower costs.

The multilevel converter requires a number of switching and control PWM signals, which cannot be generated by the available DSP because the available DSP can only at present provide about 6 pairs of PWM channels. In this instance, the field programmable gate array (FPGA) is the natural choice for medium-voltage multilevel converters. However, most of the available design techniques require special software such as the HDL coder, system generator, PSIM and ModelSim, which increases the development time and cost. In Chap. 5, a fully digital switching controller is developed for 3-phase 5-level converters. The SK 30 GH 123 IGBTs are used to develop a prototype multilevel converter, and an XC3S500E FPGA is used to develop the switching controller. The most common software such as the MATLAB/Simulink and Xilinx ISE based design technique is used which may reduce the developmental time and cost of the switching controller. The simulation results serve as a preliminary validation of the proposed design technique, which will be finally verified by the experimental results. The developed switching scheme can be used for any multilevel converter configuration with minimum changes in software environment. Moreover, the proposed design techniques and implementation issues may be useful for designing any other modern power converter.

To verify the feasibility of the high-frequency-link MMC medium-voltage converter, in Chap. 6, a scaled down 1.73 kVA laboratory prototype test platform with modular 5-level cascaded converter is developed, which converts 210 V DC (rectified generator voltage) into 3-phase 1 kV rms 50 Hz AC. The design and implementation of the prototyping, test platform, and the experimental results are analyzed and discussed. It is expected that the proposed new technology would have great potential for future renewable generation systems and smart grid applications.

The high number of levels means that medium-voltage attainability will be possible to connect the renewable generation units to the medium-voltage grid directly and it will also be possible to improve the output power quality. Since the number of components and control complexity increase linearly with the number of levels, the optimal selection of the number of converter levels is important for the best performance/cost ratio of the medium-voltage converter systems. In Chap. 7, an 11-kV system and a 33-kV system are designed and analyzed taking into account the specified system performance, control complexity, and cost and market availability of the power semiconductors. It is found that the 19-level and 43-level converters are optimal for 11- and 33-kV systems, respectively. The detailed designs and analysis of an 11-kV MMC converter and a 33-kV MMC converter systems are presented.

Chapter 8 draws the conclusions for the whole book and proposes possible future works for the further development of the technology.

References

1. Islam MR, Islam MR, Beg MRA (2008) Renewable energy resources and technologies practice in Bangladesh. Renew Sustain Energy Rev 12(2):299–343
2. Li H, Xie M, Zhang T (2013) Promote the development of renewable energy: a review and empirical study of wind power in china. Renew Sustain Energy Rev 22:101–107

3. Sahu BK, Hiloidhari M, Baruah D (2013) Global trend in wind power with special focus on the top five wind power producing countries. Renew Sustain Energy Rev 19:348–359
4. Da Z, Xiliang Z, Jiankun H, Qimin C (2011) Offshore wind energy development in China: current status and future perspective. Renew Sustain Energy Rev 15(9):4673–4684
5. Chen J (2011) Development of offshore wind power in China. Renew Sustain Energy Rev 15(9):5013–5020
6. Islam ABMS, Jameel M, Jumaat MZ, Shirazi SM, Salman FA (2012) Review of offshore energy in Malaysia and floating Spar platform for sustainable exploration. Renew Sustain Energy Rev 16(8):6268–6284
7. Zhixin W, Chuanwen J, Qian A, Chengmin W (2009) The key technology of offshore wind farm and its new development in China. Renew Sustain Energy Rev 13(1):216–222
8. Ion CP, Marinescu C (2013) Three-phase induction generators for single-phase power generation: an overview. Renew Sustain Energy Rev 22:73–80
9. Ayhan D, Saglam S (2012) A technical review of building-mounted wind power systems and a sample simulation model. Renew Sustain Energy Rev 16(1):1040–1049
10. Islam MR, Mekhilef S, Saidur R (2013) Progress and recent trends of wind energy technology. Renew Sustain Energy Rev 21:456–468
11. Iglesias RL, Arantegui RL, Alonso MA (2011) Power electronics evolution in wind turbines—a market-based analysis. Renew Sustain Energy Rev 15(9):4982–4993
12. Shafiullah GM, Oo AMT, Ali ABMS, Wolfs P (2013) Potential challenges of integrating large-scale wind energy into the power grid—a review. Renew Sustain Energy Rev 20:306–321
13. Du Z, Gu W (2009) Aerodynamics analysis of wind power. In: Proceedings of the 2009 world non-grid-connected wind power and energy conference; 24–26 Sept 2009, Nanjing, China, pp 1–3
14. Islam MR, Guo YG, Zhu JG (2010) Steady state characteristic simulation of DFIG for wind power system. In: Proceedings of the 6th international conference on electrical and computer engineering, 18–20 Dec 2010, Dhaka, Bangladesh, pp 151–154
15. Islam MR, Guo YG, Zhu JG (2014) A review of offshore wind turbine nacelle: technical challenges, and research and developmental trends. Renew Sustain Energy Rev 33:161–176
16. Anholt-offshore-wind-farm (2014) [online]. Available at: http://www.offshorewind.biz/2013/09/04/denmark-ribbon-cut-at-anholt-offshore-wind-farm/. Accessed on 13 May 2014
17. www.drmrenfrew.files.wordpress.com
18. Earth Policy Institute. Climate, energy and transportation, world cumulative wind turbine installations (online). Available at: http://www.earth-policy.org. Accessed on 25 Mar 2013)
19. IEA Wind. IEA wind task 26; the past and future cost of wind energy (online) .Available at: http://www.nrel.gov. Accessed on 20 Nov 2012
20. ABB. Wind Power Plants. Technical application papers no. 13 (Online). Available at: http://www05.abb.com. Accessed on 13 Dec 2012
21. Jeppsson J, Larsen PE, Larsson A (2013) The Swedish Energy Agency, Vattenfall. Technical description lillgrund wind power plant (online). Available at: http://www.vattenfall.se. Accessed on 23 Mar 2013
22. Pepper L (2012) Monitoring and evaluation of Blyth offshore wind farm—installation and commissioning (online). Available at: http://webarchive.nationalarchives.gov.uk. Accessed on 12 Sept 2012
23. Gerdes G, Tiedemann A, Zeelenberg S (2012) Case study: Modelling offshore wind farms—a survey for the analysis of the experiences and lessons learned by developers of offshore wind farms (Online). Available at: http://www.offshore-power.net. Accessed on 03 Dec 2012
24. Wind Turbine Installation Vessel Aeolus (2014) [Online]. Available at: http://www.vanoord.com/activities/offshore-wind-equipment. Accessed on 10 May 2014
25. Maintenance of Offshore Wind Turbine (2013) [Online]. Available at: www.power-technology.com. Accessed on 10 July 2013
26. See http://www.earth-policy.org/ for information about cumulative installed capacity of PV systems

27. See http://www.siemens.de/ for information about photovoltaic power plants
28. Du Y, Lu DDC, James G, Cornforth D (2013) Modeling and analysis of current harmonic distortion from grid connected PV inverters under different operating conditions. Sol Energy 94:182–194
29. Islam MR, Guo YG, Zhu JG (2012) A transformer-less compact and light wind turbine generating system for offshore wind farms. In: Proceedings of the 2012 IEEE international conference on power and energy, 2–5 Dec 2012, Kota Kinabalu, Malaysia, pp 605–610
30. Swiss-solar-power-plant-transformers-roof (2014) [Online]. Available at: http://www.sunwindenergy.com/news/swiss-solar-power-plant-transformers-roof. Accessed on 13 May 2014
31. Islam MR, Guo YG, Zhu JG (2014) Multilevel converters for step-up transformer-less direct grid integration of renewable generation units with medium voltage smart microgrids. Large scale renewable power generation: advances in technologies for generation, transmission and storage. Springer, Berlin, Heidelberg, pp 127–149
32. Islam MR, Guo YG, Zhu JG (2014) A multilevel medium-voltage inverter for step-up-transformer-less grid connection of photovoltaic power plants. IEEE J Photovoltaics 4(3):881–889
33. Islam MR, Guo YG, Lin ZW, Zhu JG (2014) An amorphous alloy core medium frequency magnetic-link for medium voltage photovoltaic inverters. J Appl Phys 115(17):17E710-1–17E710-3
34. Islam MR, Guo YG, Zhu JG (2013) A medium-frequency transformer with multiple secondary windings for medium-voltage converter based wind turbine generating systems. J Appl Phys 113(17):17A324-1–17A324-3
35. Islam MR, Guo YG, Zhu JG (2014) A high-frequency link multilevel cascaded medium-voltage converter for direct grid integration of renewable energy systems. IEEE Trans Power Electron 29(8):4167–4182
36. To HP, Rahman MF, Grantham C (2007). Time delay compensation for a DSP-based current-source converter using observer-predictor controller. In: Proceedings of the seventh international conference on power electronics and drive systems, 27–30 Nov 2007, Bangkok, Thailand, pp 1091–1096

Chapter 2
Power Converters for Small- to Large-Scale Photovoltaic Power Plants

Abstract This chapter presents an extensive literature survey on various different aspects of medium-voltage (MV) converter development for step-up-transformer-less direct grid integration of photovoltaic (PV) power plants. The main objective was to show how power electronic converter topologies, power electronic devices, and control complexities have affected the development of the MV converter and how to make an excellent choice of the converter topology for step-up-transformer-less grid integration through the MV converter. Besides the traditional system, which requires a step-up transformer to connect the renewable power plants to the grids, other recently proposed converter topologies for step-up-transformer-less direct grid interconnection are also introduced in detail with the aim of presenting a complete picture of power converter topologies for small- to large-scale PV power plants.

Keywords Power converters · Photovoltaic inverters · Small to large scale · Power plants · Remote area · Medium voltage

2.1 Introduction

The world's energy demand is growing remarkably due to the fast growth of population and economy in the developing countries. The energy sector is facing an accelerating amalgam crisis of the worldwide established fossil and atomic energy systems. Natural gas, coal, and crude oil are the main fossil fuels for the current world energy supply. Crude oil is the most important fossil fuel among the three main fossil flues. Figure 2.1 shows the trend of world oil consumption [1]. Coal is the second most important consumable fossil fuel. Figure 2.2 shows the world coal consumption [1]. Due to very high oil price, coal has been becoming an attractive fossil fuel in the recent years. Figure 2.3 shows the world consumption of natural gas [1]. Almost a constant growth rate has been observed for the natural gas consumption in the past decades.

M.R. Islam et al., *Power Converters for Medium Voltage Networks*,
Green Energy and Technology, DOI 10.1007/978-3-662-44529-7_2

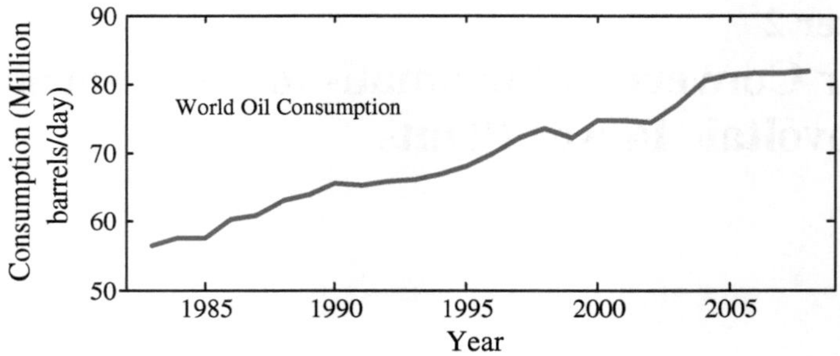

Fig. 2.1 World oil consumption [1]

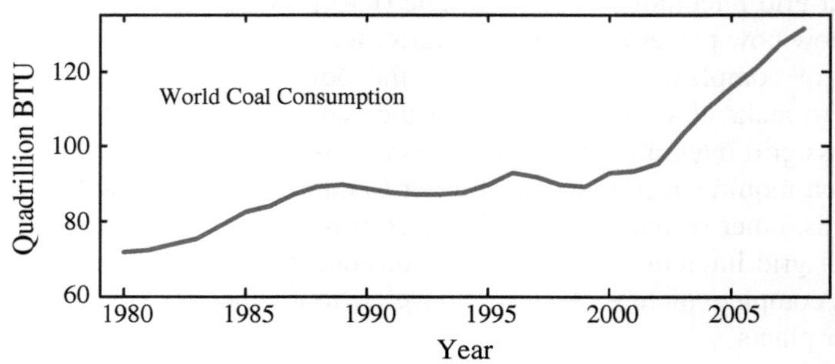

Fig. 2.2 World coal consumption [1]

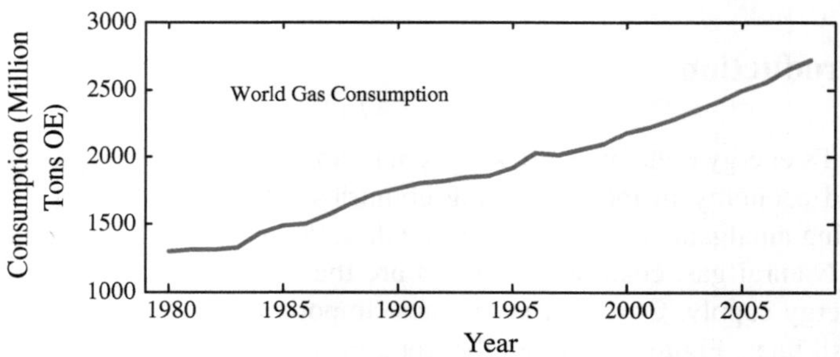

Fig. 2.3 World natural gas consumption [1]

The increasing energy demand is not only diminishing the reserve of fossil fuels, but also affecting the environment. Carbon dioxide (CO_2) gas is generated from burning of fossil fuels, which significantly contributes to the increase of average global temperatures, i.e., global warming. Figure 2.4 shows the world CO_2 emission from fossil fuel burning [1]. Scientists worldwide are now seeking

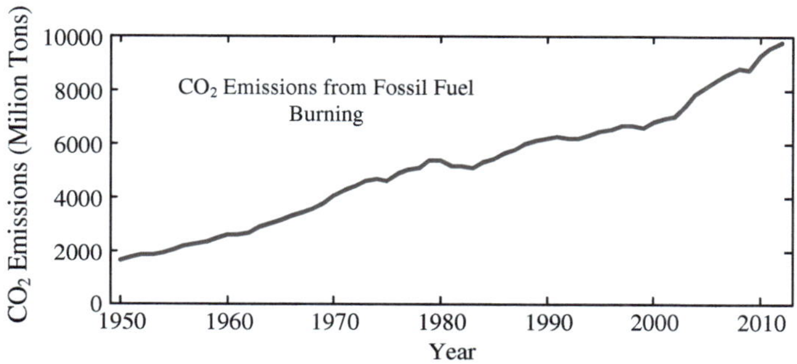

Fig. 2.4 World CO_2 emission form fossil fuel burning [1]

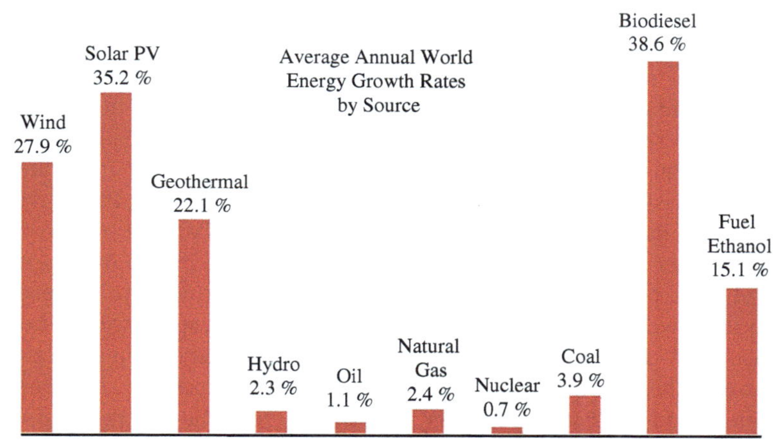

Fig. 2.5 Average annual world energy growth rates by source [1]

solutions to these two enormous challenges (energy and environment) from renewable or clean energy sources, which are richly available in almost every country.

Many countries have set targets for renewable energy use to meet the increasing energy demand and also to reduce the global warming effect. For example, the target shares of total energy from renewables by 2020 in Sweden, Finland, Austria, and Australia are 49, 38, 34, and 20 %, respectively [2]. Therefore, average annual growth rates of renewable sources are much higher than those of conventional sources in recent years. Figure 2.5 shows the average annual world energy growth rates [1]. Solar photovoltaic (PV) represents the second highest growth rate due to its abundance source and technological development of PV cell, e.g., reduction of PV module cost. Average module cost was USD 100/W and USD 29/W in 1975 and 1980, respectively, and reduced to less than USD 3.5/W in 2004. Figure 2.6 shows the average PV module cost [1].

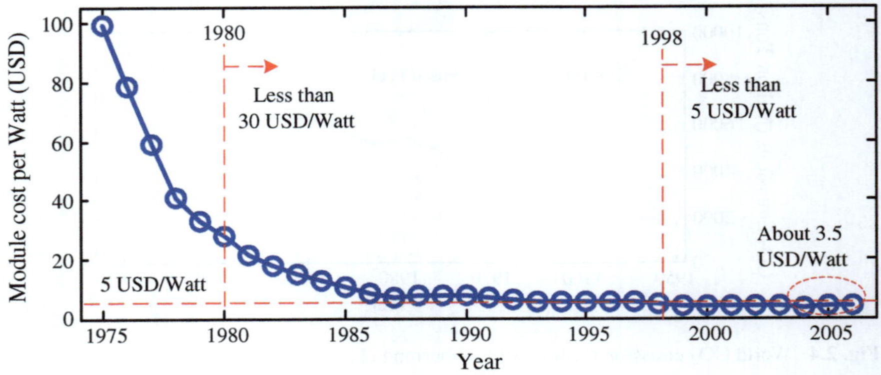

Fig. 2.6 Average photovoltaic module cost [1]

Since 2007, medium- and large-scale PV power plants have attracted a high degree of attention and the power plants of more than 10 MW in capacity have now become a reality. These multi-megawatt PV power plants require large areas of land, and thus, they are usually installed in remote areas, far from cities. For power transmission, a step-up transformer is usually used in the PV inverter system to feed in the solar energy into a medium-voltage (MV) grid (e.g., 6–36 kV). The transformer steps up the inverter output voltage from 300 V AC to grid voltage level (e.g., 6–36 kV). Although these special transformers are compact compared with conventional distribution transformers, they are still large and heavy for remote area PV applications. The large-size and heavy weight step-up transformer may increase the system weight and volume and can be expensive and complex for installation and maintenance. The MV inverter may offer the best possible solution to interconnect the PV array to the MV grid directly [3]. Moreover, it may also be possible to ensure electrical isolation through the inverter, which is important for the interconnection of MV grid and PV array [4]. Recently, advanced magnetic materials, such as amorphous and nanocrystalline alloys, have attracted significant attention to develop high-frequency magnetic links for MV inverters [5]. Compared with the power frequency transformer (operated at 50/60 Hz), the high-frequency magnetic links (in the range of a few kHz–MHz) have much smaller and lighter magnetic cores and windings and thus much lower cost. Therefore, the MV inverter for step-up-transformer-less direct grid interconnection of PV systems has become a favorable choice, since the installation of large-scale PV power plants started commercially in 2007.

This chapter incorporates PV power generation technologies, including traditional power conditioning systems, two-level low-voltage converter topologies, the limitations of power frequency step-up transformer-based grid integration of renewable generation systems in remote area applications, and advanced converter topologies for MV applications. Therefore, an extensive literature survey has been conducted focusing on many different aspects of MV converter development for step-up-transformer-less direct grid integration of PV power generation systems. The main objective of this general review on power electronic converter topologies is to show how topologies, power electronic devices, and control complexities have affected the development of the MV converter and to understand which converter topology would be a natural choice for step-up-transformer-less grid integration through the MV converter.

2.2 Solar Photovoltaic Arrays

The "solar photovoltaic" means "producing electricity from sunlight." A PV cell can be made by two semiconductor layers; one having positive charge and the other having negative charge. When the sunlight shines on a PV cell, some of the photons from the sunlight are absorbed by the semiconductor atoms, releasing electrons from the cell's negative layer. If there is an external circuit, the free electrons flow to the positive layer and produce electric current in the external circuit. Figure 2.7 shows a photograph of PV cell.

One silicon PV cell can produce about 0.5 V, which is too small to do valuable work. To increase the voltage, several individual PV cells are interconnected together in package called a module. For example, a 12-V module will have 36 individual cells connected in series. In 24 V applications, two 12-V modules can be connected in series, where the current stays unchanged. Figure 2.8 shows a photograph of 24-V PV module. Table 2.1 summarizes the electrical characteristics of solar module SX-10. When two modules are wired in parallel, their current is doubled and the voltage stays unchanged. To achieve the desired voltage and current,

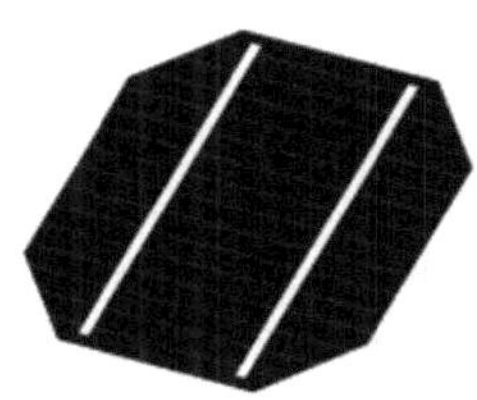

Fig. 2.7 A photograph of PV cell, which may produce 0.5 V DC

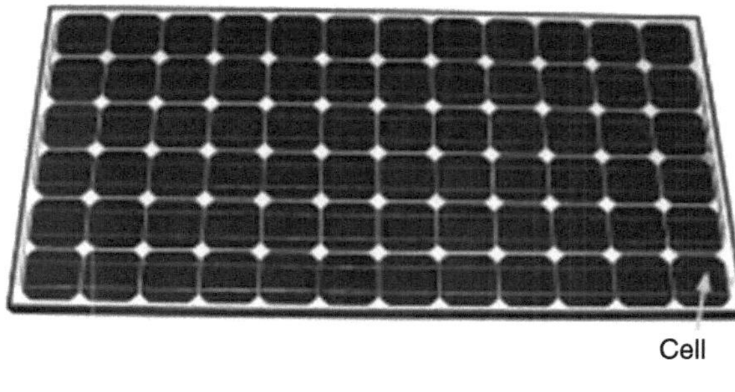

Fig. 2.8 A photograph of 24-V PV module, where 72 cells are interconnected

Table 2.1 Electrical characteristics of solar module SX-10

Parameters	Rating
Nominal voltage	12 V
Rated power	10 W
Number of PV cells	36 (connected in series)
Voltage at P_{max} (V_{mp})	16.8 V
Current at P_{max} (I_{mp})	0.59 A
Short-circuit current (I_{sc})	0.65 A
Open-circuit voltage (V_{oc})	21.0 V
Temperature coefficient of I_{sc}	(0.065 ± 0.015) %/°C
Temperature coefficient of V_{oc}	−(80 ± 10) mV/°C
Temperature coefficient of power	−(0.5 ± 0.05) %/°C

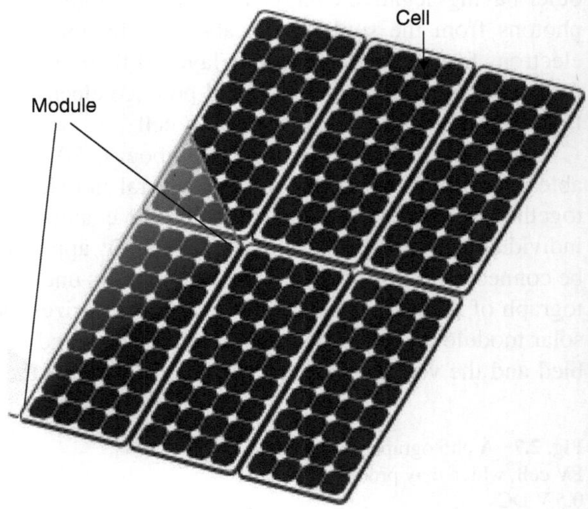

Fig. 2.9 A photograph of PV array with six 12-V modules, where 36 cells are interconnected in each module

a number of modules can be connected in series and parallel. The series–parallel combination of modules is called a PV array. Figure 2.9 shows a photograph of PV array with six 12-V modules, where 36 cells are interconnected in each module.

2.2.1 Solar Photovoltaic Array Modeling

When PV arrays are used to harvest solar energy, two important factors could limit the implementation of PV systems, i.e., high cost and low efficiency in energy conversion. The conversion efficiency of the current solar PV modules is typically only about 10–17 % [6]. In PV systems, the PV array represents about 57 % of the total cost of the system, and the battery storage system corresponds to 30 % of the cost. Other system components such as inverters and maximum power point tracker (MPPT) contribute to only 7 % of the total cost [7]. Due to the low

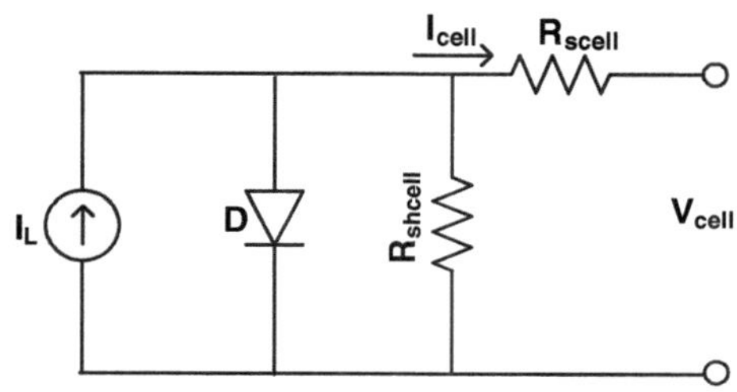

Fig. 2.10 Equivalent circuit of a PV cell

conversion efficiency and high cost of solar array, it is very desirable to operate the PV panel at the maximum power point (MPP). An ideal solar cell can be modeled by a current source in parallel with a diode. In practice, no solar cell is ideal and hence a shunt resistance and a series resistance are added to the model as shown in Fig. 2.10, where R_{scell} is the intrinsic series resistance of usually a very small value, and R_{shcell} is the equivalent shunt resistance of usually a very large value.

For a single silicon solar cell, the nonlinear *I–V* characteristic can be presented as [8, 9]

$$I_{\mathrm{cell}} = I_{\mathrm{L}} - I_{\mathrm{o}}\left[\exp\left(G(V_{\mathrm{cell}} + I_{\mathrm{cell}}R_{\mathrm{scell}})\right) - 1\right] - \frac{V_{\mathrm{cell}} + I_{\mathrm{cell}}R_{\mathrm{scell}}}{R_{\mathrm{shcell}}} \tag{2.1}$$

where
$G = \frac{q}{A_i KT}$,

q is the electronic charge ($=1.602 \times 10^{-19}$ C),
$A_i = B_i$ is the ideality factor ($=1.92$),
K is the Boltzmann's constant ($=1.38 \times 10^{-23}$ J/K),
T is the PV cell temperature,
I_{cell} is the cell output current,
V_{cell} is the cell output voltage,
I_{o} is the cell saturation current which can be presented as

$$I_{\mathrm{o}} = I_{\mathrm{or}}\left(\frac{T}{T_{\mathrm{r}}}\right)^{3} \exp\left[\frac{qE_{\mathrm{GO}}}{B_i K}\left(\frac{1}{T_{\mathrm{r}}} - \frac{1}{T}\right)\right] \tag{2.2}$$

I_{L} is the light-generated current which can be presented as

$$I_{\mathrm{L}} = \left[I_{\mathrm{SC}} + K_{I_{\mathrm{SC}}}(T_C - 28)\right] \times \frac{R_{\mathrm{ad}}}{1{,}000} \tag{2.3}$$

The array temperature T_{c} is approximately given by [9]

$$T_{\mathrm{C}} = T_{\mathrm{air}} + 0.3 \times R_{\mathrm{ad}}\ \% \tag{2.4}$$

T_{r} is the reference temperature ($=301$ K),
I_{or} is the reverse saturation current at T_{r} ($=19.9693 \times 10^{-6}$ A),
I_{os} is the cell reverse saturation current, *A*
T_{c} is the cell temperature, °C
$K_{I\mathrm{sc}}$ is the short-circuit current temperature coefficient ($=0.0017$ A/°C)

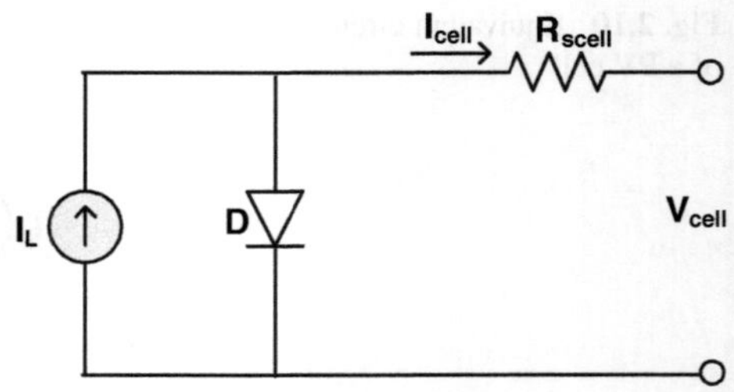

Fig. 2.11 Simplified equivalent circuit of a PV cell

R_{ad} is the cell illumination, W/m^2 (1,000 W/m^2 = 100 % illumination)
I_{sc} is the cell short-circuit current at 28 °C and 1,000 W/m^2 (=2.52 A)
EG_o is the band gap for silicon (=1.11 eV)
T_{air} is the ambient temperature, °C

Since PV arrays are built up with series and/or parallel connected combinations of solar PV cells, for an array with $n_S \times n_P$ cells, the current equation can be presented as [9, 10]

$$I_{PV} = n_p I_L - n_p I_o \left[\exp\left(G\left(\frac{V_{PV} + I_{PV} R_s}{n_s}\right)\right) - 1\right] - \frac{V_{PV} + I_{PV} R_s}{R_{sh}}. \quad (2.5)$$

where

$I_{PV} = n_P I_{cell}$ is the PV array output current,
$V_{PV} = n_S V_{cell}$ is the PV array output voltage,
n_s is the number of cells connected in series,
n_p is the number of panels connected in parallel,
$R_S = R_{Scell} \frac{n_S}{n_P}$ is the PV array series resistance, and
$R_{sh} = R_{shcell} \frac{n_S}{n_P}$ is the PV array shunt resistance.

The shunt resistance R_{shcell} is much greater than the series resistance R_{scell}, which makes the last term of (2.1) much smaller than the other terms. The simplified cell equivalent circuit is shown in Fig. 2.11.

The current–voltage relation of simplified PV cell can be expressed as

$$I_{cell} = I_L - I_o\left[\exp\left(G(V_{cell} + I_{cell} R_{scell})\right) - 1\right] \quad (2.6)$$

and the array current–voltage relation becomes

$$I_{PV} = n_p I_L - n_p I_o \left[\exp\left(G\left(\frac{V_{PV} + I_{PV} R_s}{n_s}\right)\right) - 1\right]. \quad (2.7)$$

2.2.2 Solar Photovoltaic Array Characteristics

Solar energy sources have variable daily and seasonal patterns. For example, monthly average global solar insolation at Dhaka City of Bangladesh varies between 3.92 and 7.71 kW h/m^2/day. The maximum amount of insolation

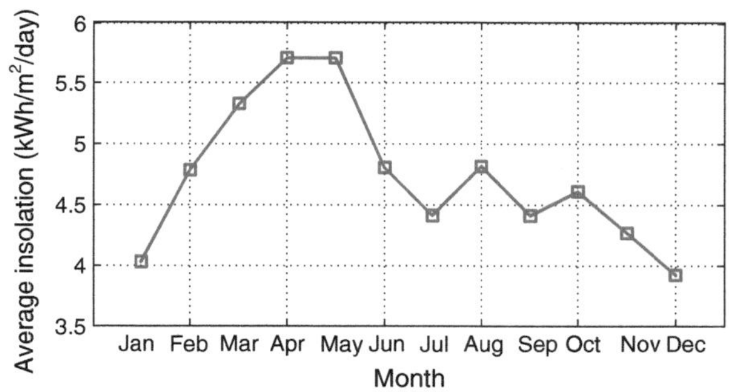

Fig. 2.12 Monthly average global solar insolation at Dhaka City of Bangladesh; recording period: 1988–1998 [11]

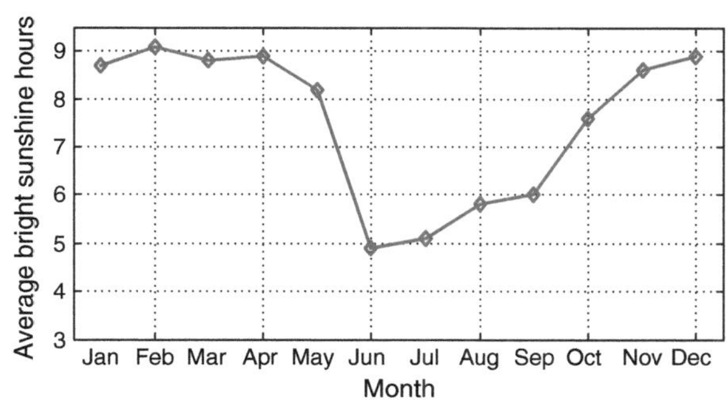

Fig. 2.13 Daily average bright sunshine hour at Dhaka City of Bangladesh; recording period: 1961–1980 [11]

is available in the months of March–May, and the minimum in December–January. Figure 2.12 shows the monthly global solar insolation at Dhaka City of Bangladesh [11]. Daily bright sunshine hours have also variable daily and seasonal patterns. For example, the daily average bright sunshine hour at Dhaka City of Bangladesh vary between 4.9 and 9.1 h. The maximum is in the months of February–April, and the minimum is in the June–September. Figure 2.13 shows the daily average bright sunshine hours at Dhaka City of Bangladesh [11]. The output characteristics of the PV array are nonlinear and critically affected by the solar radiation, temperature, and load conditions. The current versus voltage (*I–V*) and power versus voltage (*P–V*) characteristics with various irradiances at 25 °C temperature are shown in Figs. 2.14 and 2.15, respectively [12, 13]. The simulated *I–V* and *P–V* characteristics with various temperatures at 1,000 W/m^2 irradiance are illustrated in Figs. 2.16 and 2.17, respectively [12, 13].

2.2.3 Solar Photovoltaic Array Maximum Power Point

It is also observed from the power–voltage curve of the solar PV module that on the right hand side, when the voltage is almost constant, the slope of power versus

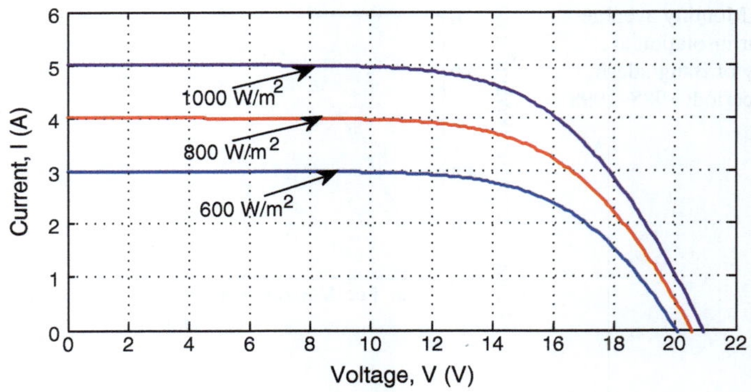

Fig. 2.14 Calculated *I–V* curves at different irradiance ranging from 600 to 1,000 W/m^2; 25 °C temperature was considered for the calculation [12, 13]

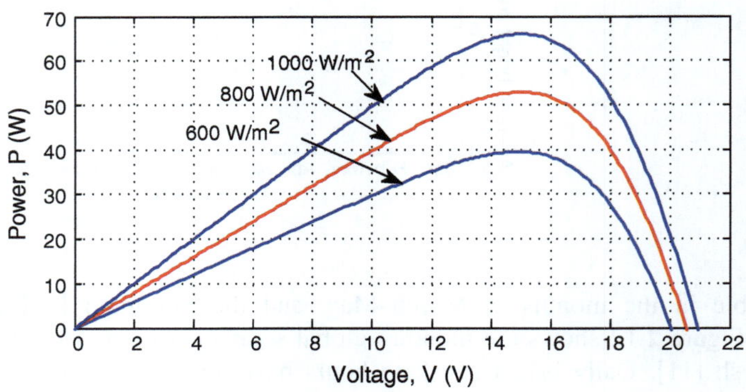

Fig. 2.15 Calculated *P–V* curves at different irradiance ranging from 600 to 1,000 W/m^2; 25 °C temperature was considered for the calculation [12, 13]

voltage is negative ($\Delta P/\Delta V < 0$), whereas on the left hand side, the slope is positive ($\Delta P/\Delta V > 0$), as illustrated in Fig. 2.18 [12, 13]. Thus, the PV array has an optimum operating point called the MPP, which varies depending on array temperature and the present insolation level. The irradiance mainly affects the output current, and the temperature mainly affects the terminal voltage of the PV array, so that the effects of both factors have to be considered when designing a PV system. However, the intermittent nature of PV sources, in terms of the power and output voltage, is a major issue when connected to the grid. Therefore, solar PV system requires a power conditioning circuit known as PV inverter that is capable of extracting the maximum power from PV source and feeds the adjusted power to the load and/or grid to their satisfaction.

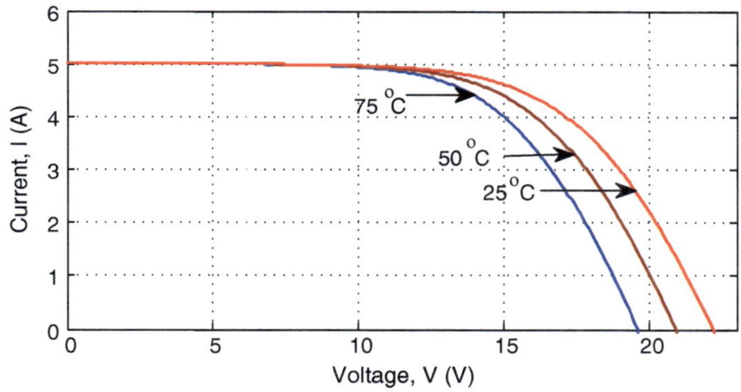

Fig. 2.16 Calculated *I–V* curves at different temperature ranging from 25 to 75 °C; irradiance 1,000 W/m^2 was considered for the calculation [12, 13]

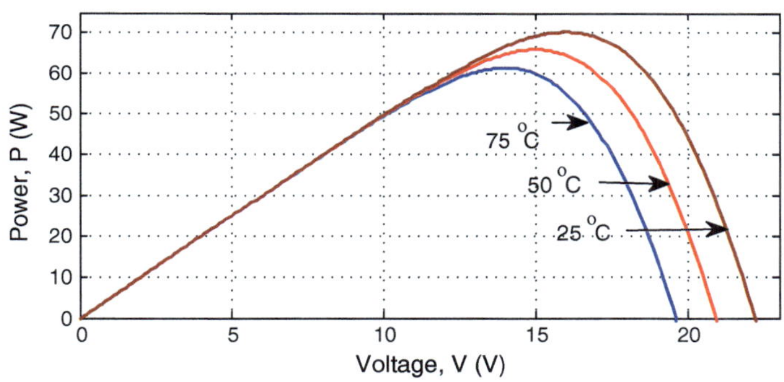

Fig. 2.17 Calculated *P–V* curves at different temperature ranging from 25 to 75 °C; irradiance 1,000 W/m^2 was considered for the calculation [12, 13]

Fig. 2.18 Maximum power point (*MPP*) determination: The slope is positive on the left side of the *P–V* curve, negative on the right side, and zero at the MPP [12, 13]

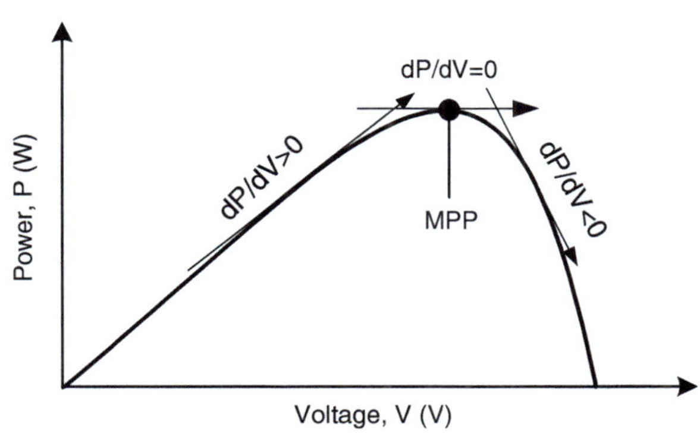

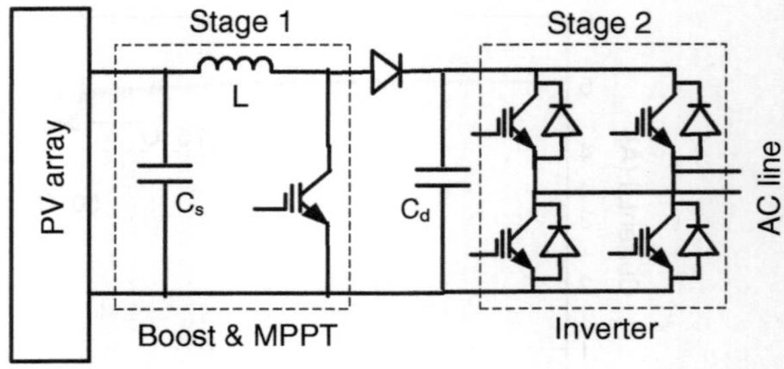

Fig. 2.19 Fundamental circuit of two-stage PV inverter [14]

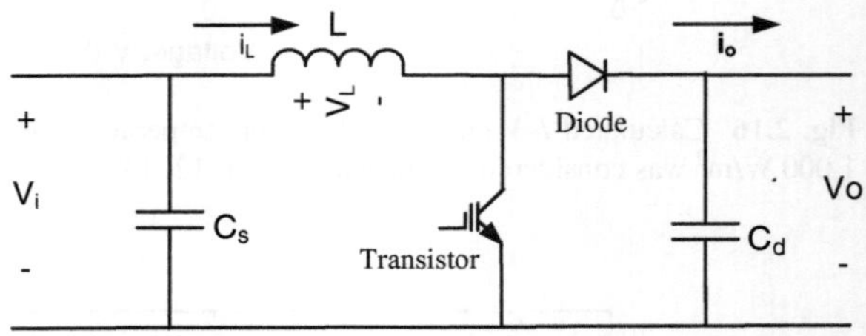

Fig. 2.20 Most commonly used DC to DC boost converter circuit

2.3 Inverters in Small-Scale Solar PV Systems

2.3.1 Two-Stage Solar PV Inverters for Small-Scale Systems

Usually, a PV inverter has two stages to shape the PV array output power for feeding into the AC load. The first stage is responsible for boosting the PV array voltage and tracking the MPP, and the second stage inverts available DC power into AC power. The most fundamental circuit topology of the two-stage PV inverter using the full-bridge as shown in Fig. 2.19 was proposed in [14].

Usually, the source-side controller tracks the maximum power and the load-side controller controls the power factor of output current and maintains constant DC voltage across capacitor C_d. When the transistor is turned on, the energy from PV array is stored in the inductor L. When transistor is turned off, the inductor stored energy is delivered to the DC-link capacitor C_d and the H-bridge inverter. The booster circuit can have two distinct modes of operation, the continuous current conduction and discontinuous current conduction, of significantly different characteristics. The booster circuit, as shown in Fig. 2.20, is also known as the step-up DC to DC converter; its typical application is to convert low input voltage to a high output voltage. When the transistor is turned on, the source voltage V_i is applied across the inductor and the rising rate of inductor current is dependent on the source voltage V_i and inductance L. Figure 2.21 shows the equivalent circuit of the DC to DC boost converter when the transistor is turned on.

When the transistor is turned off, the equivalent circuit is shown in Fig. 2.22 and the inductor voltage becomes

$$v_L = V_i - V_o \tag{2.8}$$

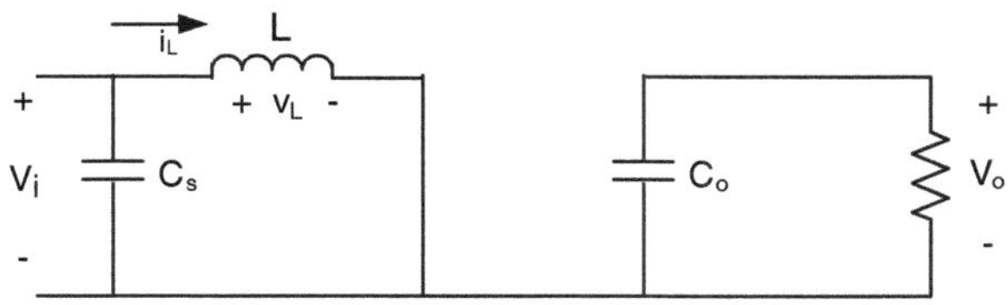

Fig. 2.21 Equivalent circuit of a DC to DC boost converter when transistor is turned on

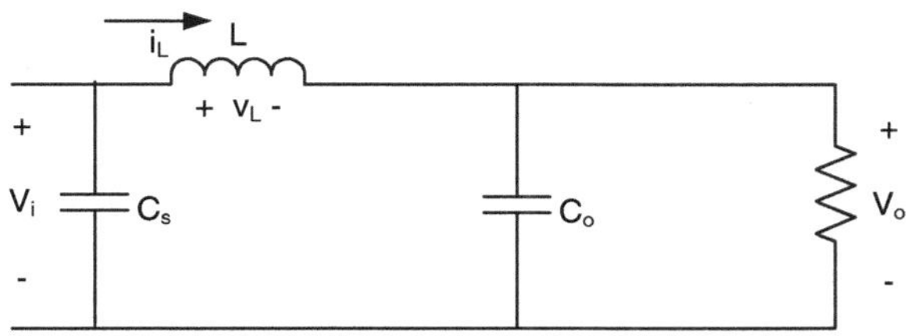

Fig. 2.22 Equivalent circuit of a DC to DC boost converter when the transistor is turned off

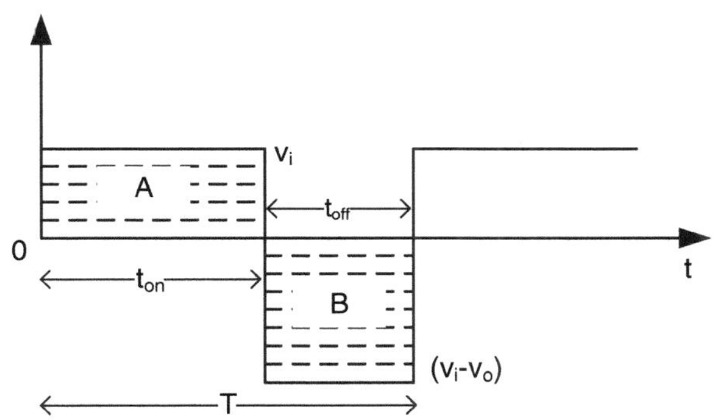

Fig. 2.23 Voltage across inductor L for time period *T*

For steady-state operation, the integral of inductor voltage v_L over one time period T must be zero, i.e., the areas A and B in Fig. 2.23 must be equal.

Therefore,

$$V_i \times t_{on} + (V_i - V_o) \times t_{off} = 0 \tag{2.9}$$

Dividing both sides by T, and rearranging all terms, we have

$$\frac{V_o}{V_i} = \frac{T}{t_{off}} = \frac{1}{1 - D} \tag{2.10}$$

Assuming the circuit is 100 % efficient, i.e., the input power (P_i) and output power (P_o) are equal, ($P_i = P_o$), or

$$V_i \times I_i = V_o \times I_o \tag{2.11}$$

one obtains

$$\frac{I_o}{I_i} = (1 - D) \tag{2.12}$$

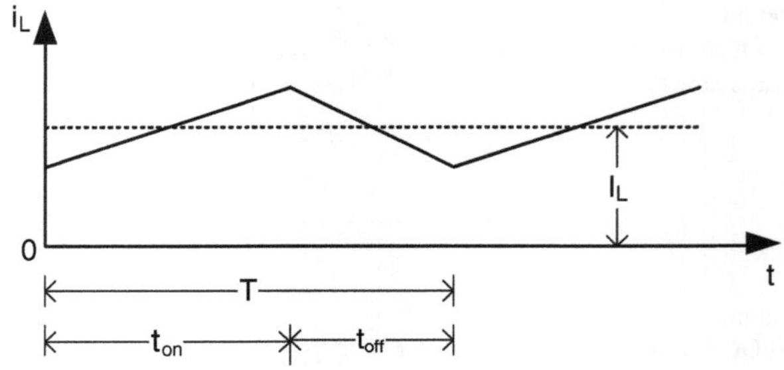

Fig. 24 Waveform of inductor current in the continuous conduction mode

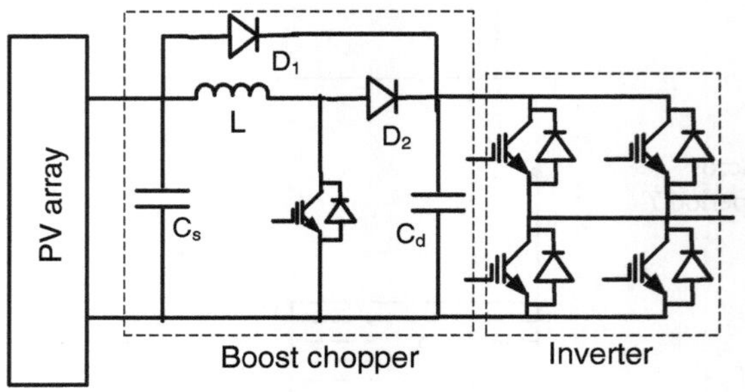

Fig. 2.25 Time-sharing chopper-based-modified circuit of two-stage PV inverter

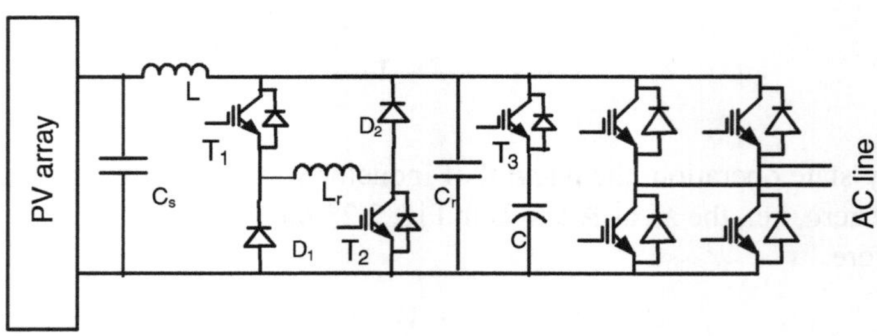

Fig. 2.26 Parallel resonant DC-link-based-modified two-stage PV inverter circuit

In the continuous operation mode, the inductor current never falls to zero in one switching cycle, i.e., either the transistor or diode is conducting. Figure 2.24 shows the inductor current in the continuous conduction mode.

Several modified topologies, such as a time-sharing dual-mode PV inverter [15] in Fig. 2.25 and a soft-switched DC to DC boost converter-based inverter [16] in Fig. 2.26, were proposed to improve the efficiency of the fundamental circuit. In the time-sharing dual-mode PV inverter, when the PV array output voltage is larger than the DC-link voltage, the transistor is always in the off state and the input current flows through the bypass diode D_1 but does not flow through the boost inductor

L and free-wheeling diode D_2. In this way, the proposed circuit can avoid the conduction losses of boost inductor, L, and the free-wheeling diode D_2. When the input voltage is smaller than the DC-link voltage, the bypass diode D_1 behaves as open circuit and the circuit operates as a fundamental two-stage PV inverter circuit. In the soft-switched DC to DC boost converter-based PV inverter, there are two stages: the converter stage and inverter stage. All the switches in both the converter and inverter stages can be turned on and off with zero voltage switching, which can reduce the switching losses significantly. The operating principle of the proposed soft switching inverter was presented with a few switching modes [16] as the following:

Mode 1 When the transistor in DC-link capacitor C is turned on with zero voltage switching, the DC-link capacitor begins to discharge and linearly decreases the main inductor L current.

Mode 2 When the transistors T_1, T_2, and T_3 are turned on with zero voltage switching, the DC-link capacitor begins to discharge.

Mode 3 When T_1 and T_2 are turned on with zero voltage switching, the resonance between resonance inductor, L_r, and resonance capacitor, C_r, is started and the main inductor current is minimized. When the resonance is finished, the current of resonant inductor, L_r, flows through diodes, D_1 and D_2.

Mode 4 When T_1 and T_2 are turned off with zero voltage switching, resonance is started. The resonant capacitor C_r is charged by the current flowing through inductors, L and L_r. The resonance stops when the voltage of resonant capacitor, C_r, equals the output voltage. The DC-link capacitor is charged by L and L_r through the parallel diode of T_3.

2.3.2 Multiple-Stages Solar PV Inverters for Small-Scale Systems

Due to the high value of stray capacitance of PV array, a galvanic connection between the ground of the grid and the PV array may exist. In this case, dangerous leakage current called common-mode current may appear through the stray capacitance between the PV array and the ground. The capacitance between PV cells and the ground can reach very high value under certain conditions, such as the PV array structure, and weather there are humidity and dust covering the PV array. This capacitance can reach up to 150 nF/kWp for crystalline-silicon cells and 1 μF/kWp for thin-film cells. The common-mode leakage current may increase the system power loss, reduce the grid current quality, and cause personnel safety problems [17, 18]. Therefore, galvanic isolation is essential because it permits easy array grounding, array isolation from grid in case of fault, and safety of personnel, and can strongly reduce the leakage current between the PV array and the ground. The connection of grid neutral and middle points of the DC link is one of the options to reduce the ground leakage currents. In this context, Rahman and Zhong [19] proposed a half-bridge topology-based PV inverter as shown in Fig. 2.27. The step-up

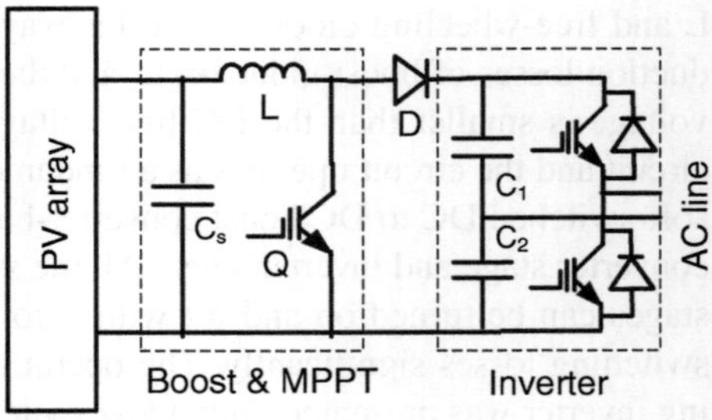

Fig. 2.27 Basic circuit of half-bridge topology-based PV inverter

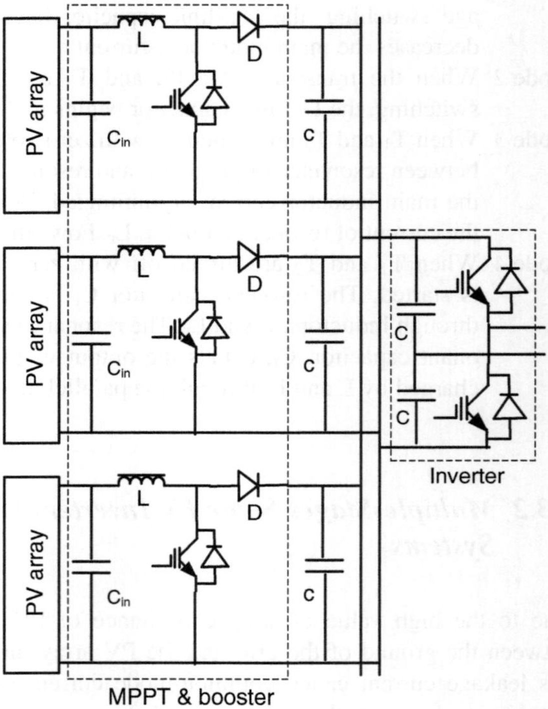

Fig. 2.28 Sunny Boy 5000TL model half-bridge topology-based PV inverter

(boost) converter raises the voltage of the PV array and also serves the operation of MPPT. Driven by the proper reference signals generated by the control algorithm to modulate the pulse width of the switching signal of switch Q, the half-bridge inverter inverts the DC power to 50 or 60 Hz AC power. Figure 2.28 illustrates the half-bridge topology-based three-string PV inverter, Sunny Boy 5000TL, commercially developed by SMA [20]. The half-bridge inverter-based topology requires less switching devices, but its DC-link voltage needs to be twice the grid voltage peak. The inverter topologies do not have electrical isolation between the array and the grid, which is critical in case of fault and safety of personnel.

The insulation can be achieved through a power frequency transformer at the grid side of the PV inverter which may also serve voltage step-up operation.

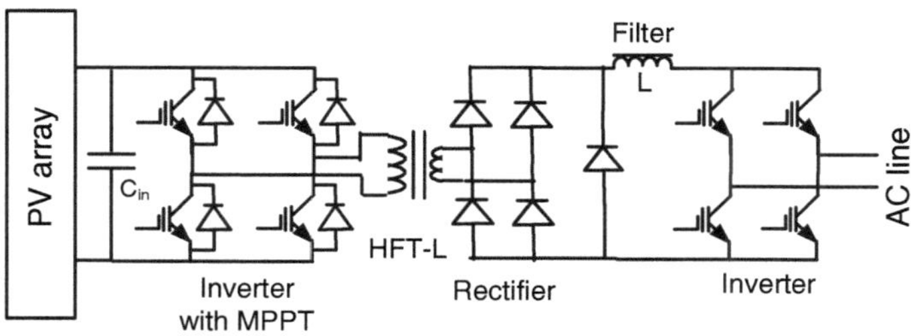

Fig. 2.29 High-frequency transformer-link (*HFT-L*)-based PV inverter power circuit

This power frequency transformer is heavy and large, increasing the PV inverter installation cost and requiring regular monitoring and maintenance. Increasing the operating frequency will lead to a compact and lightweight magnetic component of isolation transformers [4, 5, 21]. Several medium- and high-frequency (HF) transformer-based inverter topologies were developed and made available commercially. Figure 2.29 shows a high-frequency link-based inverter system [10]. The DC PV array power is converted to 50 or 60 Hz AC line power through an isolated high-frequency transformer link. The DC voltage of the PV array is firstly converted to high-frequency AC by a high-frequency inverter, which is then transformer-coupled and converted to 50 or 60 Hz AC through a high-frequency rectifier, filter, and a full-bridge inverter. The multistage power conversion may increase the cost of the converter and decrease the efficiency of the system, but it can significantly reduce the weight and volume of the power conversion system and minimize grid isolation issues. In 2010, Lu et al. [22, 23] proposed a planar high-frequency transformer-based PV inverter system as shown in Fig. 2.30. Higher output power from multiple PV arrays can be achieved by connecting each PV array to its own DC to AC converter and single-phase transformer. The primaries of the high-frequency transformer links are connected in an open delta energized by high-frequency voltage from DC to AC converters, where the converter

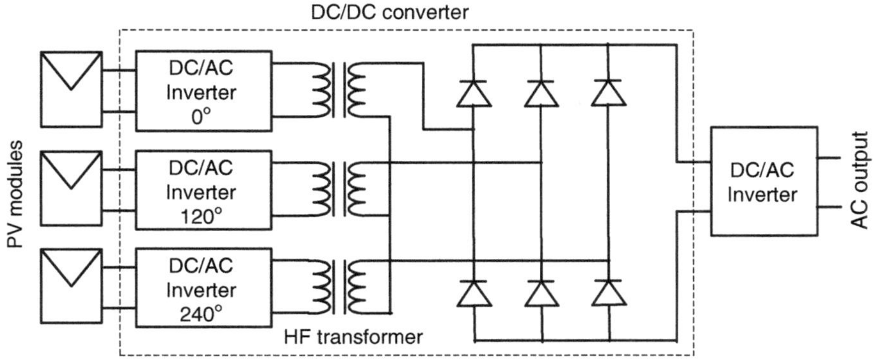

Fig. 2.30 High-frequency (*HF*) transformer-based PV inverter power circuit

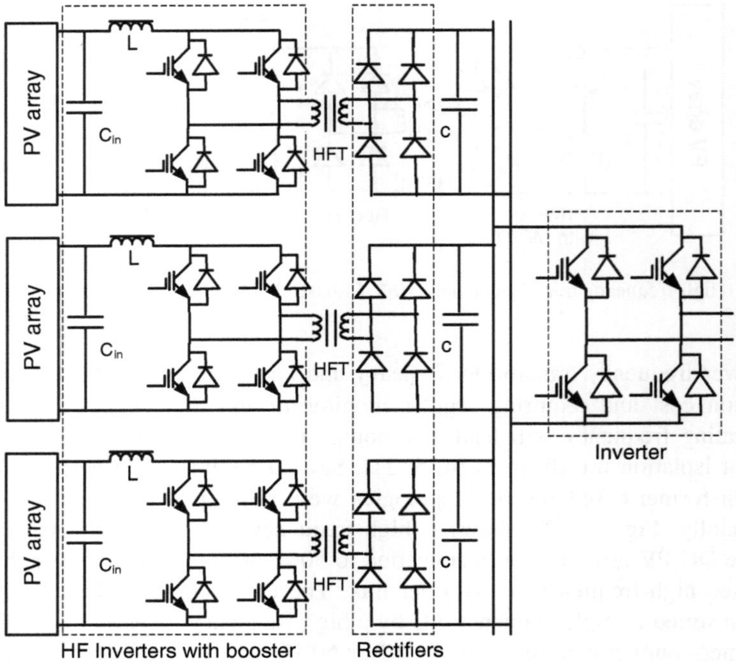

Fig. 2.31 High-frequency transformer (*HFT*)-based PowerLynx Powerlink PV 4.5 kW inverter

output voltages are phase shifted each other by 120°. The secondaries of the high-frequency transformer links are connected in a wye configuration, and the output is connected to a 50 or 60 Hz DC to AC converter through a high-frequency three-phase rectifier. The detailed design and analysis of the HF planar transformer were reported in [24, 25]. The Original Equipment Manufacturer commercially developed the high-frequency transformer-based three-string PV inverter, PowerLynx Powerlink PV 4.5 kW, as shown in Fig. 2.31 [26].

2.3.3 Single-Stage Solar PV Inverter for Small-Scale Systems

Compared to the single-stage one, the multistage power conversion is somewhat more expensive and affects the efficiency of the PV inverter. In order to reduce the volume and weight as well as the power conversion loss and cost, a hybrid PV-battery-powered DC bus system was proposed in 2009 [27]. The DC to AC conversion stage-less DC bus system is very applicable to electronic equipment and appliances with high system efficiencies. The PV-battery-powered DC bus system

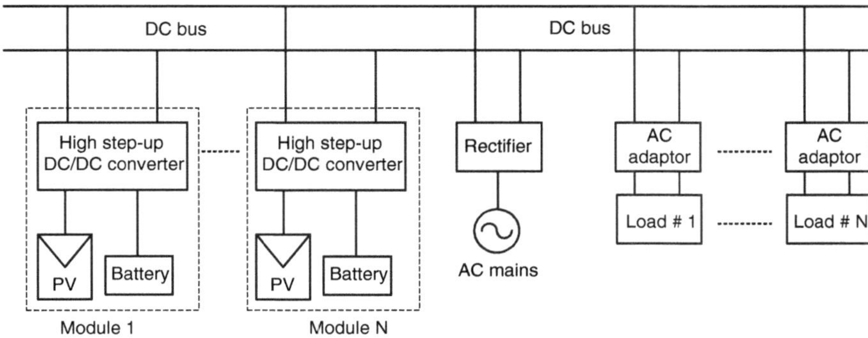

Fig. 2.32 PV-battery-powered DC bus system

Fig. 2.33 Full-bridge with MPPT-based circuit topology of single-stage inverter

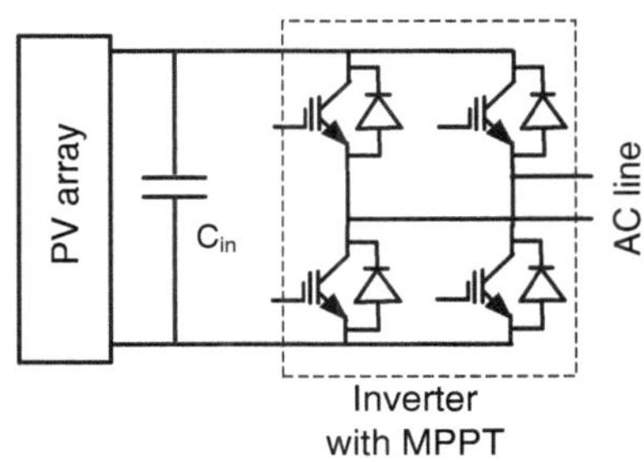

is shown in Fig. 2.32. For AC systems, a single-stage PV inverter was proposed in [28], and the circuit topology of single-stage inverter is shown in Fig. 2.33.

The proposed inverter performs a dual function: MPPT and outputting a sinusoidal current, which makes the control circuit complex. In [29], an alternative control technique was developed to reduce the complexity of the control circuit. However, the common-mode issue was not considered in the proposed single-stage inverter systems. The neutral point clamped (NPC) converter topology has the opportunity to connect the grid neutral point to middle point of the DC link, reducing the ground leakage currents. In this context, an NPC topology-based single-phase PV inverter as shown in Fig. 2.34 was presented in [30] and a three-phase PV inverter system in Fig. 2.35 was implemented in [31]. Since the presented circuits are run as buck converters, the PV array voltages should be greater than the peak values of the output AC voltages. If V_{rms} is the inverter output AC voltage and R *is* the reservation factor, the minimum array voltage can be calculated as

$$V_{\mathrm{A}} = \sqrt{2} V_{\mathrm{rms}} R. \tag{2.13}$$

Therefore, a few PV arrays in series connection are necessary to obtain the desired voltage. From the available literature, several single-stage topologies have been proposed based on either boost or buck–boost configurations. An integrated

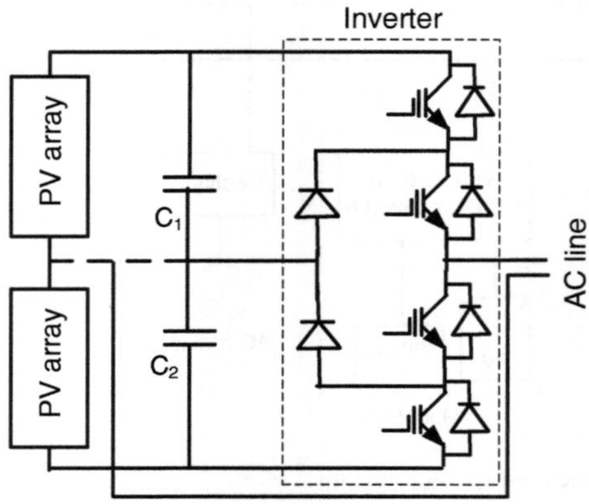

Fig. 2.34 Circuit topology of single-stage inverter: NPC with grid neutral connected to the middle point of DC link

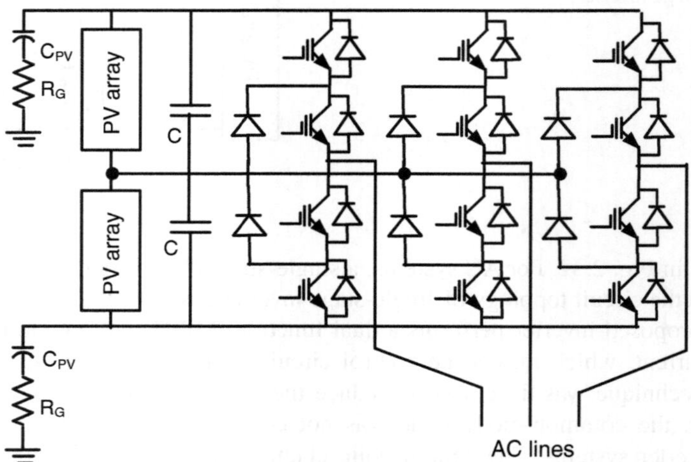

Fig. 2.35 NPC topology-based three-phase PV inverter

(boost converter and full-bridge inverter) PV inverter circuit topology shown in Fig. 2.36 was presented in [32]. The output power quality and the efficiency of the inverter are limited by the fact that the boost converter cannot generate the output voltage lower than the input voltage. A universal single-stage PV inverter shown in Fig. 2.37 was presented in [33] that can operate as a buck, boost, or buck–boost converter. This inverter can operate with a wide range of input voltage, improving the power quality and the efficiency. Using the integrated buck–boost and inversion functions, several modified configurations have been presented in [34, 35]. However, these topologies are only suitable for small-scale (e.g., <100 kW) PV systems, where the PV array normally interconnects with a low-voltage public network.

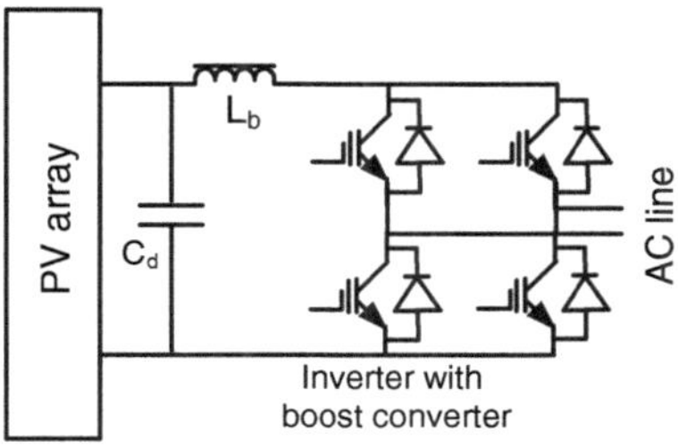

Fig. 2.36 Single-stage power circuit with boost converter

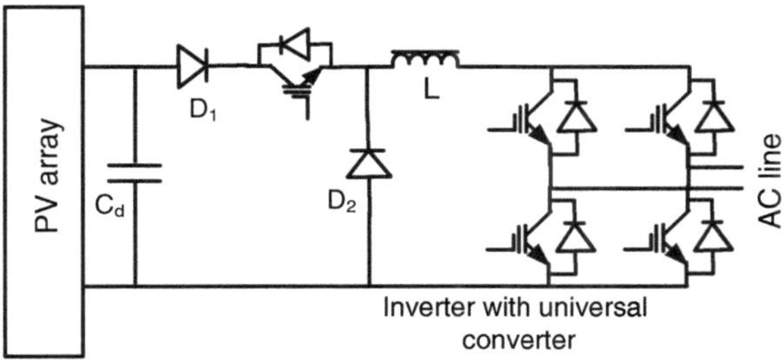

Fig. 2.37 Single-stage power circuit with universal converter

2.3.4 ABB and Siemens Solar PV Inverters for Small-Scale Systems

Although the galvanic isolation is not a requirement in Germany in the low-voltage public network, a residual current monitoring unit needs to be installed for the grid connection of transformer-less PV inverters. Most standards (e.g., IEEE 1547, EN 61000-3-2, and IEC 61727) allow the DC current injection to vary between 0.5 and 1 % of the rated current [36]. Therefore, the transformer-less inverter system should have some extra protective devices to ensure the safety issues. ABB is one of the leading manufacturers of PV inverters. In order to interconnect the PV array to the public electricity network, ABB developed the transformer-less single-phase string PV inverter model PVS300, suitable for small-scale PV systems (i.e., 3.3–8.0 kW). The PVS300 inverter converts the DC generated by PV arrays at a voltage ranging 335–800 V into single-phase 230 V AC that can be fed into the public network through protective devices as shown in Fig. 2.38 [37]. Due to the high content of harmonics in the output power and the use of high switching frequency of switching devices, the system consists of a heavy filter circuit and electromagnetic interference (EMI) filter, which may increase the losses and cost of the system.

The SINVERT PVM inverters are transformer-less, 3-phase PV inverters developed by Siemens, to convert solar DC energy at a voltage ranging in 380–850 V into 3-phase 400 V AC voltage for feeding into public electricity network. SINVERT PVM inverters are available with power outputs from 10 to 20 kW. The PV array

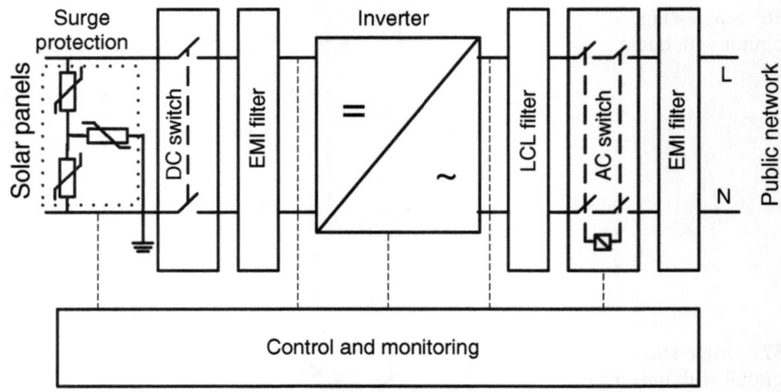

Fig. 2.38 ABB PVS300 string inverter design and grid connection

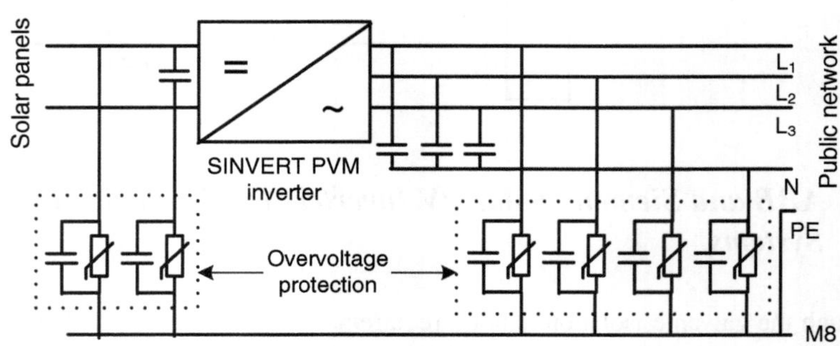

Fig. 2.39 Siemens SINVERT PVM string inverter design and grid connection

interconnection to utility grid through SINVERT inverter with safety devices is shown in Fig. 2.39 [38].

2.4 Medium- and Large-Scale Solar PV Systems

For medium- and large-scale solar PV electricity generation, there are two well-established inverter technologies: the centralized and string technologies as shown in Figs. 2.40 and 2.41, respectively.

In the centralized PV inverter technology, each string consists of a series of PV modules to reach the voltage requirement without amplification and then a few strings are parallel connected to a common inverter circuit. The number of strings mainly depends on the power levels. This technology eliminates the amplification stage, but possesses some major limitations, such as voltage mismatch loss

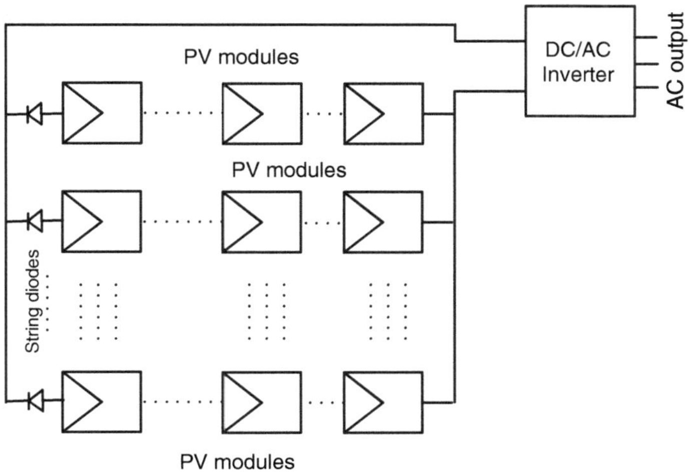

Fig. 2.40 Centralized PV inverter topology

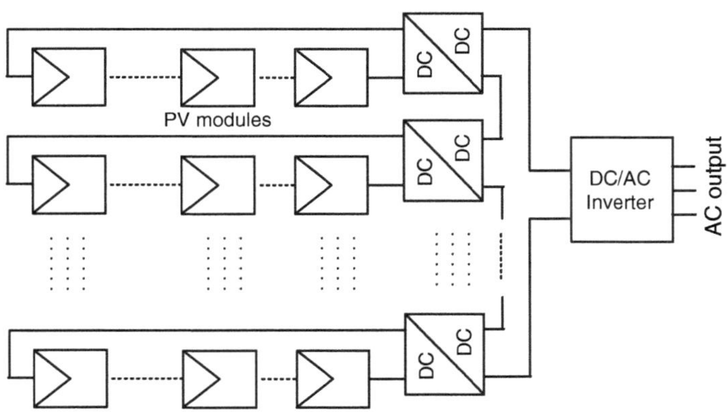

Fig. 2.41 String PV inverter topology

between PV modules and power losses due to common MPPT [26]. In order to control every string individually, the string technology uses a DC to DC converter for each string, which improves the system efficiency. In order to minimize the voltage mismatch in the strings, a battery-integrated boost converter was proposed to eliminate the voltage regulation stage [39]. The block diagram of battery-integrated boost converter is shown in Fig. 2.42, and the proposed converter-based PV inverter system is shown in Fig. 2.43. The detailed analysis of the proposed converter is presented in [40].

The ABB central inverters are especially designed for medium-scale PV power plants. The PVS800 version is a 3-phase inverter with a power capacity in the

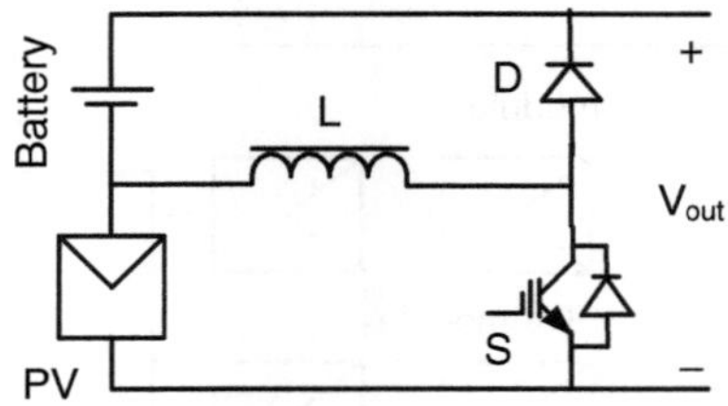

Fig. 2.42 Battery-integrated boost converter

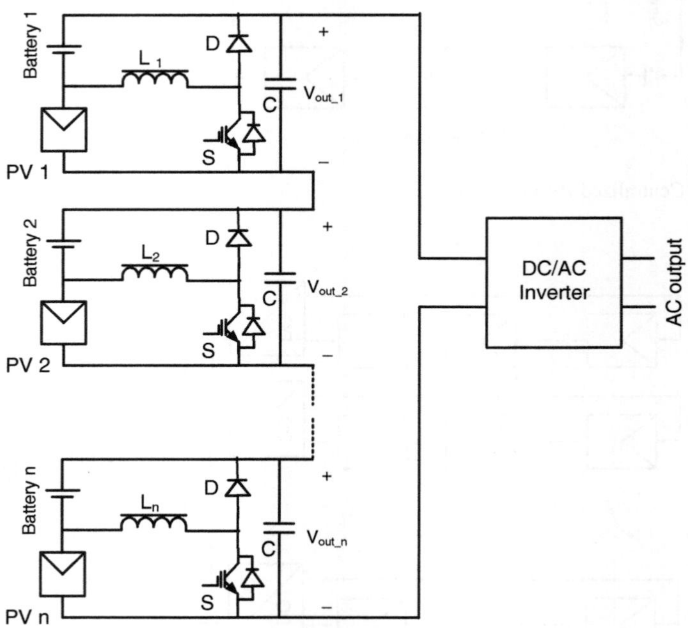

Fig. 2.43 Battery-integrated boost converter-based PV inverter system

range of 100–500 kW. The PVS800 inverter topology allows a parallel connection directly on the AC side, for grid connection through a step-up power frequency transformer [41]. The transformer steps up the inverter output voltage from 300 V AC to grid voltage level (e.g., 6–36 kV). The central inverter design and grid connection are depicted in Fig. 2.44. ABB has been delivering worldwide vacuum cast coil dry-type transformers for PV applications. The cast coil dry-type transformers are non-flammable and moisture proof. They feature a solid isolation system which is discharge-free. However, the volume and weight of a 0.4/36 kVA, 1 MVA vacuum cast coil transformer are about 4.3 m^3 and 3,250 kg, respectively, and no-load and full-load losses are 3.1 and 11.5 kW, respectively [42]. Moreover, dry-type transformers can be sensitive to water, microcracks, temperature variations,

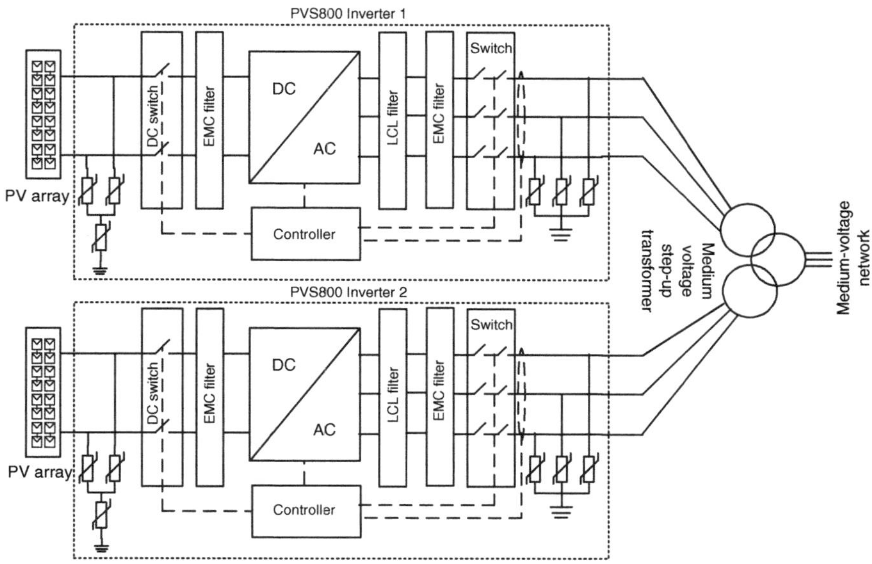

Fig. 2.44 ABB central inverter design and medium-voltage (*MV*) grid connection

and pollution which can block cooling ducts. Due to the use of traditional two-level inverter, the harmonic content in the output power is high. Usually, high-frequency pulse-width-modulated (PWM) gate signals are used to drive switching devices to reduce the harmonic content. Due to high-harmonic content and high-frequency stitching, the system requires a heavy LCL filter circuit and an electromagnetic compatibility (EMC) filter, which may increase the losses and cost of the system.

Besides its low-voltage system, Siemens also developed the SINVERT PVS inverter-based system for medium-scale PV plants. The AC output voltage and power capacity of PVS version inverters are in the range of 288–370 V and 500–630 kW, respectively, as summarized in Table 2.2. The 1–2.52 MW central inverters were designed by paralleling 2–4 PVS inverters through transformer and switchgear at the grid side. The design and grid connection of 2-inverters-based system is illustrated in Fig. 2.45 [43]. Siemens developed GEAFOL cast–resin transformers for grid connection of PV arrays. With GEAFOL, it is possible to avoid the limitations associated with liquid-filled transformers while retaining the

Table 2.2 Capacity of PVS version inverters

Versions	PVS500	PVS585	PVS600	PVS630
AC output voltage (V)	288	340	370	370
Power capacity (kW)	500	585	600	630
Min. input voltage (V)	450	530	570	570

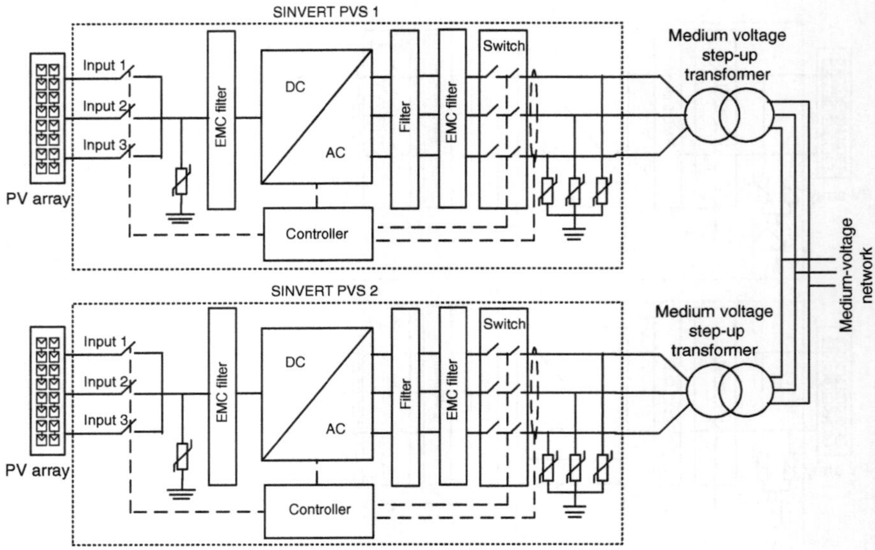

Fig. 2.45 Siemens central inverter design and medium-voltage grid connection

properties such as operational safety, resistance to humidity, mechanical strength, and compact design. However, no-load and full-load losses of a 0.4/30 kV, 1 MVA GEAFOL transformer are about 3.1 and 10 kW, respectively, and the volume and weight are about 3.5 m^3 and 2,990 kg, respectively [44].

Although these special transformers are compact compared with conventional distribution transformers, they are still large and heavy for PV applications. The large-size and heavy-step-up transformer may increase the system weight and volume and can be expensive and complex for installation and maintenance. The medium-voltage inverter may be the best possible solution to interconnect the PV array to the medium-voltage grid directly. Moreover, electrical isolation is important for the interconnection of the medium-voltage grid and PV array. Since the installation of large-scale PV power plants started commercially in 2007, the medium-voltage inverter for interconnection of PV systems has been attracting great attention.

In 2011, different MV multilevel inverter topologies were compared for the possible medium-voltage grid connection of wind turbine and PV systems [45, 46]. The modular multilevel cascaded (MMC) topology was considered as a possible candidate for medium-voltage inverter systems. The MMC inverter requires multiple isolated and balanced DC sources. In [47], a medium or HF link was proposed to generate multiple isolated and balanced DC sources for the MMC from a single source and in [48], a medium-frequency transformer link was developed to verify the feasibility of the new concept of voltage step-up using the MMC inverters. Compared with the conventional transformers operated at the power frequency (50

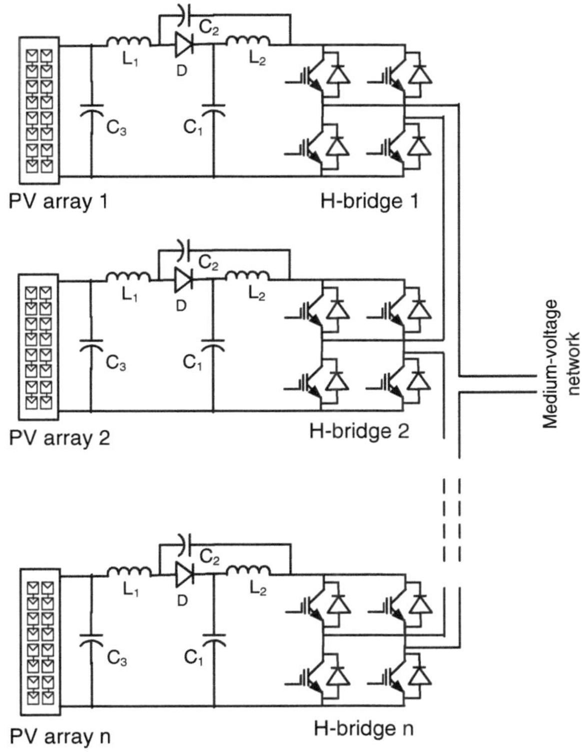

Fig. 2.46 Quasi-Z source converter-based medium-voltage PV inverter (one of 3 phases)

or 60 Hz), the medium-frequency transformer link has much smaller and lighter magnetic cores and windings and thus much lower costs.

In 2012, by combining a quasi-Z source inverter into an MMC as shown in Fig. 2.46, a medium-voltage PV inverter was proposed in [49].

The proposed inverter does not have isolation between PV array and medium-voltage grid. Multiphase isolated DC to DC converter-based MMC inverter topology as shown in Fig. 2.47 was proposed in [50, 51]. In the proposed configuration, the voltage balancing is the challenging issue, since each H-bridge cell is connected to a PV array through a DC to DC converter. A common DC link may be one of the possible solutions to minimize the voltage imbalance problem, and a single DC-link-based inverter in Fig. 2.48 was presented in [52, 53]. Although this design may reduce the voltage balancing problem in the grid side, the generation of common DC-link voltage from different PV arrays makes the inverter operation complex and limits the range of MPPT operation.

As an alternative approach to minimize the voltage imbalance problem with a wide range of MPPT operation, a common magnetic link was proposed [3]. The boost converter is considered for the MPPT operation. The array DC power is converted to a high-frequency AC through a high-frequency inverter. The inverter also

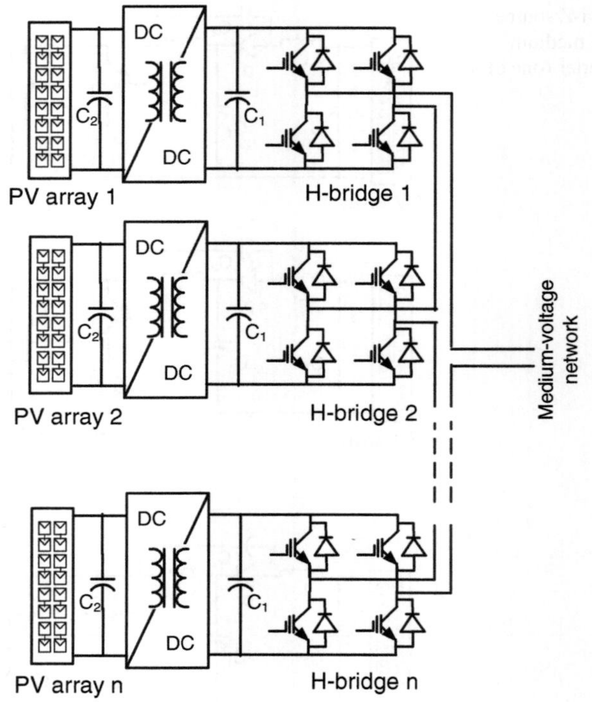

Fig. 2.47 Multiphase isolated converter-based medium-voltage PV inverter (one of 3 phases)

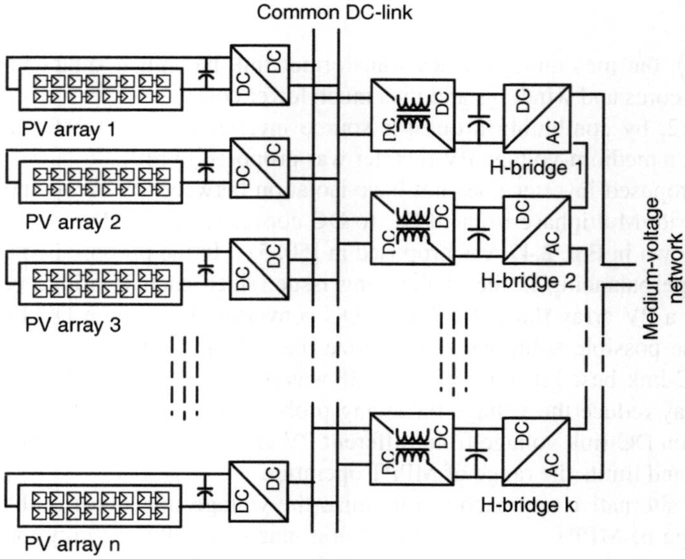

Fig. 2.48 Common DC-link-based medium-voltage PV inverter (one of 3 phases)

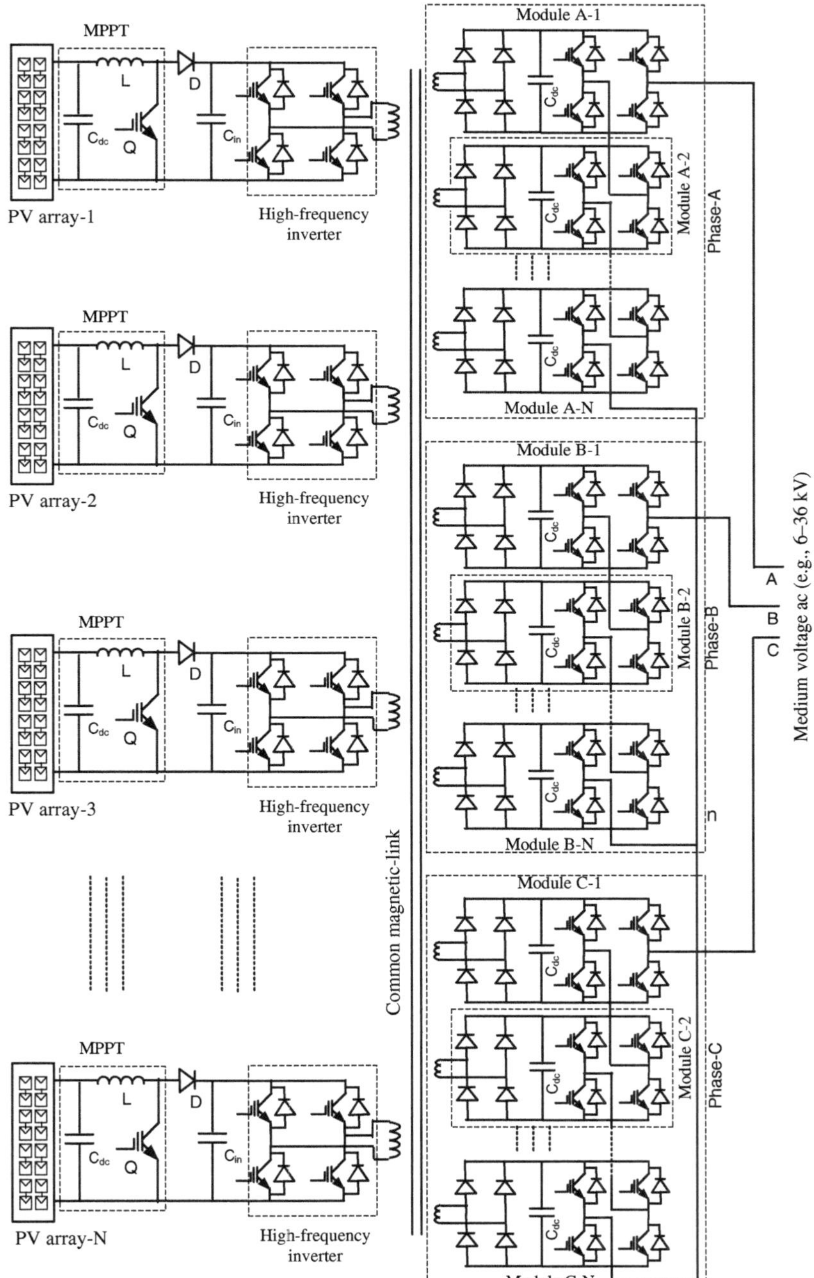

Fig. 2.49 Common magnetic-link-based 3-phase medium-voltage PV inverter

ensures constant output voltage. The inverter is connected to a primary winding of a multiwinding high-frequency magnetic link. Each secondary winding works as an isolated source and is connected to an H-bridge cell through a bridge rectifier. The number of primary windings depends on the number of PV arrays, and the number of secondary windings depends on number of levels of the inverter. The detailed power circuit of common magnetic-link-based PV inverter system is shown in Fig. 2.49 [3]. In medium- or large-scale PV power plants, several PV arrays are operated in parallel. The multiple-input multiple-output magnetic link can incorporate the parallel operation of multiple PV arrays, where each PV array is connected to a primary winding through a booster and high-frequency inverter [54]. The magnetic link also provides electrical isolation between the PV array and the grid, which can thus inherently overcome the common-mode and voltage imbalance problems and ensure a wide range of MPPT operation and safety of operating personnel.

2.5 Summary

Besides the traditional systems, which require a step-up transformer to connect the renewable generation systems to the grid, many other recently proposed converter topologies for step-up-transformer-less direct grid interconnection are also reviewed in detail with the aim of presenting a complete picture of power converter systems. In order to develop a common converter for both the wind and PV power generation systems, appropriate converter topologies should be investigated. In this chapter, most of the existing power converters are reviewed. In order to reduce the system volume and weight as well as to improve the efficiency, a number of transformer-less single-stage inverter topologies have been proposed in the past decades for small-scale PV power generation systems. The common-mode or leakage current and personal safety are really critical issues with these inverter systems. Most of the commercially available system uses the traditional two-level inverter in which the switching devices are driven by high-frequency PWM gate signals. Due to high-harmonic contents and high-frequency switching, the system requires a heavy LCL filter circuit and an EMC filter, which may increase the losses and cost of the system.

Today, the industrial trend is to move away from these heavy and large-size passive components to compact and lightweight systems that use more and more semiconductor devices in modular construction. It is found that multilevel converters with medium-frequency links would be a feasible option to develop a MV converter for the direct grid integration of PV power plants. Although different multilevel converter topologies have been developed in the last few decades, most of them are not suitable for medium-voltage applications. Because of some special features, such as the number of components scaling linearly with the number of levels, and the identical individual modules of completely modular construction enabling high-level number attainability, the MMC converter topology is considered as a highly feasible candidate for medium-voltage applications. The MMC converter requires multiple isolated and balanced DC sources. The medium-frequency magnetic link may be the best option to generate multiple isolated and balanced DC sources for the MMC inverter from a single power source. Compared with the power frequency transformers, the medium-frequency links have much smaller and lighter magnetic cores and windings and thus much lower costs.

References

1. Earth policy institute Climate, energy and transportation, world cumulative wind turbine installations [online]. Available at: http://www.earth-policy.org. Accessed on 25 Mar 2013
2. Islam MR, Guo YG, Zhu JG (2012) 11-kV series-connected H-bridge multilevel converter for direct grid connection of renewable energy systems. J Int Conf Elec Mach Syst 1(2):211–219
3. Islam MR, Guo YG, Zhu JG (2014) A multilevel medium-voltage inverter for step-up-transformer-less grid connection of photovoltaic power plants. IEEE J Photovoltaics 4(3):881–889
4. Islam MR, Guo YG, Zhu JG (2014) A high-frequency link multilevel cascaded medium-voltage converter for direct grid integration of renewable energy systems. IEEE Trans Power Electron 29(8):4167–4182
5. Islam MR, Guo YG, Lin ZW, Zhu JG (2014) An amorphous alloy core medium frequency magnetic-link for medium voltage photovoltaic inverters. J Appl Phys 115(17):17E710-1–17E710-3
6. Sharma VK, Colangelo A, Spagna G (1995) Photovoltaic technology: basic concepts, sizing of a standalone photovoltaic system for domestic applications and preliminary economic analysis. Energy Convers Manage 36(3):161–174
7. Santos JL, Antunes F, Chehab A, Cruz C (2006) A maximum power point tracker for PV systems using a high performance boost converter. Sol Energy 80(7):772–778
8. Hua C, Shen C (1998) Study of maximum power tracking techniques and control of DC/DC converters for photovoltaic power system. In: 29th annual IEEE power electronics specialists conference, Fukuoka, Japan, 17–22 May 1998, pp 86–93
9. Jaboori MG, Saied MM, Hanafy AAR (1991) A contribution to the simulation and design optimization of photovoltaic systems. IEEE Trans Energy Conversion 6(3):401–406
10. Bose BK, Szczesny PM, Steigerwald RL (1985) Microcomputer control of a residential photovoltaic power conditioning system. IEEE Trans Ind Appl 21(5):1182–1191
11. Islam MR, Islam MR, Beg MRA (2008) Renewable energy resources and technologies practice in Bangladesh. Renew Sustain Energy Rev 12(2):299–343
12. Islam MR, Guo YG, Zhu JG, Jin J (2011) Design and fabrication of a microcontroller based maximum power point tracker for renewable energy systems. J Appl Supercond Electromagnet 2(1):17–23
13. Islam MR, Guo YG, Zhu JG, Rabbani MG (2010) Simulation of PV array characteristics and fabrication of microcontroller based MPPT. In: 6th international conference on electrical and computer engineering (ICECE), Dhaka, Bangladesh, 18–20 Dec, pp 155–158
14. Sugimoto H, Dong H (1997) A new scheme for maximum photovoltaic power tracking controls. In: Power conversion conference, Nagaoka, Japan, 3–6 Aug 1997, pp 691–696
15. Ogura K, Nishida T, Hiraki E, Nakaoka M, Nagai S (2004) Time-sharing boost chopper cascaded dual mode single-phase sinewave inverter for solar photovoltaic power generation system. In: IEEE 35th annual power electronics specialists conference, Aachen, Germany, 20–25 June 2004, pp 4763–4767
16. Young-Ho K, Jun-Gu K, Young-Hyok J, Chung-Yuen W, Yong-Chae J (2010) Photovoltaic parallel resonant DC-link soft switching inverter using hysteresis current control. In:

Twenty-fifth annual IEEE applied power electronics conference and exposition, Palm Springs, Canada, 21–25 Feb 2010, pp 2275–2280
17. Nema S, Nema RK, Agnihotri G (2011) Inverter topologies and control structure in photovoltaic applications: a review. J Renew Sustain Energy 3(1):012701-1–012701-23
18. Patrao I, Figueres E, Espin FG, Garcera G (2011) Transformerless topologies for grid-connected single-phase photovoltaic inverters. Renew Sustain Energy Rev 15(7):3423–3431
19. Rahman MF, Zhong L (1997) A new, transformerless, photovoltaic array to utility grid interconnection. In: 1997 international conference on power electronics and drive systems, Singapore, 26–29 May 1997, pp 139–143
20. SMA Solar Technology AG. Sunny Boy 5000TL multi-string inverter [online]. Available at: http://www.sma.de. Accessed on 25 Feb 2013
21. Agheb E, Bahmani MA, Hoidalen HK, Thiringer T (2012) Core loss behavior in high frequency high power transformers-II: arbitrary excitation. J Renew Sustain Energy 4(3):033113-1–033113-11
22. Lu J, Stegen S, Butler D (2010) High frequency and high power density transformers for DC/DC converter used in solar PV system. In: 2010 2nd IEEE international symposium on power electronics for distributed generation systems, Hefei, China, 16–18 June 2010, pp 481–484
23. Lu J, Butler D (2011) High frequency transformers for DC/DC converter used in solar PV system. J Energy Power Eng 5:536–541
24. Lu J, Dawson F (2011) Characteristics for high frequency planar transformer with a novel comb-shaped shield. IEEE Trans Magn 47(10):4493–4496
25. Panchal C, Lu J (2011) High frequency planar transformer for universal contact-less battery charging platform. IEEE Trans Magn 47(10):2764–2767
26. Kjaer SB, Pedersen JK, Blaabjerg F (2005) A review of single-phase grid-connected inverters for photovoltaic modules. IEEE Trans Ind Appl 41(5):1292–1306
27. Lu DDC, Agelidis VG (2009) Photovoltaic-battery-powered DC bus system for common portable electronic devices. IEEE Trans Power Electron 24(3):849–855
28. Liang TJ, Kuo YC, Chen JF (2001) Single-stage photovoltaic energy conversion system. IEE Proc Electric Power Appl 148(4):339–344
29. Chen Y, Smedley KM (2004) A cost-effective single-stage inverter with maximum power point tracking. IEEE Trans Power Electron 19(5):1289–1294
30. Gonzalez R, Gubia E, Lopez J, Marroyo L (2008) Transformerless single-phase multilevel-based photovoltaic inverter. IEEE Trans Ind Electron 55(7):2694–2702
31. Cavalcanti MC, Farias AM, Olivetra KC, Neves FAS, Afonso J (2012) Eliminating leakage currents in neutral point clamped inverters for photovoltaic systems. IEEE Trans Ind Electron 59(1):435–443
32. Junior LG, Brito MAGD, Sampaio LP, Canesin CA (2011) Single stage converters for low power stand-alone and grid-connected PV systems. In: 2011 IEEE international symposium on industrial electronics, Gdansk, Poland, 27–30 June 2011, pp 1112–1117
33. Prasad BS, Jain S, Agarwal V (2008) Universal single-stage grid-connected inverter. IEEE Trans Energy Conversion 23(1):128–137
34. Wang C (2004) A novel single-stage full-bridge buck-boost inverter. IEEE Trans Power Electron 19(1):150–159
35. Patel H, Agarwal V (2009) A single-stage single-phase transformer-less doubly grounded grid-connected PV interface. IEEE Trans Energy Conversion 24(1):93–101
36. Meneses D, Blaabjerg F, García O, Cobos JA (2013) Review and comparison of step-up transformerless topologies for photovoltaic AC-module application. IEEE Trans Power Electron 28(6):2649–2663
37. ASEA Brown Boveri (ABB). Solar photovoltaic string inverter [online]. Available at: http://www.abb.com. Accessed on 13 May 2013
38. SIEMENS. SINVERT PVM operating instruction 07/2010 [online]. Available at: http://www.siemens.com/. Accessed on 13 April 2014
39. Du Y, Lu DDC (2010) Analysis of a battery-integrated boost converter for module-based series connected photovoltaic system. In: 2010 international power electronics conference, Sapporo, Japan, 21–24 June 2010, pp 694–698

40. Du Y, Lu DDC (2011) Battery-integrated boost converter utilizing distributed MPPT configuration for photovoltaic systems. Sol Energy 85(9):1992–2002
41. ASEA Brown Boveri (ABB). Solar photovoltaic central inverter [online]. Available at: http://www.abb.com. Accessed on 10 Mar 2014)
42. Islam MR, Guo YG, Zhu JG (2014) A review of offshore wind turbine nacelle: technical challenges, and research and developmental trends. Renew Sustain Energy Rev 33:161–176
43. SIEMENS. SINVERT PVS solar photovoltaic inverter [online]. Available at: http://www.siemens.com/. Accessed on 13 Feb 2014)
44. ASEA Brown Boveri (ABB). GEAFOL transformer [online]. Available at: http://www.abb.com. Accessed on 02 July 2013)
45. Islam MR, Guo YG, Zhu JG (2011) Performance and cost comparison of NPC, FC and SCHB multilevel converter topologies for high-voltage applications. In: 2011 international conference on electrical machines and systems, Beijing, China, 20–23 Aug 2011, pp 1–6
46. Islam MR, Guo YG, Zhu JG, Dorrell (2011) Design and comparison of 11 kV multilevel voltage source converters for local grid based renewable energy systems. In: Proceedings of the 37th annual conference on IEEE industrial electronics society, Melbourne, Australia,7–10 Nov 2011, pp 3596–3601
47. Islam MR, Guo YG, Zhu JG (2011) H-bridge multilevel voltage source converter for direct grid connection of renewable energy systems. In: 2011 IEEE PES innovative smart grid technologies Asia (ISGT), Perth, Australia, 13–16 Nov 2011, pp 1–7
48. Islam MR, Guo YG, Zhu JG (2013) A medium-frequency transformer with multiple secondary windings for medium-voltage converter based wind turbine generating systems. J Appl Phys 113(17):17A324-1–17A324-3
49. Sun D, Ge B, Peng FZ, Haitham AR, Bi D, Liu Y (2012) A new grid-connected PV system based on cascaded H-bridge quasi-Z source inverter. In: 2012 IEEE international symposium on industrial electronics, Hangzhou, China, 28–31 May 2012, pp 951–956
50. Choi H, Zhao W, Ciobotaru M, Agelidis VG (2012) Large-scale PV system based on the multiphase isolated DC/DC converter. In: 3rd IEEE international symposium on power electronics for distributed generation systems, Aalborg, Denmark, 25–28 June 2012, pp 801–807
51. Zhao W, Choi H, Konstantinou G, Ciobotaru M, Agelidis VG (2012) Cascaded H-bridge multilevel converter for large-scale PV grid-integration with isolated DC-DC stage. In: 3rd IEEE international symposium on power electronics for distributed generation systems, Aalborg, Denmark, 25–28 June 2012, pp 849–856
52. Kouro S, Fuentes C, Perez M, Rodriguez J (2012) Single DC-link cascaded H-bridge multilevel multistring photovoltaic energy conversion system with inherent balanced operation. In: 38th annual conference on IEEE industrial electronics society, Montreal, QC, Canada, 25–28 Oct 2012, pp 4998–5005
53. Rivera S, Wu B, Kouro S, Wang H, Zhang D (2012) Cascaded H-bridge multilevel converter topology and three-phase balance control for large scale photovoltaic systems. In: 3rd IEEE international symposium on power electronics for distributed generation systems, Aalborg, Denmark, 25–28 June 2012, pp 690–697
54. Islam MR, Lei G, Guo YG, Zhu JG (2014) Optimal design of high-frequency magnetic-links for power converters used in grid connected renewable energy systems. IEEE Trans Magn. doi:10.1109/TMAG.2014.2329939

Chapter 3
Power Converter Topologies for Grid-Integrated Medium-Voltage Applications

Abstract Although several converter topologies have been used in low-voltage applications, most of the topologies are not suitable in medium-voltage applications. The selection of converter topology in medium-voltage applications is really a critical problem and highly affects the converter performance and cost. The main aim of this chapter was to find out a suitable converter topology, which can interconnect the renewable generation units directly to the medium-voltage grid with mature semiconductor devices. Different multilevel converter topologies, such as neutral point clamped (NPC), flying capacitor (FC), and modular multilevel cascaded (MMC), have been considered and compared for the design of an 11 kV converter system. The comparison is made in terms of the number of semiconductors, semiconductor cost and availability, total harmonic distortions (THDs), filter size, and control complexity of the converters. The performance is analyzed and compared in the MATLAB/Simulink environment. To ensure quality performance, a level-shifted carrier-based switching scheme is used for the NPC topologies and a phase-shifted carrier-based switching scheme is used for the FC and MMC converter topologies with a carrier frequency of 1–2 kHz and modulation index of 0.8–0.9.

Keywords Power converters · Neutral point clamped (NPC) · Modular multilevel cascaded (MMC) · Flying capacitor (FC) · Topology selection · Medium-voltage · Two-level · Switching and control · MATLAB/Simulink

3.1 Introduction

In terms of technology development, a continuous race to develop higher voltage and higher current power semiconductors for utilization in high-power systems still goes on. Many recent generations of devices are suitable for low-voltage applications while medium-voltage semiconductors are still under development. The highest voltage rating of the commercially available insulated gate bipolar transistor (IGBT) is 6.5 kV, which is suitable for 2.88 kV or lower-voltage

M.R. Islam et al., *Power Converters for Medium Voltage Networks*, Green Energy and Technology, DOI 10.1007/978-3-662-44529-7_3

converter system with traditional 2-level converter topology. In order to interconnect the 2-level converter-based renewable generation system to medium-voltage grid, a power frequency step-up transformer is commonly used [1, 2]. Due to the high percentage of harmonic distortion in the output-generated power, the 2-level converter-based system requires a large-size line filter. The use of high-voltage-rated devices may increase the switching and conduction losses of the 2-level converter. Moreover, the price of power semiconductor devices increases rapidly with their voltage as well as power ratings. Therefore, the traditional 2-level converter topologies are not suitable for medium- and high-voltage networks.

The cascade connection of lower-rated semiconductors may be a cost-effective solution for medium-voltage applications. The multilevel converter topologies represent the simplest and most efficient way to connect the semiconductor devices in cascade form. The cascade connection gives the ability to use low-voltage-rated matured and low-cost semiconductor devices. Medium- and high-voltage high-power converters can be designed by using available semiconductor devices with the multilevel converter topology. In comparison with conventional 2-level converters, multilevel converters present lower switching losses, lower-voltage stress on switching devices, and better harmonic performance. These remarkable features allow the direct connection of renewable energy systems to the grid without using large, heavy, and costly power transformers. In addition to this, they serve to minimize the input and output filter requirements. Although several multilevel converter topologies have been used in low-voltage applications, most of them are not suitable in medium-voltage networks as the number of auxiliary components scales quadratically with the number of levels. Because of their special features (e.g., the number of components scales linearly with the number of levels, and individual modules are identical and completely modular in construction hence enabling high-level number attainability), the modular multilevel cascaded (MMC) converter topology can be considered as a possible candidate for medium-voltage applications [3]. Several power semiconductor vendors such as Semikron, ABB, IXYS, and Mitsubishi Electric produce switching devices in modular forms, and all the devices of an H-bridge inverter in a single pack with a gate drive circuit, which reduces not only the complexity and overall size of the converter, but also the converter developmental time and semiconductor costs. The modularity also provides a means to modify or rectify the converter circuit easily without changing other sections of the converter.

The main aim of this chapter was to find out a suitable converter topology, which can interconnect the renewable generation units directly (without using the traditional step-up transformer and line filter) to the medium-voltage grid with low-voltage low-cost matured semiconductor devices. Besides the multilevel converter topologies, the traditional 2-level converter topologies have also been considered with their switching schemes, power handling capacity, and performance. Different multilevel converter topologies, such as neutral point clamped (NPC), flying capacitor (FC), and MMC, have been considered for the design of an 11 kV converter system. The comparison is made in terms of the number of semiconductors, semiconductor cost and availability, total harmonic distortions (THDs), filter size, and control complexity of the converters. The performance is analyzed and compared in the MATLAB/Simulink environment. To ensure quality performance, a level-shifted carrier-based switching scheme is used for the NPC topologies and a phase-shifted carrier-based switching scheme is used for the FC and MMC converter topologies with a carrier frequency of 1–2 kHz and modulation index of 0.8–0.9. The modeling of circuit configurations, modulation schemes, and switching techniques in MATLAB/Simulink are presented in detail.

3.2 Two-Level Converters

3.2.1 Single-Phase Converter

3.2.1.1 Half-Bridge Converter

Figure 3.1 illustrates the circuit diagram of half-bridge converter consisting of two switching devices, e.g., insulated gate bipolar device (IGBT) and a three-wire DC power supply. The switching devices can be controlled in a smart way for progressively wider pulses at the positive peak and narrower pulses at the negative peak, which may help to filter out high-frequency switching harmonics. The unipolar pulse width modulation (PWM) scheme with a sinusoidal reference and a triangular carrier can be used to operate switching devices. The comparator of the modulation scheme compares the reference signal with the carrier signal and generates Boolean output of 0 or 1. This generated gate pulse may operate the top IGBT of the half-bridge converter. The bottom IGBT can be operated by the inverted gate pulse of the Boolean output. Figure 3.2 shows the modulation scheme of half-bridge converter. The freewheeling diodes permit current flow even if all switches are open and lagging currents flow in case of inductive loads.

The two switching devices of a half-bridge converter should not be on at the same time, to avoid the direct short circuit between the terminals of DC power supply. The peak load voltage, V_L, of half-bridge converter can be expressed as

$$V_L = m_i \frac{V_s}{2} \tag{3.1}$$

where m_i is the modulation index and V_s the DC power supply voltage. Table 3.1 summarizes the operation of half-bridge converter circuit. The output voltage waveform is shown in Fig. 3.3.

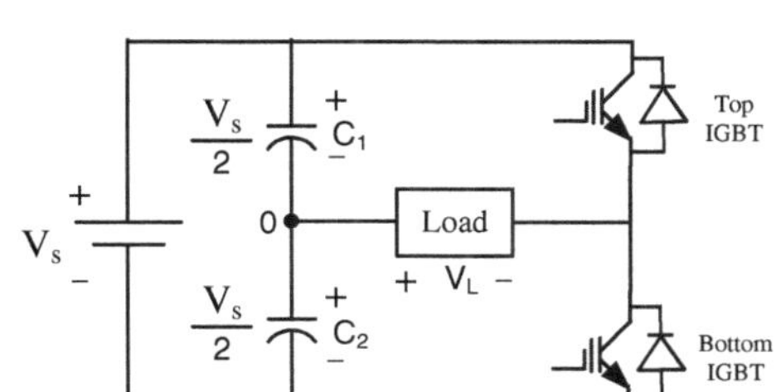

Fig. 3.1 Circuit diagram of half-bridge voltage source converter

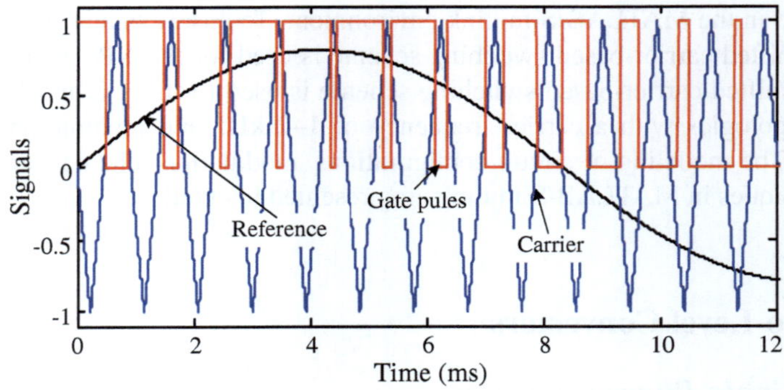

Fig. 3.2 Modulation scheme of half-bridge PWM converter

Table 3.1 Switching scheme of half-bridge converter

Top IGBT	Bottom IGBT	Source voltage	Load voltage
On	Off	V_s	$V_L = +m_i \frac{V_s}{2}$
Off	On		$V_L = -m_i \frac{V_s}{2}$
Off	Off		$V_L = 0$
On	On		Must be avoided

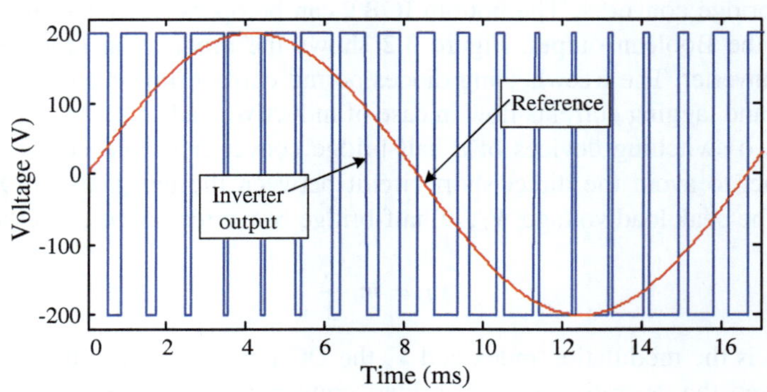

Fig. 3.3 Output voltage waveform of half-bridge PWM converter

If T is the time period and T_p the time when the top switch is in the on state during a cycle, the rms output voltage, $V_{L(\text{rms})}$, of the half-bridge PWM converter can be deduced as

$$
\begin{aligned}
V_{L(\text{rms})} &= \sqrt{\frac{1}{T}\int_0^T V_L^2 \mathrm{d}t} \\
&= \sqrt{\frac{1}{T}\left(\int_0^{T_p} m_i^2 \frac{V_s^2}{4}\mathrm{d}t + \int_{T_p}^{T} m_i^2 \frac{V_s^2}{4}\mathrm{d}t\right)} \\
&= \sqrt{\frac{m_i^2 V_s^2}{4T}\left(\int_0^{T_p} \mathrm{d}t + \int_{T_p}^{T} \mathrm{d}t\right)} \qquad (3.2)\\
&= \frac{m_i V_s}{2}\sqrt{\frac{1}{T}[t]_0^{T_p} + \frac{1}{T}[t]_{T_p}^{T}} \\
&= \frac{m_i V_s}{2}\sqrt{\frac{1}{T}(T_p - 0 + T - T_p)} \\
&= \frac{m_i V_s}{2} \qquad (3.3)
\end{aligned}
$$

Figure 3.4 plots the frequency spectrum of output voltage, which contains about 146 % THD with a carrier frequency of 1 kHz. Although the half-bridge converter requires less number of switching devices, the requirements of double voltage-rated DC power supply and large-size line filter may limit the use of this converter. The peak inverse voltage (PIV) of the switching devices is equal to the total DC power supply voltage, V_s. Maximum 1.44 kV converter can be designed with the commercially available IGBTs. Table 3.2 summarizes the possible maximum output voltages of half-bridge converters with commercially available switching devices.

3.2.1.2 Full-Bridge Converter

Figure 3.5 illustrates the circuit diagram of full-bridge converter consisting of four switching devices with two on each half-bridge arm. Due to the similarity of

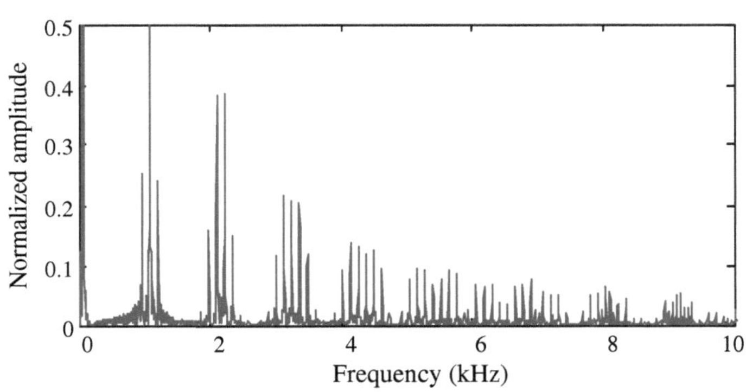

Fig. 3.4 Frequency spectrum of output voltage of half-bridge PWM converter

Table 3.2 Maximum rms output voltage of half-bridge converters with modulation index at 0.8

Voltage rating of IGBT (V)	$V_{com@100FIT}$ (V)	Voltage rating of DC supply (V)	Converter rms output voltage (V)
6,500	3,600	3,600	1,440
4,500	2,250	2,250	900
3,300	1,800	1,800	720
2,500	1,200	1,200	480
1,700	900	900	360

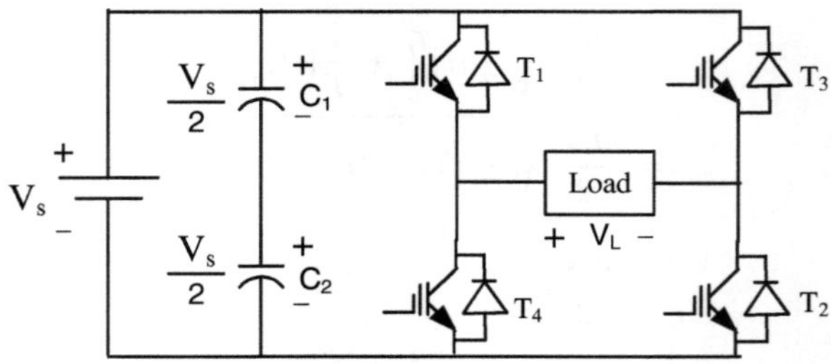

Fig. 3.5 Circuit diagram of full-bridge PWM converter

Table 3.3 Switching scheme of full-bridge converter

T_1	T_2	T_3	T_4	Source voltage	Load voltage
On	On	Off	Off	V_s	$V_L = +m_i V_s$
Off	Off	Off	Off		0
Off	Off	On	On		$V_L = -m_i V_s$
On	On	On	On		Must be avoided

structure with capital letter "H," it is also called H-bridge converter. The switching device pairs $T_1 - T_2$ and $T_3 - T_4$ can be switched on and off alternately with each pair providing the opposite polarity of voltage across the load. The two switching devices of the same arm should not be on at the same time, to avoid direct short circuit between the terminals of the DC power supply. The switching devices can be controlled in a smart way for progressively wider pulses at the center and narrower pulses at the edges of half-cycle output voltage waveform, which may help to filter out high-frequency switching harmonics. The modulation scheme as depicted in Fig. 3.2 can also be used to switch the switching pairs. The Boolean output by comparison of carrier signal and positive half-cycle of the reference signal may switch T_1 and T_2 simultaneously. The Boolean output by comparison of carrier signal and negative half-cycle of the reference signal may switch T_3 and T_4 simultaneously. Table 3.3 shows the switching scheme of a full-wave converter. The output voltage waveform of full-bridge converter has three voltage levels, e.g., 200, 0, and −200 V in the voltage waveform as shown in Fig. 3.6. Due to the higher number of voltage levels, the THD is much lower than that of half-bridge

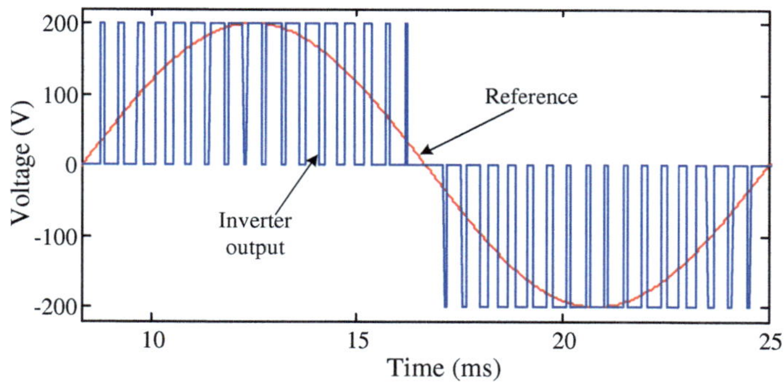

Fig. 3.6 Output voltage waveform of full-bridge PWM converter

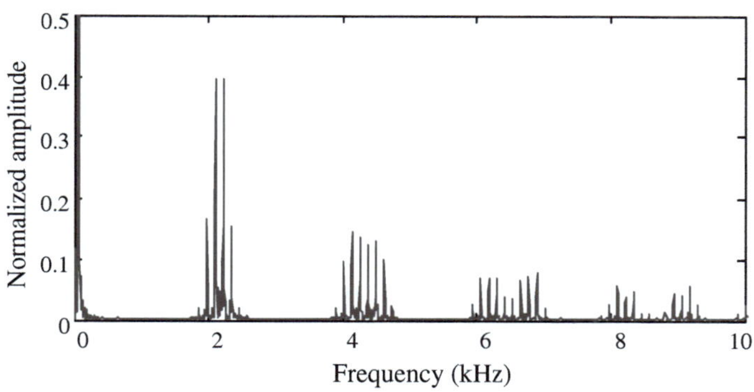

Fig. 3.7 Frequency spectrum of output voltage of full-bridge PWM converter

converter. At 1 kHz PWM switching frequency, the full-bridge converter may provide an output voltage with less than 80 % THD. Figure 3.7 shows the frequency spectrum of output voltage of the full-bridge converter.

If the output PWM voltage waveform as shown in Fig. 3.6 is approximated by square wave as shown in Fig. 3.8, the rms value of the output voltage can be calculated as

$$\begin{aligned} V_{L(\mathrm{rms})} &= \sqrt{\frac{1}{T}\int_0^T V_L^2 \mathrm{d}t} \\ &= \sqrt{\frac{1}{T}\left(\int_0^{\frac{T}{2}} m_i^2 V_s^2 \mathrm{d}t + \int_{\frac{T}{2}}^T m_i^2 V_s^2 \mathrm{d}t\right)} \\ &= \sqrt{\frac{2}{T}\int_0^{\frac{T}{2}} m_i^2 V_s^2 \mathrm{d}t} \end{aligned} \tag{3.4}$$

$$= m_i V_s \tag{3.5}$$

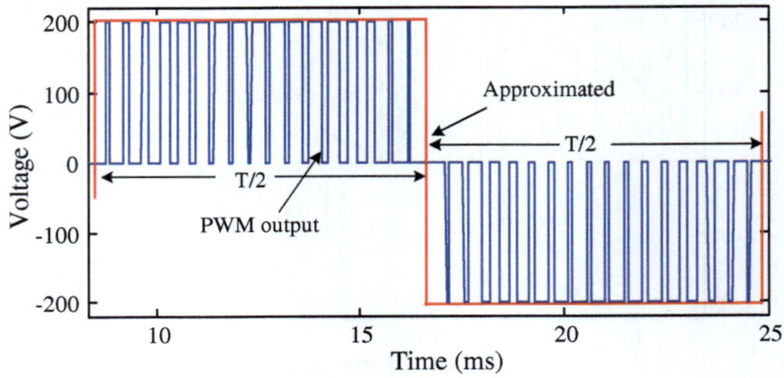

Fig. 3.8 Output voltage of full-bridge PWM converter with square wave approximation

Table 3.4 Maximum rms output voltage of full-bridge converters with modulation index at 0.8

Voltage rating of IGBT (V)	$V_{com@100FIT}$ (V)	Voltage rating of DC supply (V)	Converter rms output voltage (V)
6,500	3,600	3,600	2,880
4,500	2,250	2,250	1,800
3,300	1,800	1,800	1,440
2,500	1,200	1,200	960
1,700	900	900	720

Compared with the half-bridge converter, almost double-rated voltage converter can be designed by using the full-bridge converter topology with same voltage-rated IGBTs. Table 3.4 summarizes the possible maximum converter ratings that can be designed with the available IGBTs. Although the full-bridge converter requires two more switching devices, which may increase converter power loss and cost, the smaller THD can reduce the size of line filter significantly, which may reduce the converter power loss and cost and make the converter compact and lightweight.

3.2.2 Three-Phase Converter

Three-phase converters are more common than single-phase half- or full-bridge converters. Figure 3.9 illustrates the power circuit of a 3-phase bridge converter, which consists of six switching devices with two for each phase. Large capacitors are usually connected at the input terminal to make the supply voltage constant and also minimize the harmonics fed back to the source especially when the DC

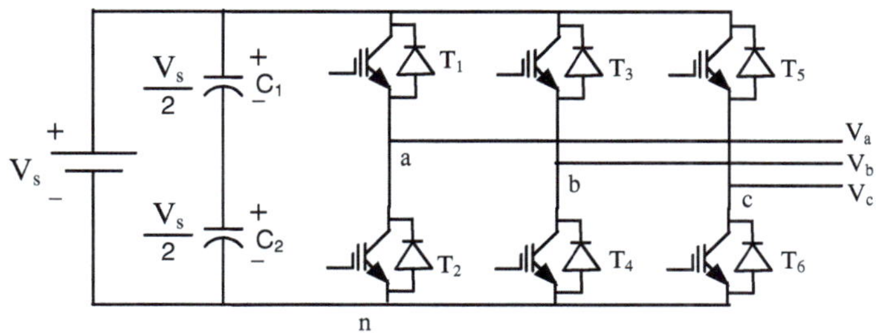

Fig. 3.9 Circuit diagram of 3-phase 2-level converter

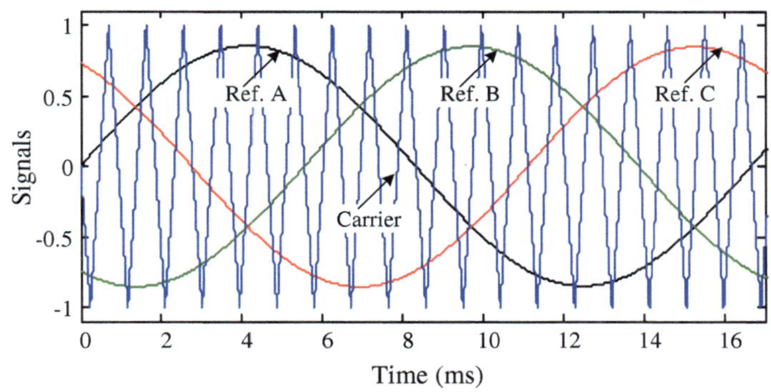

Fig. 3.10 Modulation scheme of 3-phase 2-level converter

supply is through a rectifier circuit. The triangular carrier signal is compared with the sinusoidal modulating signal. When the modulating signal is greater than the carrier, pulse 1 is high (1) and pulse 2 is low (0). Single carrier and three modulating or reference signals (120° phase shift between each other) may generate six gate pulses to drive the converter. Figure 3.10 shows the PWM switching scheme of the 3-phase bridge converter. The Boolean output through comparison of Ref A with carrier signal may switch T_1, and its inverted output may switch T_2. The gate pulse generated from Ref B and carrier may switch T_3, and its inverted gate pulse may switch T_4. Similarly, Ref C and carrier may generate gate pulses for T_5 and T_6. The line-to-line voltage waveform of 3-phase bridge converter is almost the same as that of the full-bridge converter. Figure 3.11 shows the line-to-line output voltage waveform of 3-phase bridge converter. Similar to the single-phase full-bridge converter, the line-to-line voltage of the 3-phase bridge converter contains about 80 % THD. Figure 3.12 shows the frequency spectrum of the output line-to-line voltage. The maximum output voltage generation capacity is also similar to the single-phase full-bridge converter, e.g., maximum 2.88 kV (rms) converter

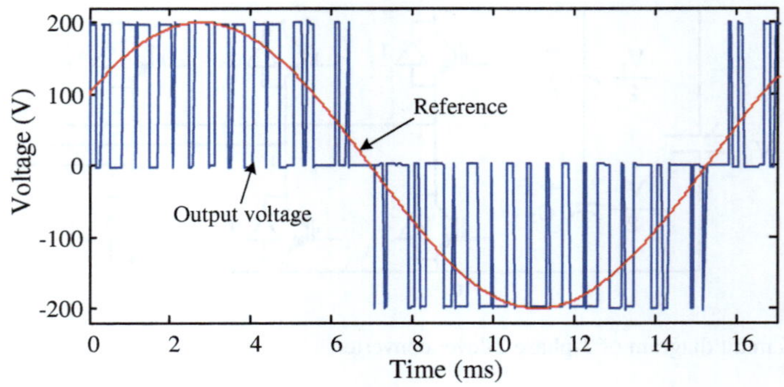

Fig. 3.11 Output line voltage of 3-phase 2-level converter

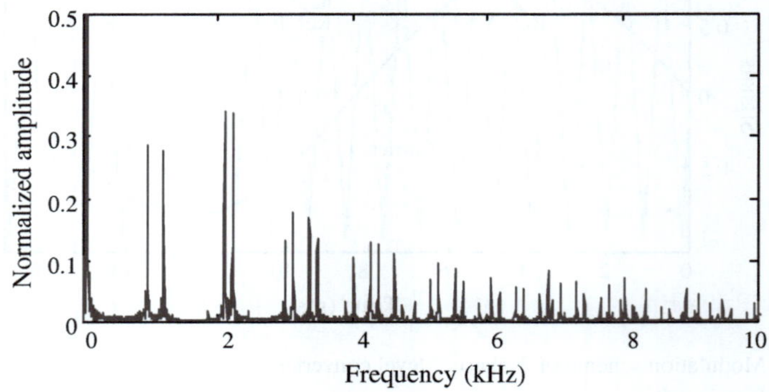

Fig. 3.12 Frequency spectrum of output line voltage of 3-phase 2-level converter

with 6.5 kV IGBTs. Although the 3-phase bridge converter requires 33 % additional switching devices, it can handle about 73 % more power than that of the single-phase full-bridge converter. Therefore, the 3-phase converter could be a smart choice for medium- and high-power applications. If V_{l-l} is the converter line-to-line voltage and I_l the line current, the power handling capacity of the 3-phase 2-level converter can be calculated as

$$P_T = \sqrt{3} V_{l-l} I_l. \tag{3.6}$$

3.2.3 *Space Vector Pulse Width Modulation (SV-PWM)*

The main aim of any modulation technique is to obtain desired output having a maximum fundamental component with minimum harmonics and less switching losses. The space vector pulse width modulation (SV-PWM) method is an advanced PWM method, and it is possibly the best among all the PWM techniques for high-power applications. Compared to sinusoidal pulse width modulation (SPWM), SV-PWM offers 15 % higher DC-voltage utilization and 33 % fewer communications per cycle. Moreover, it can obtain a better harmonic performance and less switching loss.

In each cycle of SV-PWM, the desired output voltage is approximated by a time average of three voltage vectors, in which two are nonzero voltage vectors adjacent to the reference and the third is a zero vector. The strategy of sequencing these three switching vectors within each cycle will affect the current ripple, the switching losses and the spectrum of the output voltages and currents. Moreover, the switching sequence is not unique. Since the SV-PWM offers superior performance with respect to other modulation techniques, it is important to establish the sequencing strategy which is well suited for the systems.

In SV-PWM, the converter is controlled through the concept of converter states. Each converter state corresponds to a certain combination of switches. There are two switches in each leg of a 2-level 3-phase converter. The switches in one leg cannot be both on or off at the same time (the former leads to a short circuit of the associated phase, and the latter results in an open circuit). There are three converter legs in a 3-phase system, so the total number of converter states is eight.

Figure 3.13 shows the eight possible combinations of on and off patterns of 3-phase 2-level converter. The on and off states of the lower switches are the inverted states of the upper ones. The phase voltages corresponding to the eight combinations of switching patterns can be calculated. This transformation results in six nonzero voltage vectors and two zero vectors. The nonzero vectors form the axes of a hexagon containing six sectors ($V_1 - V_6$) as shown in Fig. 3.13. The angle between any adjacent two nonzero voltage vectors is 60°. The zero vectors are at the origin and give a zero voltage vector to the output. The envelope of the hexagon formed by the nonzero vectors is the locus of the maximum output

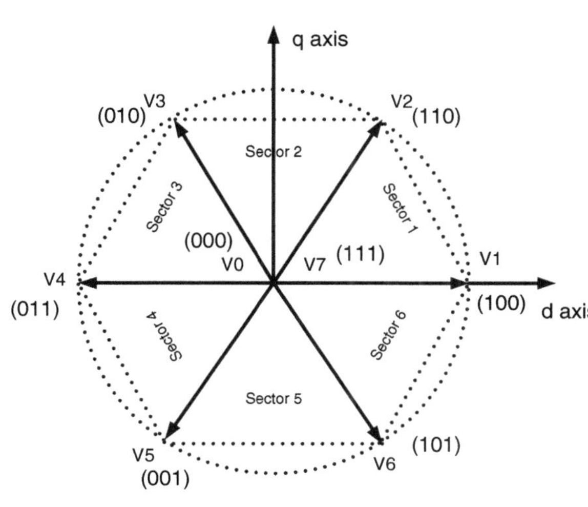

Fig. 3.13 Space vector diagram of 3-phase 2-level converter

Table 3.5 Eight switching states of 3-phase 2-level converter

Voltage vector	Switching vector			Line voltage		
	T_1	T_3	T_5	V_{ab}	V_{bc}	V_{ca}
V_0	0	0	0	0	0	0
V_1	1	0	0	V_s	0	$-V_s$
V_2	1	1	0	0	V_s	$-V_s$
V_3	0	1	0	$-V_s$	V_s	0
V_4	0	1	1	$-V_s$	0	V_s
V_5	0	0	1	0	$-V_s$	V_s
V_6	1	0	1	V_s	$-V_s$	0
V_7	1	1	1	0	0	0

voltage. If versus is the DC supply voltage of 3-phase 2-level converter, the length of six active space vectors can be calculated as

$$\frac{2}{3}V_s \tag{3.7}$$

When the upper switch on a phase leg is on, i.e., T_1, T_3, or T_5 is 1, the corresponding lower switch should be off, i.e., T_2, T_4, or T_6 is 0. In the 2-level SV-PWM, V_0 and V_7 generate the same line voltage (which is zero) and hence are redundant states. The eight possible switching states and corresponding phase as well as the line voltages are shown in Table 3.5. One can choose one or both of the zero voltage vectors, as suited, to reduce the number of switching or manipulate harmonics. These two zero vectors are the only redundant space vectors for a 2-level SV-PWM. As the number of levels of the SV-PWM implementation increases, the number of redundant space vectors increases, which can be exploited to achieve the desired performance characteristics.

Usually, when creating space vectors, the three time varying quantities are sinusoids of the same amplitude and frequency that have 120° phase shifts between each other. When this is the case, the space vector at any given time maintains its magnitude. As time increases, the angle of the space vector increases, causing the vector to rotate with frequency equal to the frequency of the sinusoids. Each of the converter states can be mapped into the dq plane by applying Park's transformation to the phase voltages as given. The vectorial representation of a given 3-phase system as shown in Fig. 3.14 can be carried out by the relation (3.9) to (3.15) [4]. Figure 3.15 describes the technique to calculate the d-axis component corresponding to 3-phase voltages. The q-axis component can be calculated from (3.12), which can be deduced from Fig. 3.16.

$$V_d = V_{an} - V_{bn}\cos 60° - V_{cn}\cos 60° \tag{3.8}$$

$$= V_{an} - \frac{1}{2}V_{bn} - \frac{1}{2}V_{cn} \tag{3.9}$$

$$V_q = 0 + V_{bn}\cos 30° - V_{cn}\cos 30° \tag{3.10}$$

$$= 0V_{\text{an}} + \frac{\sqrt{3}}{2}V_{\text{bn}} - \frac{\sqrt{3}}{2}V_{\text{cn}} \tag{3.11}$$

$$\begin{bmatrix} V_d \\ V_q \end{bmatrix} = \begin{bmatrix} 1 & -\frac{1}{2} & -\frac{1}{2} \\ 0 & \frac{\sqrt{3}}{2} & -\frac{\sqrt{3}}{2} \end{bmatrix} \begin{bmatrix} V_{\text{an}} \\ V_{\text{bn}} \\ V_{\text{cn}} \end{bmatrix} \tag{3.12}$$

$$|V_{\text{ref}}| = \sqrt{V_d^2 + V_q^2} \tag{3.13}$$

$$\alpha = \tan^{-1}\left(\frac{V_q}{V_d}\right) \tag{3.14}$$

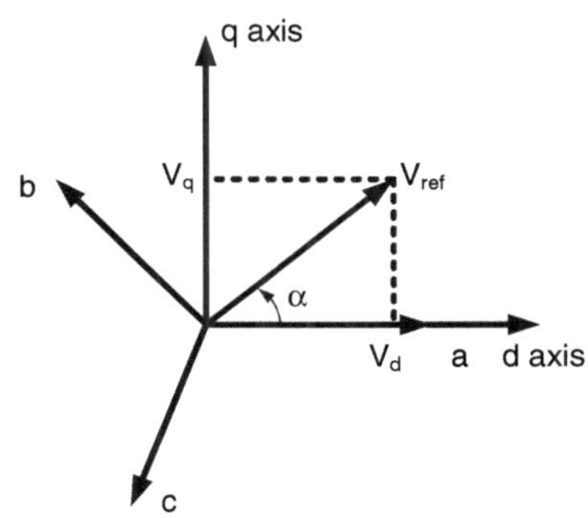

Fig. 3.14 The dq plane

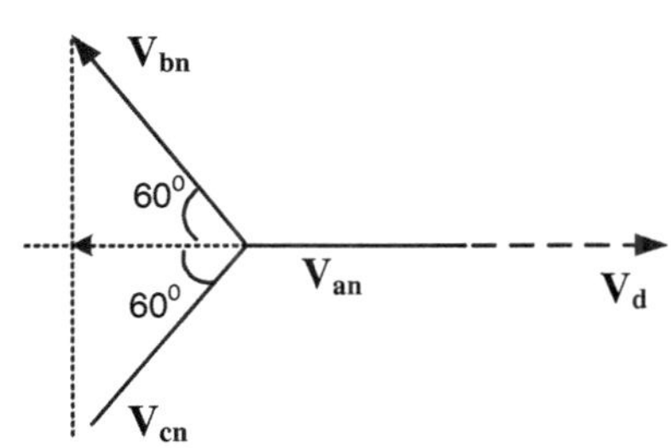

Fig. 3.15 *d*-axis component calculation

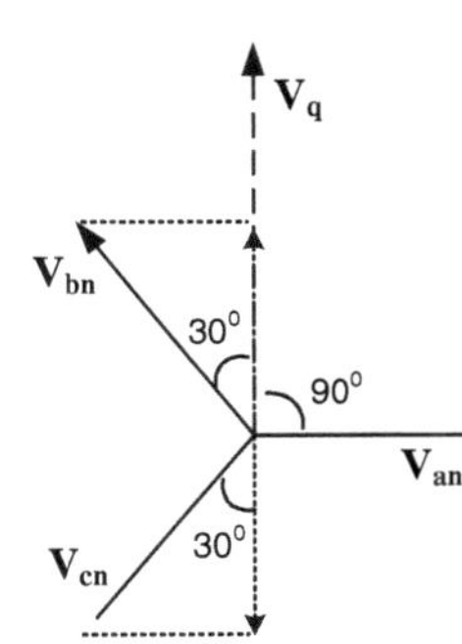

Fig. 3.16 *q*-axis component calculation

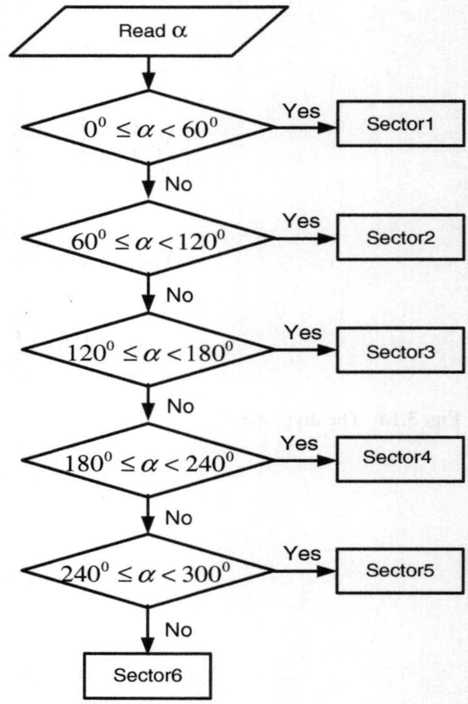

Fig. 3.17 Algorithm to select the sector

Equation (3.14) is used to calculate α, and the reference vector is located according to an algorithm as shown in Fig. 3.17.

The only difference among PWM schemes that use adjacent vectors is the selection of the zero vector(s) and the sequence in which the vectors are applied within the switching cycle. The degrees of freedom we have in the selection of a switching pattern are as follows:

- selection of the zero vector whether or not we would use V_7 (111) or V_0 (000) or both sequencing of the vectors, and
- splitting of the duty cycles of the vectors without additional commutations.

Considering the degrees of freedom, there are four most commonly used switching patterns:

- seven switching states,
- three switching states (with only one zero vector),
- three switching states (with both zero vectors), and
- four switching states.

3.2.3.1 Switching Pattern of Seven Switching States

The number of commutations and switching states in one sampling period is six or seven, respectively, as shown in Fig. 3.18. Both zero vectors (V_7 and V_0) are used

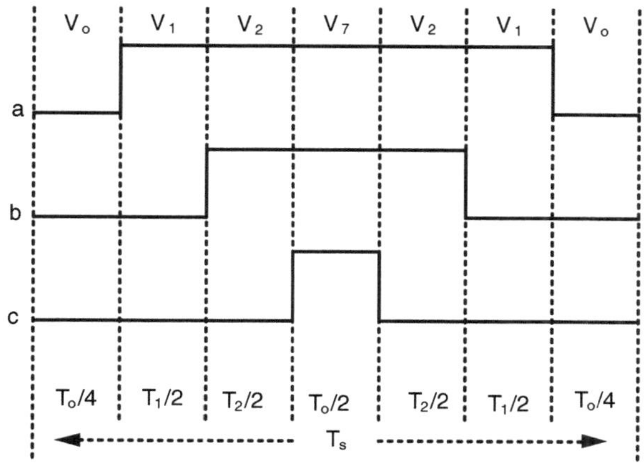

Fig. 3.18 Seven-state switching pattern at sector 1

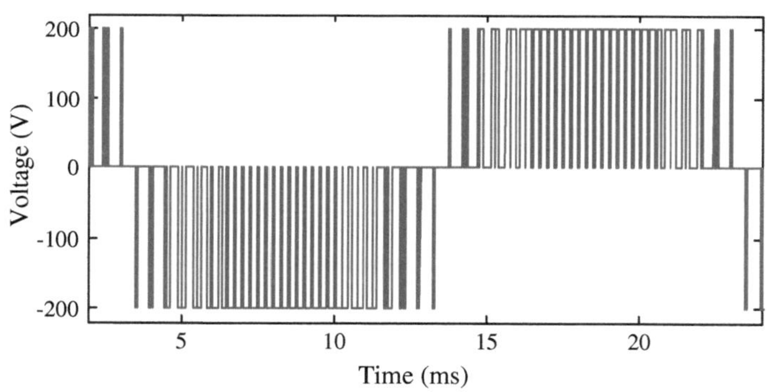

Fig. 3.19 Output line-to-line voltage with seven-state switching pattern

in each sector, where V_7 is placed in the middle and V_0 at both side of the pattern. For example, if the reference is located in the first sector, the possible switching sequence could be V_0–V_1–V_2–V_7–V_2–V_1–V_0, etc. The switching scheme has been implemented in MATLAB/Simulink environment. The output line-to-line voltage of 3-phase 2-level converter with this switching scheme is shown in Fig. 3.19. Due to the symmetry in the switching waveform this scheme shows the lowest THD. Compared with SPWM, seven switching states switching scheme gives about 45 % less harmonic distortion. Figure 3.20 shows the frequency spectrum of line-to-line voltage, where about 55 % THD was measured. Figure 3.21 plots the line current of 3-phase 2-level converter with seven switching states switching scheme-based SV-PWM.

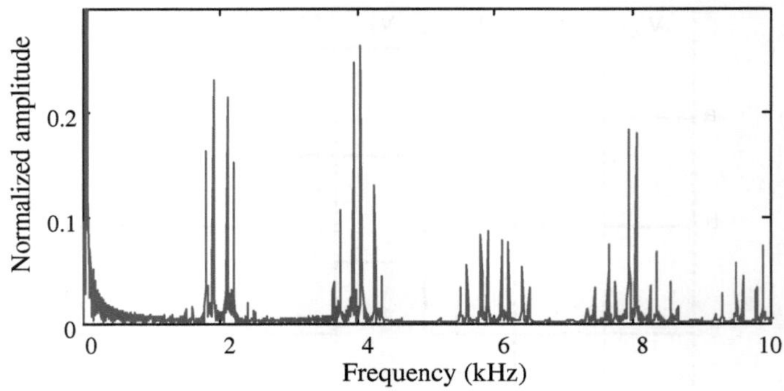

Fig. 3.20 Frequency spectrum of output voltage with seven-state switching pattern

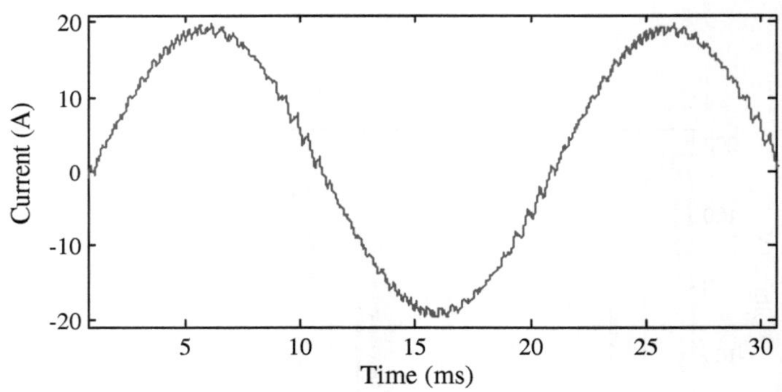

Fig. 3.21 Line current with seven-state switching pattern

3.2.3.2 Three Switching States Switching Pattern (Only One Zero Vector)

Since the selection of the nonzero vectors is based on the desired output voltage vector and the phase and magnitude of the current are determined by the load, it is not always possible to avoid switching the phase carrying the highest current. In such a case, the phase carrying the second highest current is not switched on so that the switching losses are still reduced. In this scheme, the number of commutations in one sampling period is four and there are only three switching states in T_s. With this strategy, the switching sequence is repeated in each cycle as long as the reference vector is within the same sector. For the case when a reference is in sector 1, the switching sequence is given in Fig. 3.22. If V_7 is selected as the zero vector, each switching cycle will have two commutations in the inverter leg *b* and two in leg *c* (a total of 4), while leg *a* will have the upper switch on during the entire cycle. If the zero vector is V_0, the number of commutations (i.e., transitions)

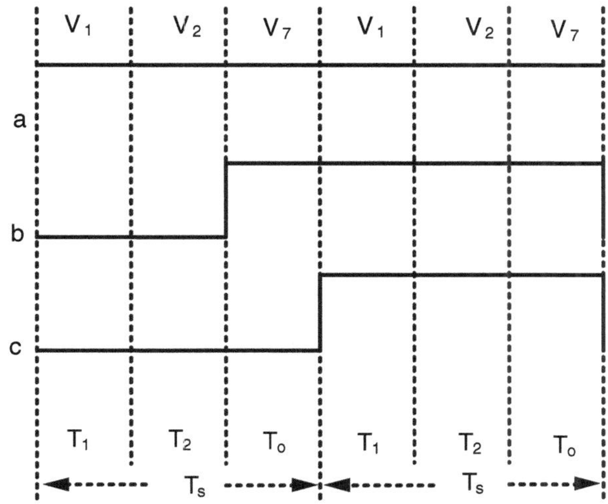

Fig. 3.22 Three-state switching pattern (only one zero vector)

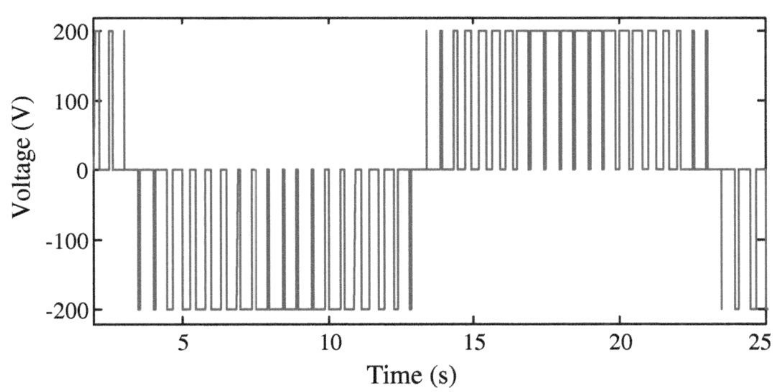

Fig. 3.23 Output line-to-line voltage with three-state switching pattern (only one zero vector)

per cycle will be the same, but the inverter leg that will cease to commutate in sector 1 will be leg *c*. This redundancy of zero vectors allows for a reduction of commutation losses. Selection of the zero vector should be done so as to avoid commutation in the inverter leg carrying the largest current. The output line-to-line voltage of 3-phase 2-level converter with three switching states switching pattern (only one zero vector) is shown in Fig. 3.23. Figure 3.24 shows the frequency spectrum of output line-to-line voltage. Compared with the seven-state switching scheme, three switching states switching pattern (only one zero vector) gives significantly higher harmonics, i.e., about 20 % more. Still it is about 15 % lower than that of the SPWM scheme. Figure 3.25 plots the line current, which is

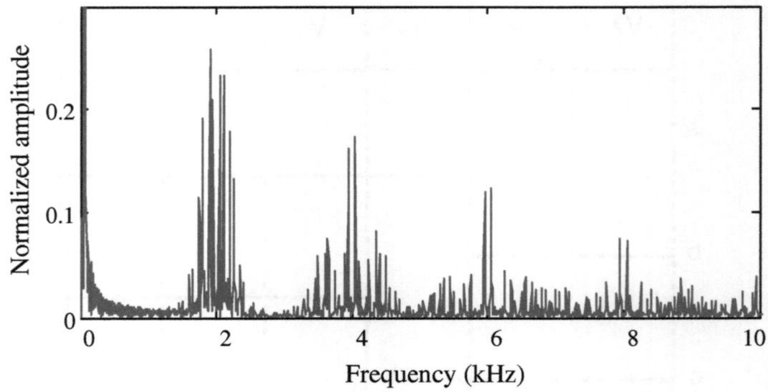

Fig. 3.24 Frequency spectrum of output line-to-line voltage with three-state switching pattern (only one zero vector)

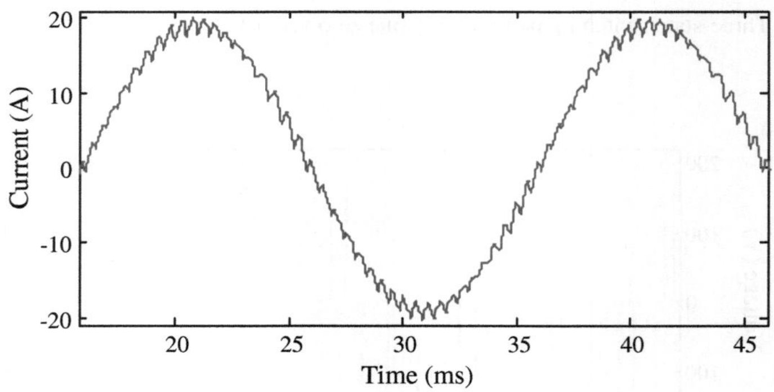

Fig. 3.25 Line current with three-state switching pattern (only one zero vector)

more distorted compared with the line current in Fig. 3.21 with the same circuit parameters.

3.2.3.3 Three Switching States Switching Pattern (Both Zero Vectors but Alternatively)

One can modify the switching sequence between the voltage vectors, so that each transfer from one state to the other involves only one commutation. Figure 3.26 shows the switching pattern of three-state switching (both zero vectors but alternatively). This requires the use of both zero vectors (V_7 and V_0) in a given sector and a reversal of the switching sequence every cycle. In this scheme, the zero vectors V_7 and V_0 are used alternatively in adjacent cycles, so that the effective switching frequency is halved. However, the sampling period is still T_s.

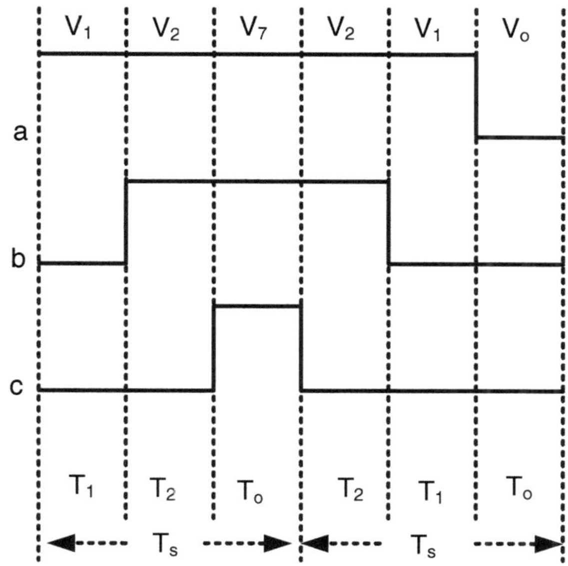

Fig. 3.26 Switching pattern of three-state switching (both zero vectors alternatively)

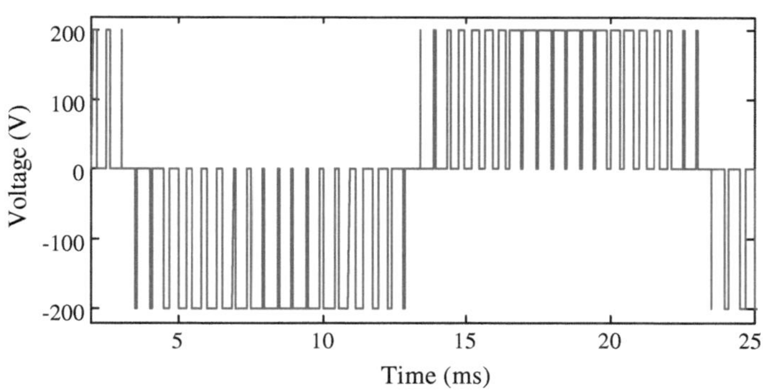

Fig. 3.27 Output line-to-line voltage of 3-phase 2-level converter with three-state switching (both zero vectors alternatively)

The benefit is a reduction of commutations (to three). The switching losses for this scheme are expected to be ideally 50 % as compared to those of the seven-state switching scheme, but the THD is significantly higher due to the existence of the harmonics at half of the sampling frequency. About 68 % THD was measured on the line-to-line output voltage. In this scheme, the number of commutations in one sampling period is three and there are only three switching states in T_s. The output line-to-line voltage of 3-phase 2-level converter with three-state switching (both zero vectors alternatively) is shown in Fig. 3.27. Although the harmonic

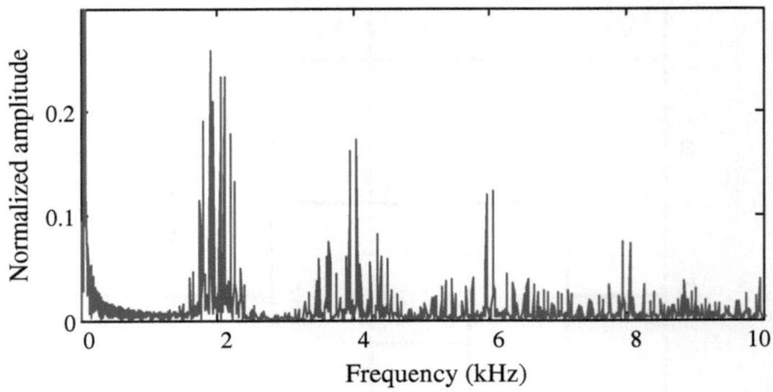

Fig. 3.28 Harmonic spectrum of line-to-line voltage with three-state switching (both zero vectors alternatively)

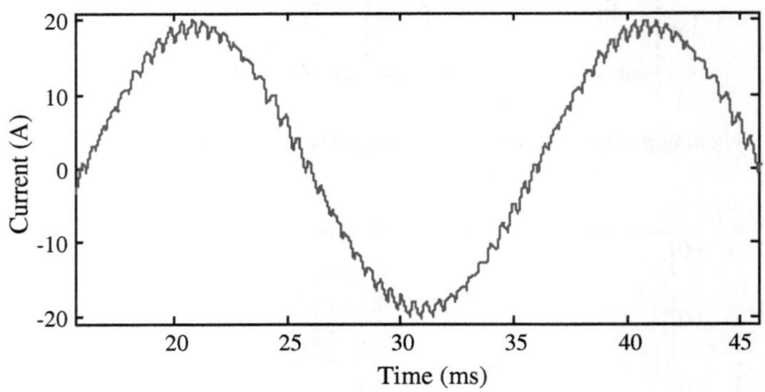

Fig. 3.29 Line current of 3-phase 2-level converter with three-state switching (both zero vectors alternatively)

performance of this switching pattern is similar with the three-state switching pattern (only one zero vector), the number of commutation is higher (3 times in a period). Figure 3.28 shows the harmonic spectrum of output line-to-line voltage with three-state switching (both zero vectors alternatively). The line current waveform is also almost the same as in Fig. 3.25. Figure 3.29 shows the line current waveform with three-state switching (both zero vectors alternatively).

3.2.3.4 Four Switching States Switching Pattern

Both zero vectors, V_7 and V_0, are used to reduce the number of commutations per switching cycle. The switching sequence is reversed after each switching cycle within a sector. For example, if the reference is in the first sector, the possible switching sequence can be V_0–V_1–V_2–V_7–V_7–V_2–V_1–V_0, as shown in Fig. 3.30.

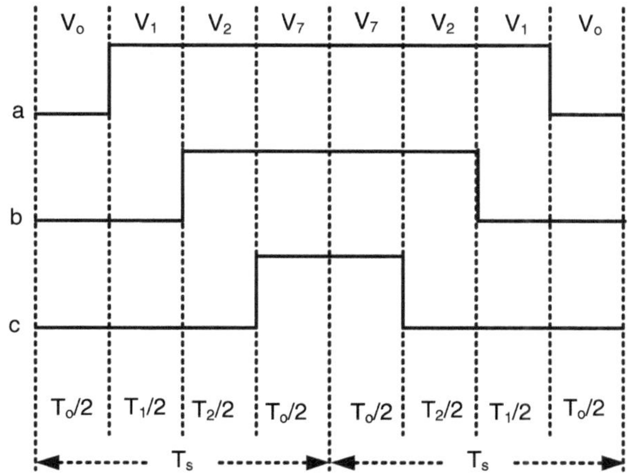

Fig. 3.30 Four-state switching pattern

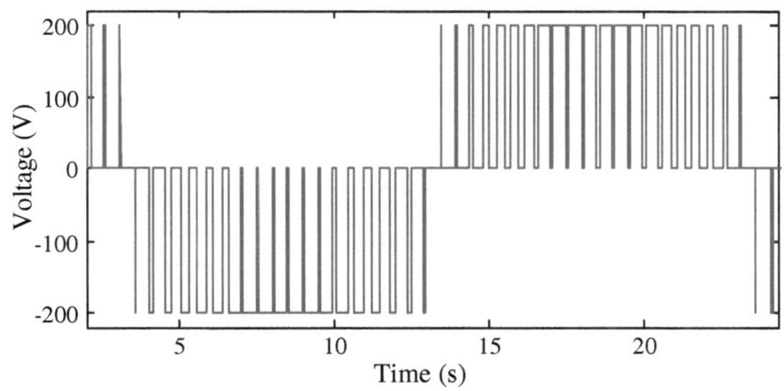

Fig. 3.31 Output line-to-line voltage of 3-phase 2-level converter with four-state switching pattern

The main advantage of this approach is that it requires only three commutations per cycle and gives symmetrical pulses.

The benefit is a reduction of commutations (to three). The switching losses for this scheme are expected to be ideally 50 % as compared to those of the seven-state switching scheme, but the THD is significantly higher due to the existence of the harmonics at half of the sampling frequency. In this scheme, the number of commutations in one sampling period is three and there are only four switching states in T_s. The output line-to-line voltage of 3-phase 2-level converter with four-state switching scheme is shown in Fig. 3.31. The harmonic performance of this switching scheme is slightly better than those of three-state switching pattern (only one zero vector) and three-state switching (both zero vectors alternatively).

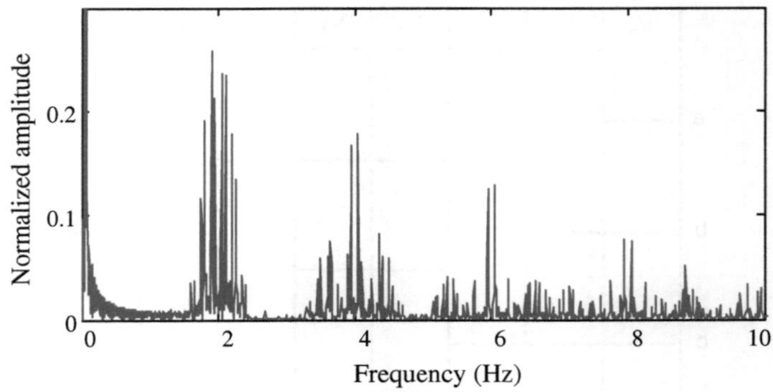

Fig. 3.32 Harmonic spectrum of line-to-line voltage with four-state switching pattern

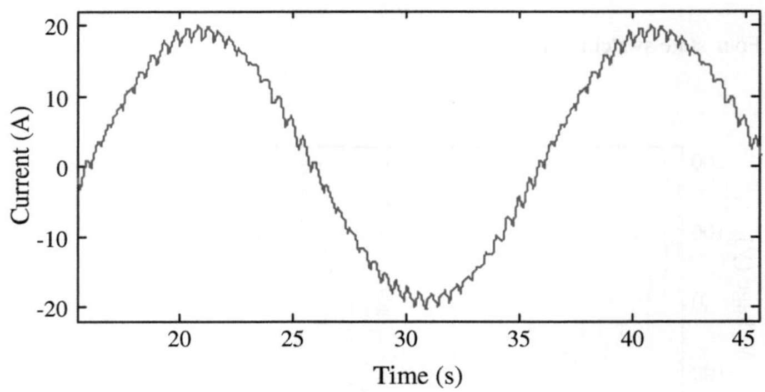

Fig. 3.33 Line current of 3-phase 2-level converter with four-state switching pattern

About 65 % THD was measured. Figure 3.32 shows the frequency spectrum of line-to-line voltage. The line current waveform is also slightly less distorted than those in Figs. 3.25 and 3.29. Figure 3.33 shows the line current waveform with four-state switching pattern.

3.2.3.5 Evaluation of Switching Patterns

Table 3.6 summarizes the performance of four SV-PWM switching patterns. It can be clearly seen that a pattern with high THD has low losses and vice versa. The switching losses are approximately proportional to the magnitude of the current being switched, and hence, it would be advantageous to avoid switching the converter leg carrying the highest instantaneous current. Compared with the seven-state switching scheme (if 100 % switching loss is considered with this scheme), the three-state switching pattern (only one zero vector), three-state switching (both zero vectors alternatively), and four-state switching pattern gives about 33, 50,

Table 3.6 Comparison of switching patterns

Switching patterns	Seven states	Three states with		Four states
		V_0 or V_7	V_0 and V_7	
Commutations in T_s	6	2	3	3
Switching loss (%)	100	33	50	50
THD (%)	55	70	68	65

and 50 % switching losses, respectively. In this way, the three-state and four-state switching patterns are the best choice for power converters.

3.3 Multilevel Converters

The multilevel converters are power electronic circuits that can provide output voltage with more than two levels and be operated in inverter or rectifier mode. The 3-level converter is the lowest level number multilevel converter. Although the number of levels of output phase voltage of 3-level converter is the same as that of line-to-line voltage of 2-level full-bridge converter, the line-to-line voltage of 3-level converter consists of five-voltage levels. There are three basic topologies in the multilevel converter: NPC, FC, and MMC. In 1975, the concept of multilevel converter topology was proposed [5], and in the last few decades, several multilevel converter topologies have been patented [6, 7]. In order to achieve high voltage using low-voltage switching devices, the multilevel converter topology uses a number of switching devices with low-voltage DC sources. The proper control of the switching devices superimposes these multiple DC sources in a staircase (time shifted) form in order to achieve high voltage at the output. The renewable energy sources, such as wind turbine generators, solar PV arrays, and fuel cells, can be used as the multiple DC-voltage sources. The NPC and FC converters require single DC supply, but the MMC converter requires multiple isolated and balanced supplies. High-frequency magnetic link could be a possible option to generate multiple DC supplies for MMC converter from single source. In multilevel converters, the rated voltage of the switching devices depends on the rating of the DC-voltage sources to which they are connected. Therefore, the rated voltage of switching devices is much lower compared with a 2-level converter. The power circuits of one phase of different 3-phase multilevel converters are shown in Fig. 3.34. According to different switching states, it is possible to achieve high-voltage level on the output voltage by adding up the DC-voltage sources compared with the 2-level inverters.

According to the output voltage waveform of two level, the quality of output voltage depends only on the sampling frequency (on X-axis), so that a higher sampling frequency means better voltage waveform, or closer to the sine wave reference. On the other hand, increasing the switching frequency would increase the switching loss and cause electromagnetic compatibility (EMC) problems. The

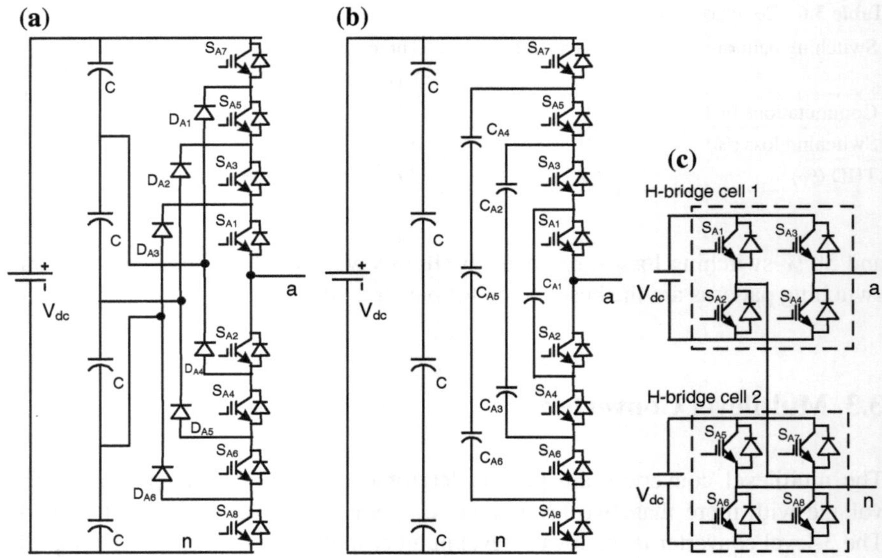

Fig. 3.34 Five-level power converter circuit topologies [2], **a** NPC converter topology, **b** FC converter topology, and **c** MMC converter topology

higher switching frequency means higher dv/dt, which would decrease the stability and reliability of the switching devices. Moreover, the semiconductor devices also have limited range of switching frequencies. The switching frequency of the most commonly available power IGBTs is at 10 kHz or lower. For example, the recommended switching frequency of Semikron 1,200 V 123 NTP, 126 Trench, 12E4HD, and 12 V IGBTs are 9, 5, 9, and 12 kHz, respectively. The 7-level converter output voltage waveform is depicted in Fig. 3.35. In the voltage waveform, there is an additional parameter to control the quality of the output voltage waveform, i.e., the number of voltage levels. Both the *X*- and *Y*-axes can be controlled to improve the converter performance effectively. The switching frequency can be reduced to a low value with an increase to the number of voltage levels, making the converter more stable and reliable and making it possible to use low-voltage-rated matured and low-cost semiconductor devices.

Synthesizing a staircase output voltage, which is closer to the sinusoidal voltage reference compared with the 2-level converter output, allows for a reduction in the harmonic content in voltage waveforms, which would lead to the size and cost reduction of the line filter circuit of power conversion systems. Each level of the staircase consists of a few PWM pulses, as shown in Fig. 3.36.

Like the 2-level converter, the carrier signal frequency of the PWM generator can be adjusted to control the quality of the output waveform. However, the number of levels of the staircase waveform mostly affects the quality of the output voltage waveform. Figure 3.37 plots the THDs of 11 and 33 kV converters with different numbers of converter levels. For an 11 kV converter, 19-level or

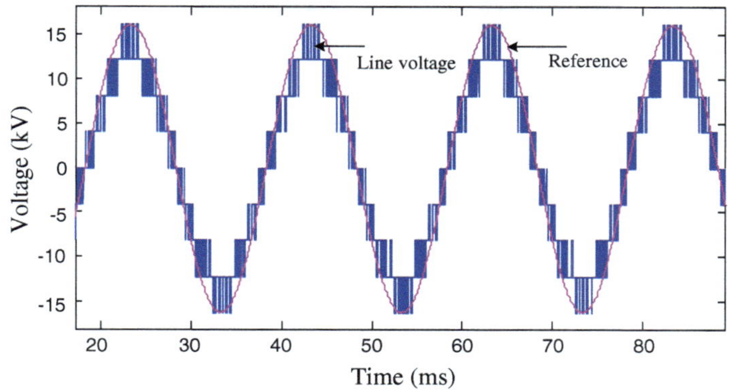

Fig. 3.35 Line-to-line output voltage waveform of 5-level converter

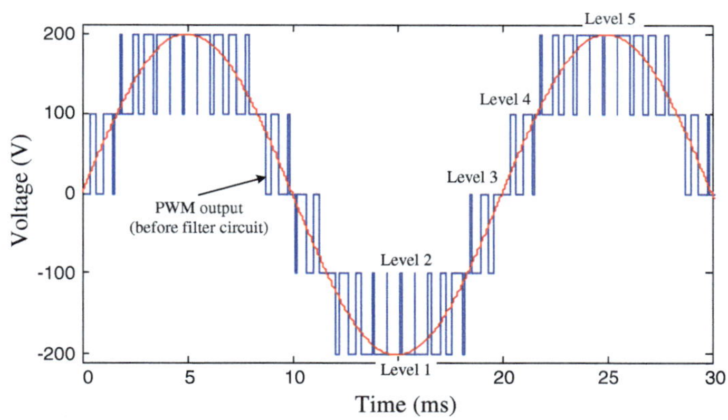

Fig. 3.36 Staircase output phase voltage waveform of 5-level converter

higher level converter may provide output power with less than 5 % THD, which inherently satisfies the IEEE and IEC standards. For a 33 kV system, 21-level or higher level converter may satisfy the 5 % THD requirements without using line filter. Although the number of semiconductor devices increases with the number of converter levels, the high-level converter enables the use of low-voltage devices, which are not only low cost but also matured forms of technology. Table 3.7 lists the market prices of some available power IGBTs.

Therefore, the semiconductor cost of multilevel converters is lower than that of the equivalent 2-level converters operated at the same power rating. Figure 3.38 plots the per-unit prices of switching devices. It can be seen from Fig. 3.38 that the 2.5 kV or lower-rated IGBTs cost much lower than the high-voltage ones. The highest voltage rating of the commercially available IGBT is 6.5 kV, and this is suitable for 2.88 kV or lower-voltage converter systems with the traditional

Table 3.7 Market prices of some available power IGBTs [8]

Voltage (V)	Current (A)	Brand	Price (AUD)
600	400	Powerex	157.91
	300	Powerex	118.28
1,200	3,600	Eupec	2,072.31
	2,400	Eupec	1,582.39
	1,400	Infineon	1,100.02
	600	Powerex	414.51
	400	Powerex	200.44
	300	Powerex	207.28
	200	Eupec	151.64
1,700	3,600	Eupec	2,624.48
	2,400	Eupec	2,017.17
	1,200	Eupec	1,106.11
	600	Infineon	425.00
	450	Eupec	339.60
	300	Eupec	238.36
	225	Eupec	195.21
	150	Semikron	168.00
3,300	1,500	Eupec	3,007.82
	1,200	Eupec	2,431.88
	400	Eupec	1,369.31
6,500	600	Eupec	3,997.60
	400	Eupec	3,078.18
	200	Eupec	1,918.98

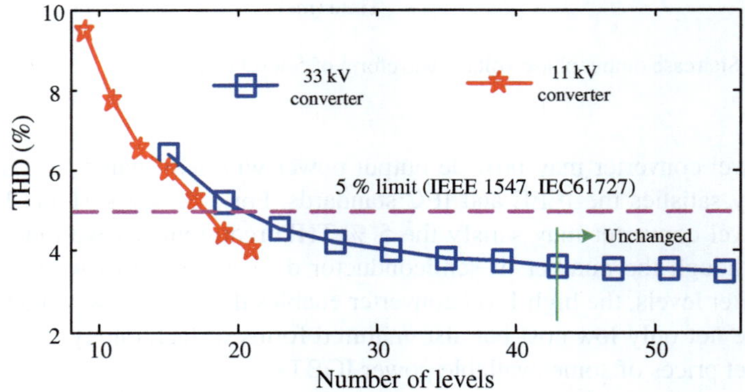

Fig. 3.37 THDs of 33 and 11 kV multilevel converters [3]

2-level converter topology. Table 3.8 summarizes the voltage ratings of commercially available IGBTs, where $V_{com@100FIT}$ is the device commutation voltage for a device reliability of 100 failures in time (FIT) due to cosmic radiation. The multilevel converter topology enables the possibility to develop medium- and

Table 3.8 Voltage handling capacity of available power IGBTs

Rated device voltage (V)	$V_{com@100FIT}$ (V)
6,500	3,600
4,500	2,250
3,300	1,800
2,500	1,200
1,700	900
1,200	800
900	600
600	300

Table 3.9 Voltage handling capacity of power converters

Device voltage rating (V)	Number of levels	Line voltage (kV, rms)
6,500	2	2.88
	5	10.18
	11	25.45
	19	45.82
	21	50.91
3,300	2	1.44
	5	5.09
	11	12.72
	19	22.91
	21	25.45
1,200	2	0.64
	5	2.26
	11	5.65
	19	10.18
	21	11.31
600	2	0.240
	5	0.848
	11	2.12
	19	3.81
	21	4.24

high-voltage converters using commercially available semiconductor devices. Table 3.9 summarizes the voltage handling capacity of 2-level and multilevel converters. With 2-level converter topology, converters of a maximum of 2.88 kV rms can be designed using commercially available devices, but this limitation can be easily overcome with multilevel topologies.

Series connection of switching devices is a well-established method to develop a medium-voltage converter with 2-level converter topology. In order to reduce the effects of non-unique device characteristics and mismatch of driver circuit operations, the method uses switching aid circuits, such as snubber and/or active gate control circuits, which increases the switching loss. The switching loss mostly depends on the communication voltage of the switching devices, which can be much lower with multilevel converter topologies than that of the 2-level converters. Moreover, multilevel converters require no switching aid circuits, which improve the commutation times. Therefore, multilevel converters give lower switching losses than those of the 2-level converters. A power loss comparison between the MMC converter and 2-level converter was analyzed, and it was reported that the switching losses of MMC converters were much lower than those of the 2-level series-connected converters operating under the same conditions [9].

Among the variety of topologies for multilevel converter structures reported in the recent decades, the NPC, FC, and MMC converter with separate DC sources are three commonly used structures. Most of the other topologies for multilevel converters have been designed through the modification or combination of these three basic topologies. A brief description of these three multilevel converter topologies is presented in Sects. 2.1–2.3.

3.3.1 Neutral Point Clamped Converter

In order to increase the converter voltage rating without increasing the voltage rating of switching devices, the NPC converter topology was introduced in 1981 [10]. A 3-phase 7-level NPC converter circuit is shown in Fig. 3.39. A phase leg of a 7-level converter consists of six pairs of switching devices and five pairs of diodes. Each switch in a pair works in complimentary mode, and the respective diode pair provides access to the midpoint. There are seven switching states to synthesize the 7-level phase voltage. If it is assumed that each auxiliary diode voltage rating is the same as the active switching device voltage rating, the number of diodes required for each phase leg will be $(2m - 4)$.

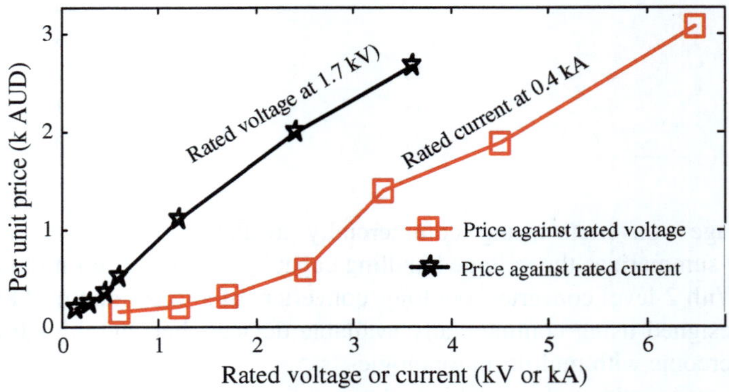

Fig. 3.38 THDs of 33 and 11 kV multilevel converters [3]

An m-level NPC converter has an m-level output phase voltage and a $(2m - 1)$-level output line voltage. Figure 3.40 shows the output line-to-line voltage waveform of 7-level converter. There are seven-voltage levels to the peak value and 13 levels to peak-to-peak value. The difference between two successive voltage levels is equal to the voltage across each DC-link capacitor. The three phases of the NPC converter share a common DC bus voltage, which has been subdivided by $(m - 1)$ capacitors into equally m-levels. Each phase leg of the NPC converter requires $(2m - 2)$ active switching devices, and the commutation voltage of each active switching device is $(m - 1)$ times lower than the total DC-bus voltage. For example, in total, there are six DC-link capacitors in a 7-level NPC converter and the voltage across each capacitor is $V_{dc}/6$.

In 1998, the NPC converter-based high-power AC motor drive system was proposed [11]. In 2002, another application of the NPC converter for high-power medium-voltage variable speed motor drives was proposed [12]. The NPC converter has also attracted significant attention because it serves as an interface between a high-voltage DC transmission line and an AC transmission line [13].

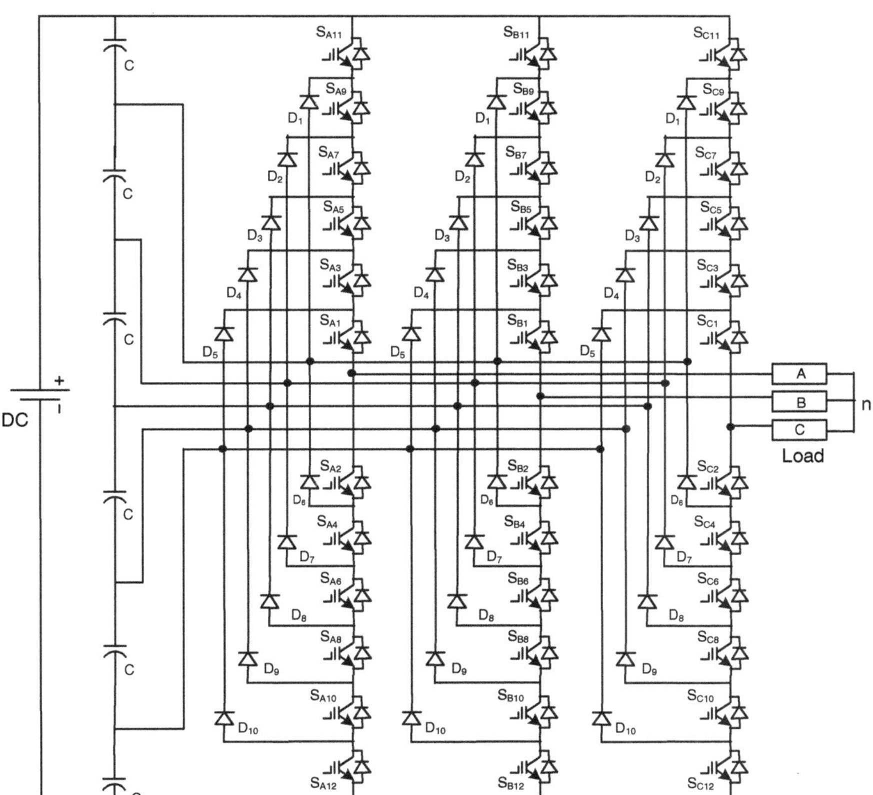

Fig. 3.39 Three-phase 7-level NPC converter circuit topology

The NPC converter topology has the ability to connect the neutral point to the middle point of the DC link thereby reducing the ground leakage currents and enabling this topology to form back-to-back connection. Accordingly, in the recent years, the back-to-back NPC converters have also been largely utilized in the grid interfacing of renewable energy sources such as wind turbine generators and solar PV arrays. In total, there are $(6m - 12)$ auxiliary diodes required for a 3-phase m-level converter.

When the number of converter levels is sufficiently high, the number of diodes required will make the system impractical to implement. This is due to that the diode reverse recovery of these huge clamping diodes becomes a critical factor. The reverse recovery problem can be reduced by using the newly developed silicon carbide diode which has negligible reverse recovery time. However, this technology is still under development. Only 40 A or lower-rated silicon carbide diodes are currently available on the market. All the converters in this chapter are designed with 250 A rated diodes, as the silicon carbide technology is still not available for these designs. On the other hand, these special diodes significantly increase the semiconductor cost of the converter (a 3-phase 43-level NPC converter requires 246 diodes). For example, the unit price of a 600 V 40 A silicon carbide diode is about 25 USD. Therefore, the reverse recovery problem is still a challenging issue for designing medium- or high-voltage converters with NPC converter topology. In addition to the requirement for several auxiliary diodes, the capacitor voltage balancing is one of the notable drawbacks of NPC topology. Unbalanced capacitor voltages may cause distortions in output waveform and damage switching devices due to the over voltage breakdown of switching devices. For this reason, many researchers are turning their attention to developing extra circuits and control algorithms [14], but these circuits and algorithms are for no more than five levels. Moreover, the boost switch may limit the voltage and power rating of the converter. Therefore, this topology is not suitable for medium- or high-voltage applications, where a higher number of levels are required.

3.3.2 Flying Capacitor Converter

In 1992, FC converter was proposed [15]. The overall structure of the FC converter is comparable with that of the NPC converter except that the converter uses auxiliary capacitors instead of auxiliary diodes. The circuit topology of a 3-phase 7-level FC converter is shown in Fig. 3.41.

A phase leg of the 7-level converter consists of six pairs of switching devices and five branches of capacitors. An m-level FC converter has an m-level output phase voltage and a $(2m - 1)$-level output line voltage. The three phases of the FC converter may share a common DC-bus voltage similar to the NPC converter, which has been subdivided by $(m - 1)$ capacitors into equal m-levels. Each phase leg of the FC converter requires $(2m - 2)$ active switching devices, and the commutation voltage of each active switching device is $(m - 1)$ times lower than

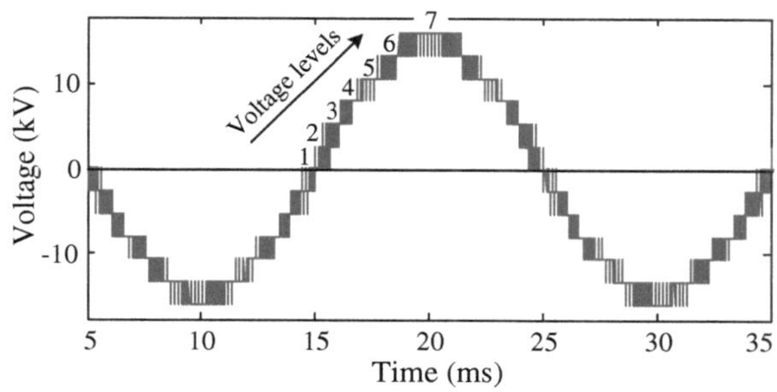

Fig. 3.40 Output line-to-line voltage of 7-level NPC converter

the total DC-bus voltage. In addition to the DC-link capacitors, the FC converter requires large number of auxiliary capacitors. If the voltage rating of the capacitors is equal to that of the active switching devices, the number of auxiliary capacitors required for each phase leg will be $(m^2 - 3m + 2)/2$ [16]. There are several possible switching states to synthesize a desired output. The optimal switching states may ensure good performance and reduce power losses of the converter as well.

Due to some features (e.g., the control of real and reactive power and the capability to ride through short-duration outage and voltage sags), the FC-voltage source converter has found its application in power systems. In 2002, an FC multilevel-voltage source converter-based unified power-flow controller was proposed [17]. In 2007, a 3-level FC multilevel converter-based high-voltage direct current power transmission system was presented [18]. However, the number of auxiliary capacitors required is quadratically related to the number of levels. When the number of converter levels is sufficiently high, the number of capacitors required makes the system impractical to implement. The large number of capacitors not only increases the converter size and weight but also significantly reduces the lifetime of the converter. Due to the regulation of the capacitor voltages, the FC multilevel converter requires complex control strategies. Moreover, the pre-charging of clamping capacitors with the required voltage level may increase the complexity and cost, and decrease the performance of the converter.

3.3.3 Modular Multilevel Cascaded Converter

In the NPC and FC converters, modification of circuit configuration, e.g., increase or decrease of the number of levels is complicated, which affects the entire circuit of the converter. On the other hand, the MMC converter is completely modular in structure, where it is easy to increase or decrease the number of levels of the converter. The modification of number of converter levels does not affect the entire circuit. In MMC converters, each phase leg consists of a series of modular single-phase full-bridge or H-bridge inverter cells connected in series. A 3-phase 7-level

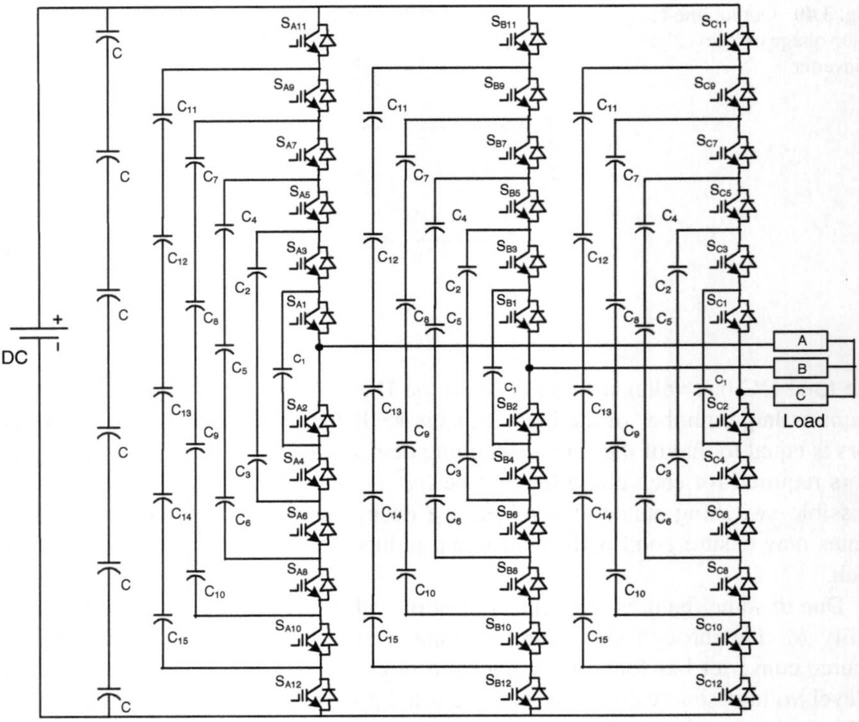

Fig. 3.41 Three-phase 7-level FC converter circuit topology

MMC converter circuit is illustrated in Fig. 3.42, which is Wye connected. The 3-phase MMC converter can also be connected in delta.

Each isolated and balanced DC source is connected to a single-phase full-bridge or H-bridge inverter cell. The requirement of multiple isolated and balanced DC supplies may limit the application of MMC converter. For example, a 3-phase 7-level MMC converter requires nine DC supplies. Recent research has proposed advanced magnetic material-based high-frequency magnetic link toward the solution of this problem [19, 20]. The four switching devices of each H-bridge inverter cell can generate three different voltage outputs: $+V_{dc}$, 0, and $-V_{dc}$ by connecting the isolated DC source to the AC output. The proper control of the switching devices superimposes the voltage outputs of all the H-bridge cells of a phase in a staircase form in order to achieve high voltage at the converter output. Each H-bridge inverter cell contributes two voltage levels in the output voltage. Similar to the NPC and FC converters, an m-level MMC converter also has an m-level output phase voltage and a $(2m - 1)$-level output line-to-line voltage. As the NPC and FC multilevel converters, the MMC converter requires $(2m - 2)$ active switching devices per phase leg. The commutation voltage of each active switching device is $(m - 1)$ times lower than the total DC-bus voltage. There are several possible switch combinations to synthesize the 7-level phase voltage.

Because of their special features (e.g., the number of components scales linearly with the number of levels, and individual modules are identical and completely modular in construction hence enabling high-level number attainability), the MMC converter topology has been considered the best possible candidate for medium- to high-voltage high-power applications. In 1990, a single-phase MMC converter structure was presented for plasma stabilization applications [21]. In 1997, Robicon Corporation presented a medium-voltage high-power 3-phase MMC converter for motor drive applications [22]. The complicated multi-winding phase-shifted power-frequency transformer was used to deliver electric power to all the floating H-bridge inverter cells. In 1996, an MMC-based static synchronous compensator for reactive power control was presented [23]. In 1999, the converter was proposed for heavy-duty electric and hybrid-electric vehicles that have large electric motor drives [24]. A series of isolated batteries were used to deliver electric power to all the floating H-bridge inverter cells.

3.3.4 Switching Schemes of Multilevel Converters

A phase shift or level shift can be introduced between the carrier signals of contiguous cells to produce a phase-shifted or level-shifted switching pattern between them. In this way, a stepped multilevel waveform is generated. For the m-level converter, $(m - 1)$ phase-shifted or line-shifted carriers with the same frequency f_{ca} are required. If f_m is the frequency of the reference signal, the frequency modulation index of the converter can be calculated as

$$m_f = \frac{f_{\text{ca}}}{f_m}. \tag{3.15}$$

3.3.4.1 Phase-Shifted Carrier-Based Switching Scheme

The phase-shifted carriers are specially conceived for FC [25] and MMC [22] converters. Since each FC cell is a 2-level converter, and each H-bridge cell is a 3-level inverter, the traditional bipolar (using one carrier signal that is compared to the reference to decide between two different voltage levels, typically the positive and negative busbars of a voltage source converter) and unipolar PWM techniques can be used, respectively. Due to the modularity of these topologies, each cell can be modulated independently using the same reference signal in a phase. If A_c the peak-to-peak amplitude of the carrier signals and A_m the peak-to-peak amplitude of the reference signal, the amplitude modulation index can be calculated as

$$m_{\text{ap}} = \frac{A_m}{A_c}. \tag{3.16}$$

Figure 3.43 shows the block diagram of the phase-shifted switching scheme for the 7-level converter. This modulation scheme requires three reference signals;

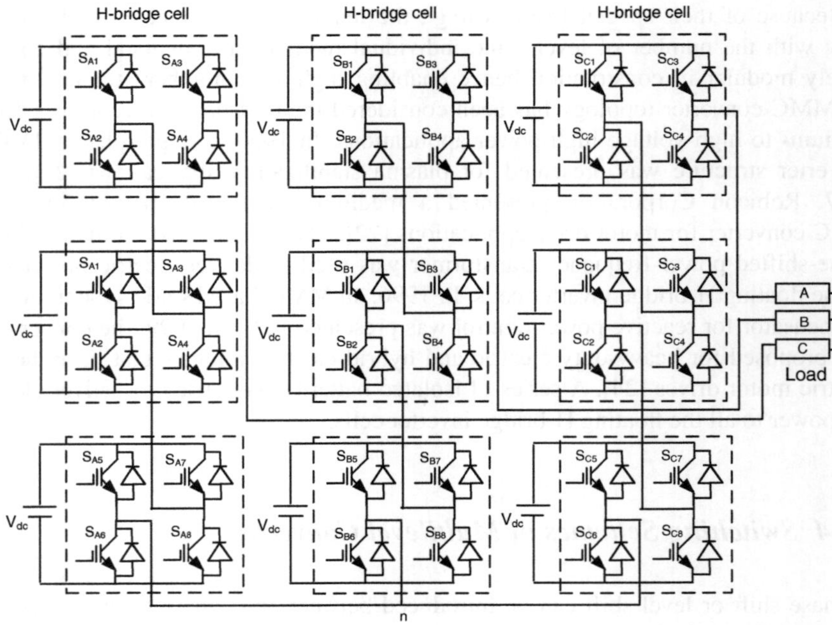

Fig. 3.42 Three-phase 7-level modular multilevel cascaded converter topology

120° shifted by each other and one for a particular phase leg. The number of carrier signals depends on the number of converter levels; in total, there are $(m - 1)/2$ carrier signals in an m-level converter. The carrier phase shifting for that particular cell or pair can be calculated as [26]

$$\theta_{\text{ps}} = \frac{360^\circ}{(m-1)}. \tag{3.17}$$

The MATLAB/Simulink model to generate 3-phase-shifted carrier signals for a 3-phase 7-level converter is illustrated in Fig. 3.44. These three-generated carriers may be used to drive the left-side H-bridge arms of three H-bridge inverter cells on a phase leg. Figure 3.45 plots the three-generated carrier signals. The inverted carrier signals may be used to generate gate pulses to drive the right-side H-bridge leg of three H-bridge inverter cells on a phase leg. Figure 3.46 plots the inverted carrier signals. In total, there are six carrier signals and three reference signals in the modulation scheme of 3-phase 7-level converter. Figure 3.47 plots all the carrier and reference signals of a 3-phase 7-level converter.

Each compare unit generates one switching signal for the top switching device of a half-bridge cell or a pair. The inverted form of this switching signal drives the bottom switching device. Figure 3.48 shows the gate pulse generation technique for the top switching device. The technique to generate gate pulse for the bottom switching device is illustrated in Fig. 3.49. For the left half-bridge cell, one

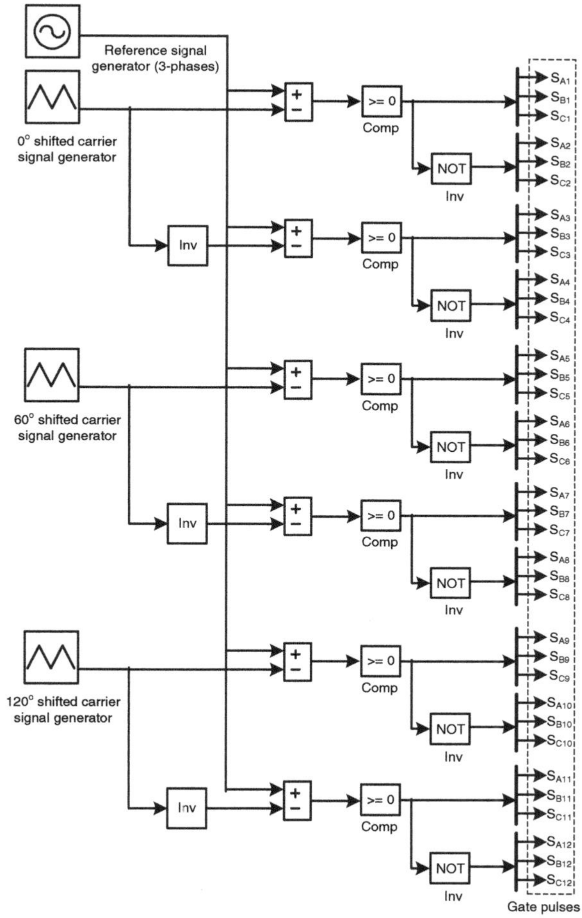

Fig. 3.43 Phase-shifted switching scheme for the 3-phase 7-level FC and MMC multilevel converters

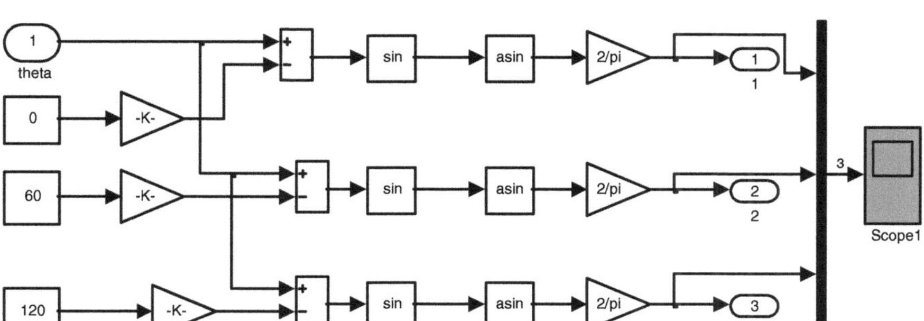

Fig. 3.44 MATLAB/Simulink model to generate phase-shifted carrier signals for 7-level converter

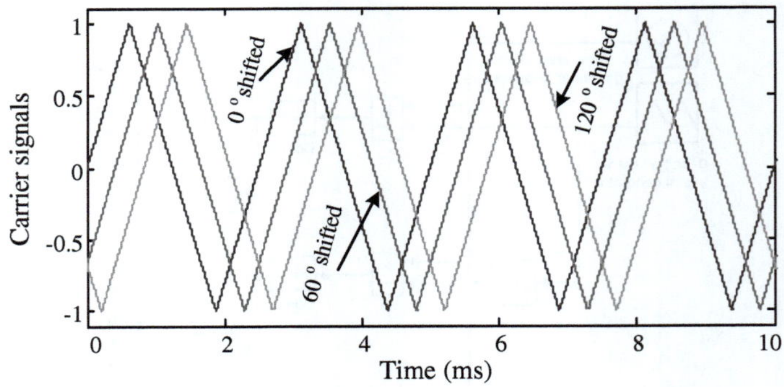

Fig. 3.45 Three-generated phase-shifted carrier signals

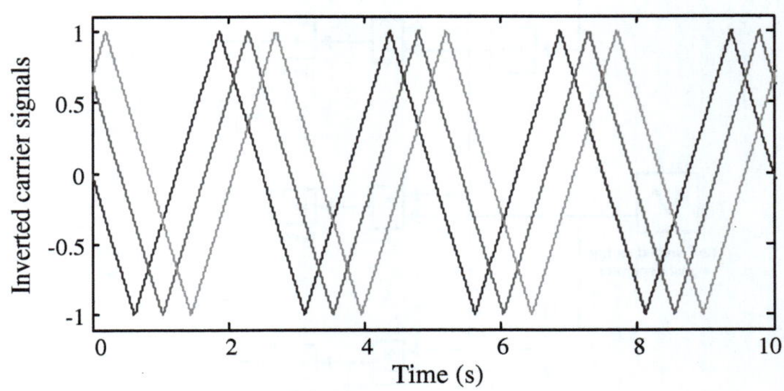

Fig. 3.46 Three-inverted phase-shifted carrier signals

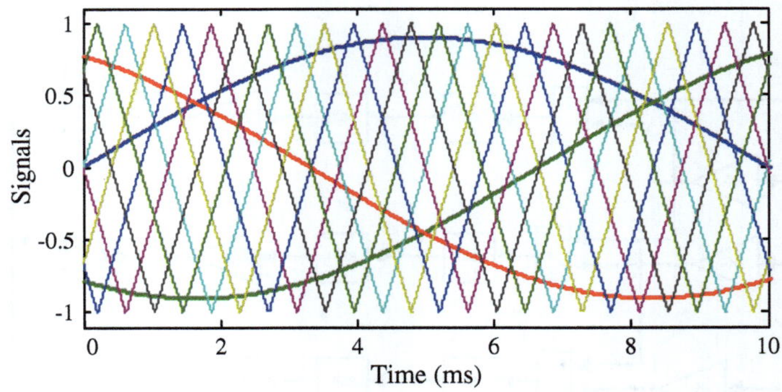

Fig. 3.47 Three reference signals and six phase-shifted carrier signals-based modulation scheme of 3-phase 7-level converter

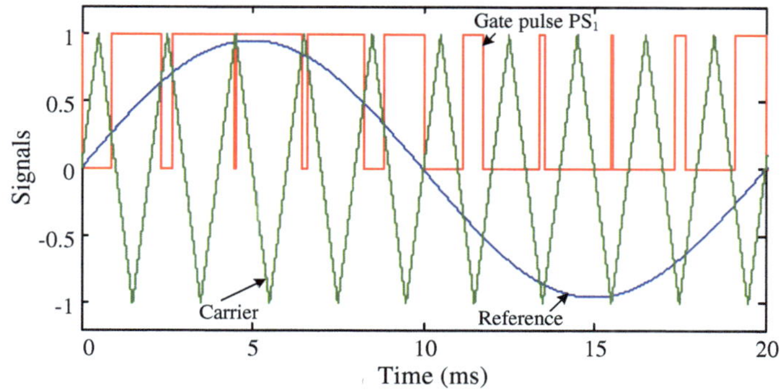

Fig. 3.48 Generation of *gate pulse* PS_1 with phase-shifted carrier

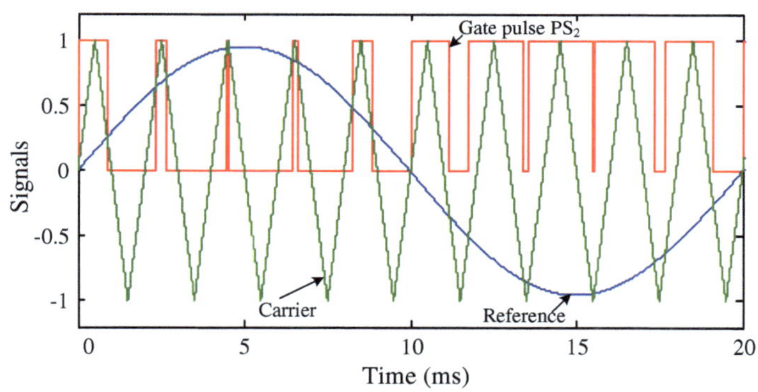

Fig. 3.49 Generation of *gate pulse* PS_2 with phase-shifted carrier

is asserted when the reference signal value is greater than or equal to the carrier signal value, and the other is asserted when the reference signal value is less than the carrier signal value. For the right half-bridge cell, one is asserted when the inverted carrier signal value is greater than or equal to the reference signal value, and the other is asserted when the inverted carrier signal value is less than the reference signal value. Figure 3.50 shows the MATLAB/Simulink model to generate the gate pulses for a 3-phase 7-level converter.

For the FC converter, the advantage of the even power distribution is that once the FCs are properly charged (initialized to their corresponding values), no imbalance will be produced due to the self-balancing property of this topology and as a result there is no need to control the DC-link voltages [27, 28]. Another interesting feature is that the total output voltage has a switching pattern with k (number of the power cells) times the frequency of the switching pattern of each cell. This multiplicative effect is produced by the phase shifts of the carriers. Hence, lower THD is obtained at the output, using the k time's lower frequency carriers. With the phase-shifted carrier-based modulation scheme, the control signal assignment to the appropriate semiconductor of the MMC converter is easy and this remains simple even when the level number increases to higher values.

3.3.4.2 Level-Shifted Carrier-Based Switching Scheme

In the level-shifted scheme, the carriers are arranged in vertical shifts instead of the horizontal shift. Each carrier is set between two voltage levels, and hence, it is named as level shifted. Since each carrier is associated with two levels, the same principle of bipolar PWM can be applied which takes into account the fact that the control signal has to be directed to the appropriate semiconductors in order to generate the corresponding levels [29]. The carriers span the whole amplitude range that can be generated by the converter.

The level-shifted carrier-based modulation scheme is especially useful for NPC converters, since each carrier can be easily associated with two power switches of the converter [30]. Moreover, the level-shifted carrier-based modulation scheme provides less distorted line voltages because all carriers are in phase in contrast to the phase-shifted carrier-based modulation scheme. Although the level-shifted carrier-based modulation scheme may provide reasonable performance, the complexity of control signal assignment to the appropriate semiconductors makes it unpopular for MMC multilevel converter applications. Further, the level-shifted carrier-based modulation scheme generates more distorted input current in the MMC converter and a capacitor voltage imbalance in the FC converter [30]. Figure 3.51 plots the reference and carrier signals for a 3-phase 7-level NPC converter. In the level-shifted scheme, the carriers on a particular phase leg are level shifted by

$$\theta_{ls} = \frac{\pm A_m}{(m-1)}. \tag{3.18}$$

The peak-to-peak amplitude of carrier signals can be calculated by

$$A_{\mathrm{cl}} = \frac{A_m}{(m-1)}. \tag{3.19}$$

Figure 3.52 shows the block diagram of the level shifted switching scheme for the 3-phase 7-level converter.

Each comparator unit generates one switching signal for the top switching device of a pair. The inverted form of this switching signal drives the bottom switching device. Figure 3.53 shows the gate pulse generation technique for the top switching device. The technique to generate gate pulse for the bottom switching device is illustrated in Fig. 3.54. For the top pair, one is asserted when the reference signal value is greater than or equal to the carrier signal value, and the other is asserted when the reference signal value is less than the carrier signal value. For the second top pair, one is asserted when the shifted carrier signal value is

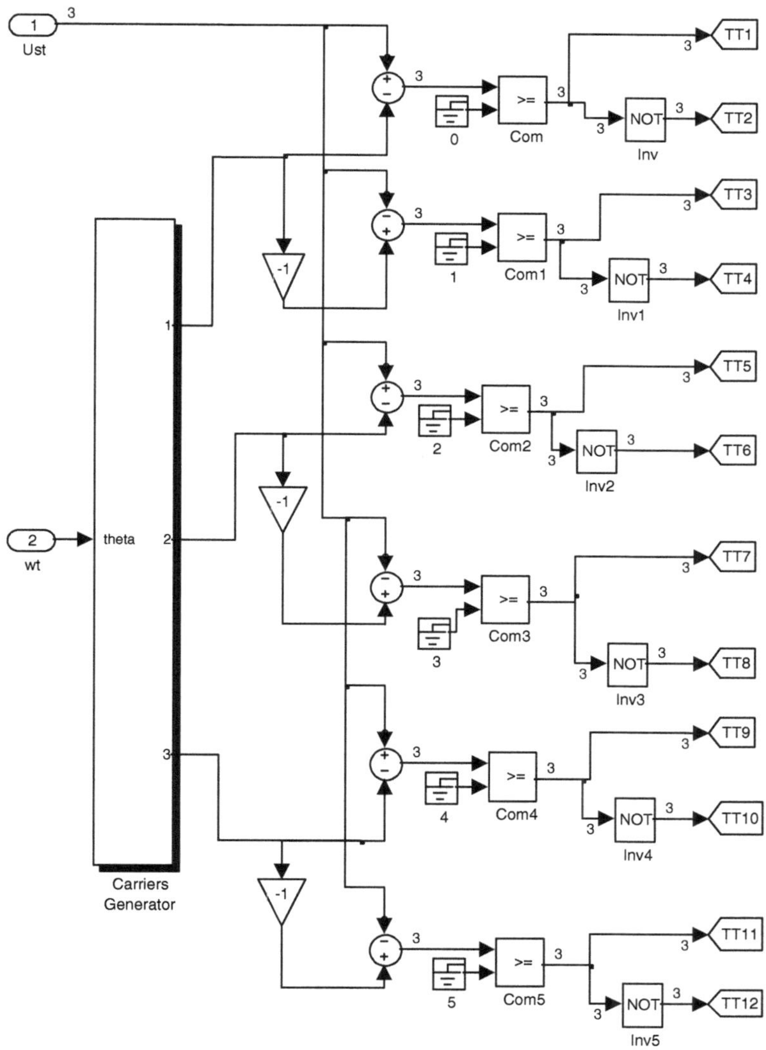

Fig. 3.50 MATLAB/Simulink model to generate gate pulses for 3-phase 7-level converter

greater than or equal to the reference signal value, and the other is asserted when the shifted carrier signal value is less than the reference signal value. In multilevel converters with level-shifted carriers, the amplitude modulation index can be calculated as

$$m_{al} = \frac{A_m}{(m-1)A_{\mathrm{cl}}}. \tag{3.20}$$

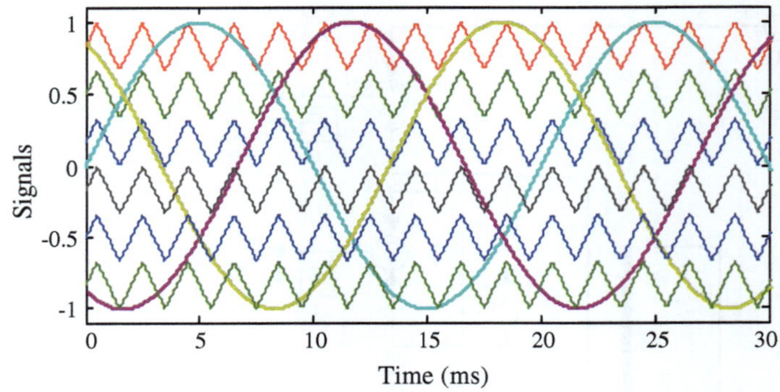

Fig. 3.51 Three reference signals and four level-shifted carrier signals

3.3.4.3 Reference Signals in Switching Schemes

Different types of major reference signals used in the traditional converters can also be used in the multilevel converter system, e.g., sinusoidal, third-harmonic injected sinusoidal, 60° modulated sinusoidal and trapezoidal. Each has its unique advantages and disadvantages. It is possible to reduce the switching losses by reducing the number of switching occurrences in each period. Flatting the top of the reference signal waveform in the third-harmonic injected sinusoidal, 60° modulated sinusoidal and the trapezoidal not only allows the possibility of switching loss reduction but also increased the range of linear modulation. These four reference signals with phase-shifted carrier signals generate four modulation schemes: (a) the phase-shifted carriers with sinusoidal references known as the sine pulse width modulation (SPWM), (b) the phase-shifted carriers with third-harmonic injected sinusoidal references known as the third-harmonic injected pulse width modulation (THPWM), (c) the phase-shifted carriers with 60° modulated sinusoidal references known as the 60° pulse width modulation (SDPWM), and (d) the phase-shifted carriers with trapezoidal type references known as the trapezoidal pulse width modulation (TRPWM). Figure 3.55 shows the MATLAB/Simulink model to generate 120° shifted sinusoidal three reference signals. Parameters such as modulation index, frequency, and phase displacement can be controlled through this block. Figure 3.56 plots the sinusoidal reference and carrier signals.

The MATLAB/Simulink model as shown in Fig. 3.57 can be used to generate the third-harmonic injected sinusoidal reference signals. The amplitude of the third-harmonic component can be controlled to obtain the desired quality output. The flatting top of the reference signal reduces the number of commutation in a cycle, which helps reduce the switching losses of the converter. Figure 3.58 shows the reference and carrier signals of THPWM scheme.

The amplitude-limited triangular signals can be used as the trapezoidal reference signal, which can easily be generated by the MATLAB/Simulink model as

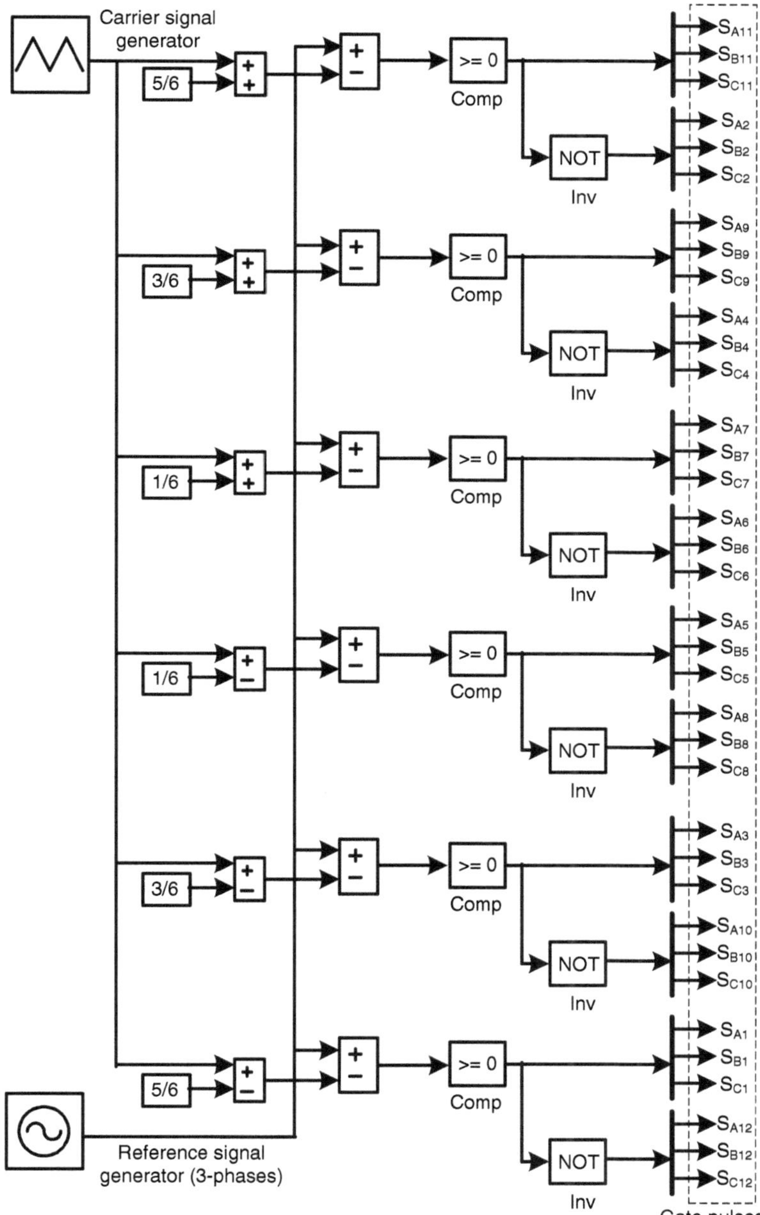

Fig. 3.52 Level-shifted modulation technique for a 3-phase 7-level NPC converter

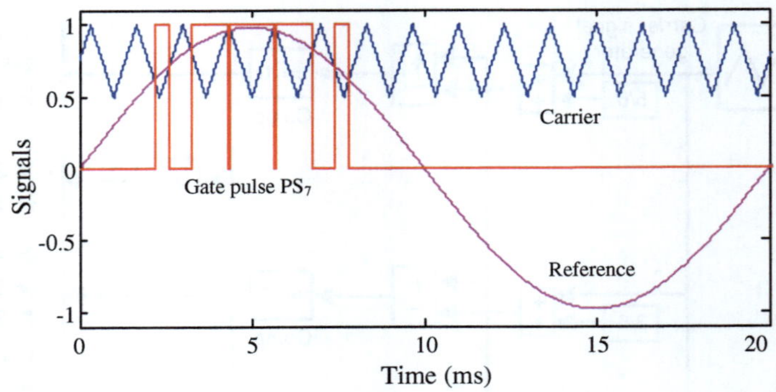

Fig. 3.53 Generation of *gate pulse* PS_7 with level-shifted carrier

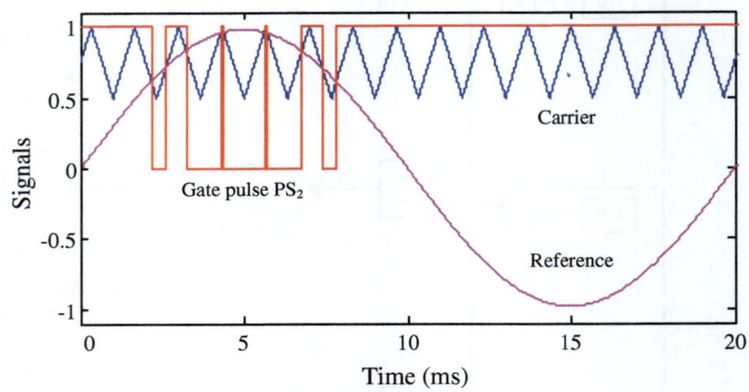

Fig. 3.54 Generation of *gate pulse* PS_2 with level-shifted carrier

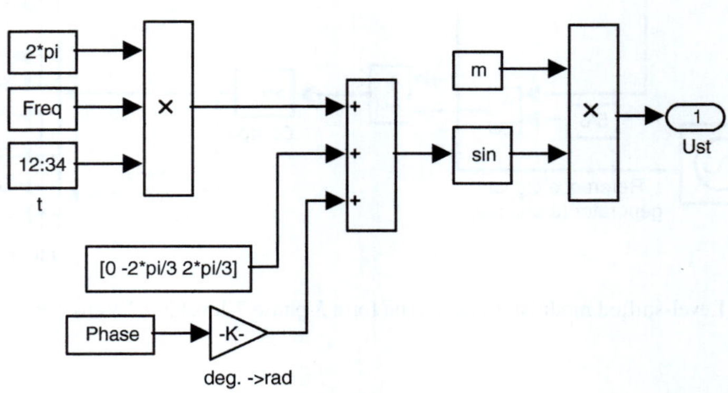

Fig. 3.55 MATLAB/Simulink model to generate sinusoidal reference signals

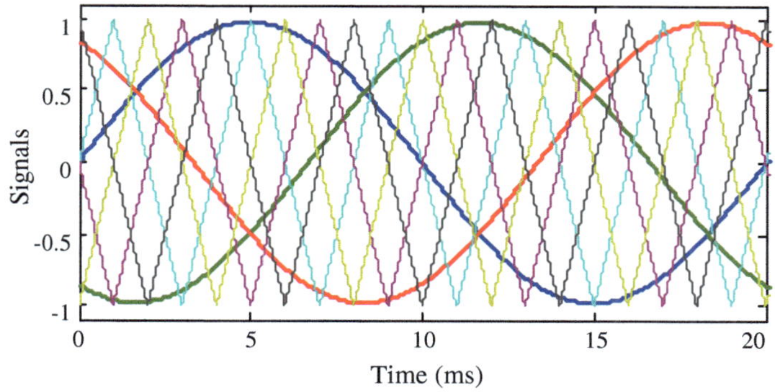

Fig. 3.56 SPWM modulation scheme for 3-phase 5-level converter

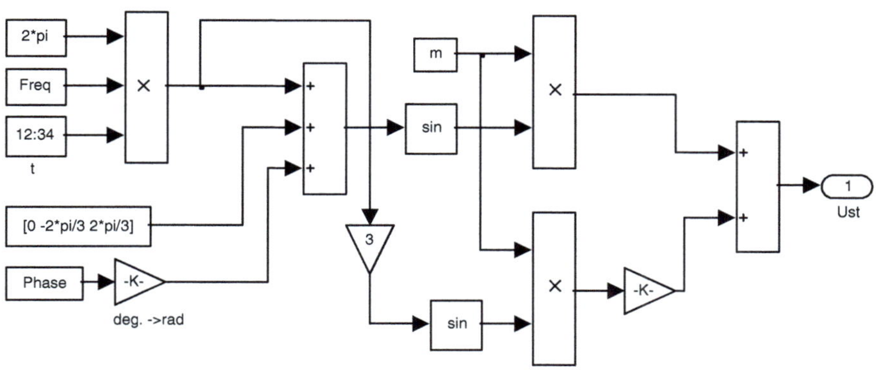

Fig. 3.57 MATLAB/Simulink model to generate third-harmonic injected sinusoidal reference signals for 3-phase 5-level converter

shown in Fig. 3.59. The parameters of the reference signals such as the frequency, maximum amplitude, and phase displacement can easily be controlled in the Simulink environment. This modulation scheme also uses flatting top references, which significantly reduce the switching losses of the converter. Figure 3.60 plots the trapezoidal reference signals.

Sixty degree pulse width modulation scheme also use sinusoidal references but their amplitudes are limited to 60°, clamped the magnitude of the reference signal, $\sin\theta$ at a constant value while θ is higher than 60°. The limiting of the peak value actually makes the top of the reference signals flattened. The MATLAB/Simulink model to generate sinusoidal reference signals can also be used to generate 60° reference signals. An extra amplitude limiter may be added in the model as shown in Fig. 3.61. Figure 3.62 plots the reference and carrier signals for SDPWM scheme.

The modulation schemes are applied on the 7-level to 19-level MMC converter systems, and performance is analyzed in the MATLAB/Simulink environment.

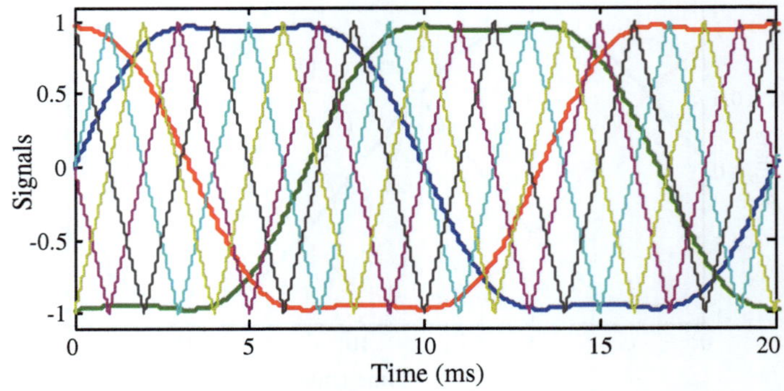

Fig. 3.58 Third-harmonic injected pulse width modulation schemes for 3-phase 5-level converter

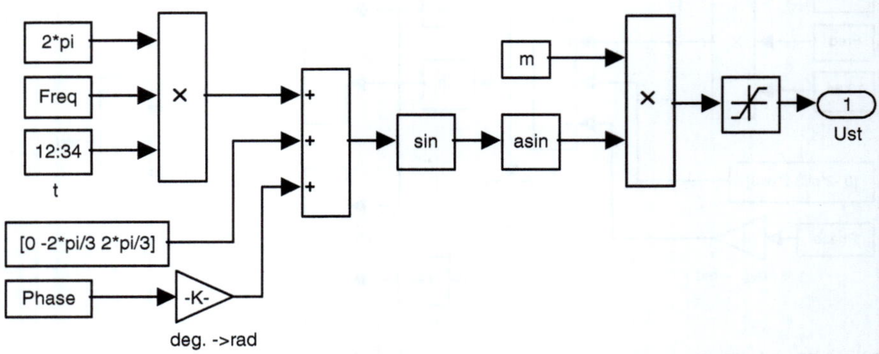

Fig. 3.59 MATLAB/Simulink model to generate trapezoidal reference signals for 3-phase 5-level converter

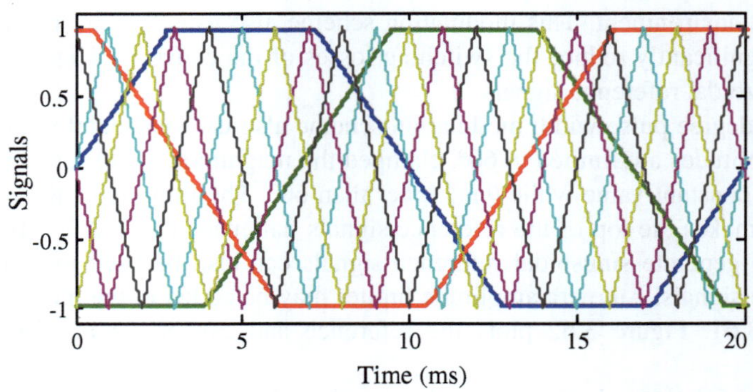

Fig. 3.60 Trapezoidal pulse width modulation schemes for 3-phase 5-level converter

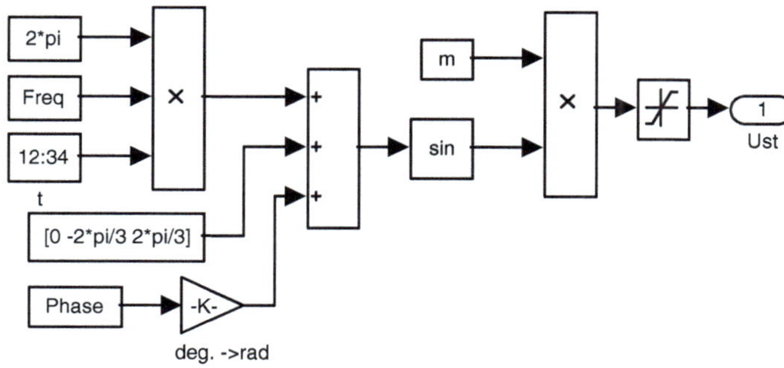

Fig. 3.61 MATLAB/Simulink model to generate 60° sinusoidal reference signals for 3-phase 5-level converter

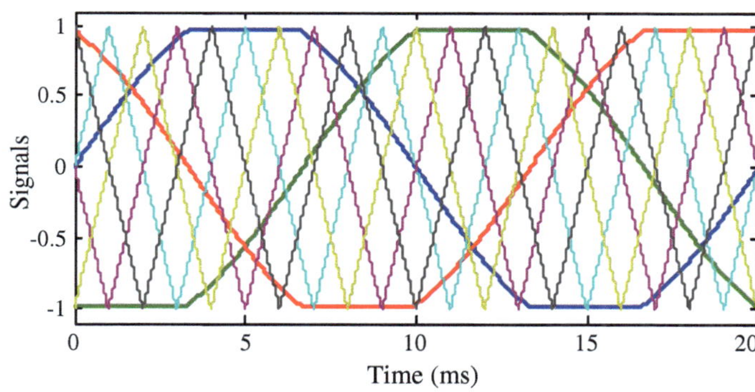

Fig. 3.62 Sixty degree pulse width modulation schemes for 3-phase 5-level converter

Of these four modulation schemes, the THPWM-based SPWM scheme gives the best harmonic performance, as shown in Fig. 3.63. The SDPWM and TRPWM schemes have higher lower-order harmonic content than the SPWM and THPWM schemes. In addition, the SPWM has shown a higher reduction rate for the high-level number.

3.4 Selection of Multilevel Converter Topology for Medium-Voltage Applications

The objective of this section is to find a suitable multilevel converter topology to design a medium-voltage converter using matured low-voltage semiconductor technology for renewable generation systems. Parameter identification is a vital part of the medium-voltage converter design process. The operating conditions and basic converter data are summarized in Table 3.10. If $V_{ll(\mathrm{rms})}$ is the grid line-to-line

Table 3.10 Converter specification

Technical data	Five-level	11-level
Converter line voltage	11 kV	11 kV
Minimum DC-link voltage	3,882 V	1,553 V
Nominal DC-link voltage	4,044 V	1,618 V
Phase current	250 A	250 A
Apparent output power	4.76 MVA	4.76 MVA
Converter carrier frequency	1–2 kHz	1–2 kHz
Output frequency	50 Hz	50 Hz

voltage and m the number of levels of the converter, the minimum DC-link voltage of each H-bridge inverter cell or each DC-link capacitor can be calculated as

$$V_{\mathrm{dc(min)}} = \sqrt{2}\frac{V_{ll(\mathrm{rms})}}{(m-1)}. \tag{3.21}$$

To determine the nominal DC-link voltage of each H-bridge inverter cell or each DC-link capacitor, a voltage reserve of 4 % is assumed, i.e.,

$$V_{\mathrm{dc(nom)}} = 1.04\,\mathrm{V}_{\mathrm{dc(min)}}. \tag{3.22}$$

If $I_{p(\mathrm{rms})}$ is the converter phase current, the apparent output power (for Wye connection) can be calculated from

$$S_c = \sqrt{3}V_{ll(\mathrm{rms})}I_{p(\mathrm{rms})}. \tag{3.23}$$

Semiconductor utilization is a very important issue in the design process of a medium- or high-voltage converter system as this is a considerable cost. The DC-link voltage needs to be considered when selecting IGBTs and diodes voltage ratings, and a cosmic ray effect assessment may also be necessary. In addition to the output capacity and voltage ratings of the converter, the availability of IGBT and diode modules in the market needs to be considered in the design process. For the 5-level converter, the commutation voltage of each H-bridge cell or DC-link capacitor is 4,044 V. The highest-rated IGBT in the market is 6.5 kV, which recommends the use of the 3.6 kV or lower-voltage applications. Therefore, two 4.5 kV series-connected IGBTs are assumed as a single switch for all the 5-level converter topologies. Table 3.11 summarizes the design of the power semiconductors for the converter specification in Table 3.10, with a carrier frequency of 1–2 kHz. To enable a converter output phase current of 250 A, a 400 A current rating is chosen for the power semiconductors. Assuming, the maximum capacitor voltage ripple, ΔV_c to be 5 % of the DC-link voltage and the DC output current, I_{dc} to the maximum amplitude of the phase current, $I_{p(\mathrm{rms})}$ [31], we can calculate the capacitance of the FC as

$$C = \frac{I_{p(\mathrm{rms})}}{n_c \Delta V_c f_c}. \tag{3.24}$$

where n_c is the number of series-connected FC cells.

Table 3.11 Components of different multilevel converters [16]

Number of levels	5			11		
Topology	NPC	FC	MMC	NPC	FC	MMC
IGBTs	48	48	48	60	60	60
Rated voltage (kV)	4.5	4.5	4.5	3.3	3.3	3.3
V_{com} (V)	2,022	2,022	2,022	1,618	1,618	1,618
$V_{com@100FIT}$ (V)	2,250	2,250	2,250	1,800	1,800	1,800
DVUF (%)	90	90	90	90	90	90
Diodes	18	–	–	54	–	–
Capacitors	–	18	–	–	135	–

In order to select a suitable converter topology for a medium-voltage converter, the following basic multilevel converter topologies are considered: a 5-level NPC, a 5-level FC, a 5-level MMC, an 11-level NPC, an 11-level FC, and an 11-level MMC. The comparison is made in terms of the number of semiconductors, semiconductor cost, THDs, filter size, and control complexity of the converters. The performance is analyzed and compared in the MATLAB/Simulink environment. To generate switching pulses, a level-shifted carriers-based switching scheme is used for the NPC converter topologies and a phase-shifted carriers-based switching scheme is used for the FC and MMC converter topologies with a carrier frequency of 1–2 kHz and a modulation index of 0.8–0.98.

3.4.1 *Detailed Design and Analysis of 5-Level and 11-Level NPC Converters*

With a 5-level converter topology, the commutation voltage of each DC-link capacitor is 4,044 V, i.e., each voltage level contributes 4,044 V to the output voltage waveform. The output line voltage waveform of a 5-level NPC converter consists of nine voltage levels and moderately matches the reference sine signal. Figure 3.64 plots the line voltage of a 5-level NPC converter without using the line filter circuit. The THDs of the line voltage are calculated at 17.26 %. The frequency spectrum is depicted in Fig. 3.65.

With the 11-level converter topology, the communication voltage of each DC-link capacitor is 1,618 V, i.e., each voltage level contributes 1,618 V to the output voltage waveform. The output line voltage waveform of an 11-level NPC converter consists of twenty-one voltage levels, and this aligns strongly with the reference sine signal. Figure 3.66 plots the line voltage of an 11-level NPC converter without using the line filter circuit. The THDs of the line voltage are calculated at 7.07 %. The frequency spectrum is depicted in Fig. 3.67.

The output voltage waveform of an 11-level NPC converter coincides to a large degree with the reference sine wave as compared to other 5-level converter

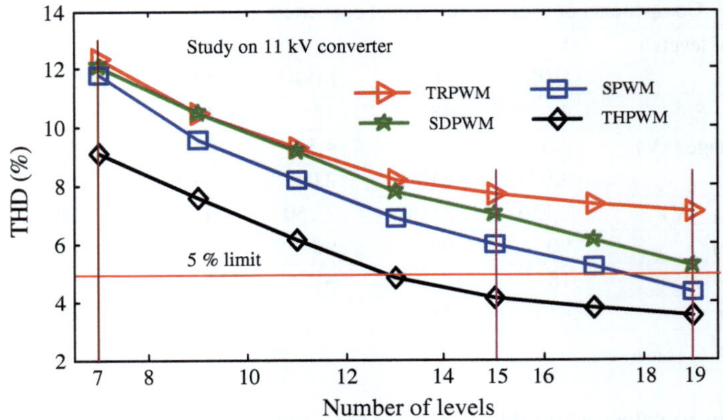

Fig. 3.63 Calculated THD at different level numbers ranging from 7-level to 19-level [1]

Fig. 3.64 Line voltage of 5-level neutral point clamped converter

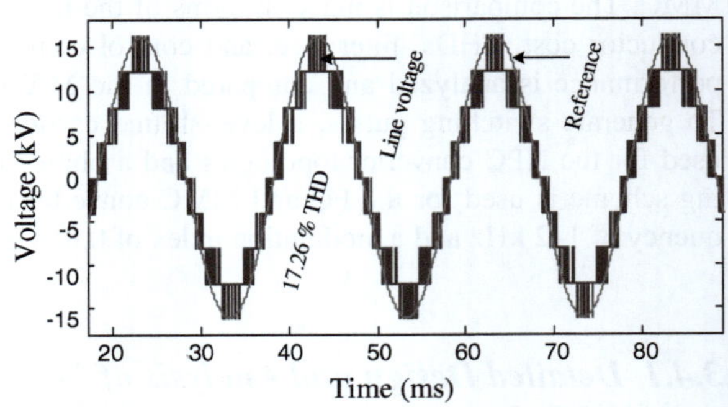

Fig. 3.65 Frequency spectrum of 5-level neutral point clamped converter

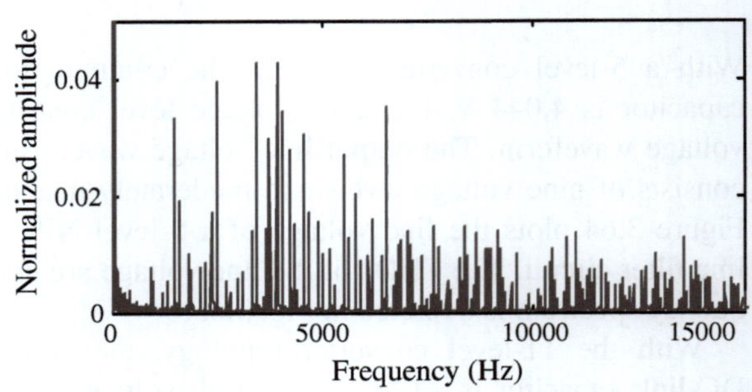

outputs. All 11-level converter output voltage waveforms are very close to the reference sine wave while the NPC converter performance is better than the others. Furthermore, from the output figures, it is clear that increasing the number of levels means improving the converter performance. It is assumed that each blocking diode voltage rating is the same as the active device voltage rating. A total of 54 diodes are required for an 11-level NPC converter. The large number of diodes affects

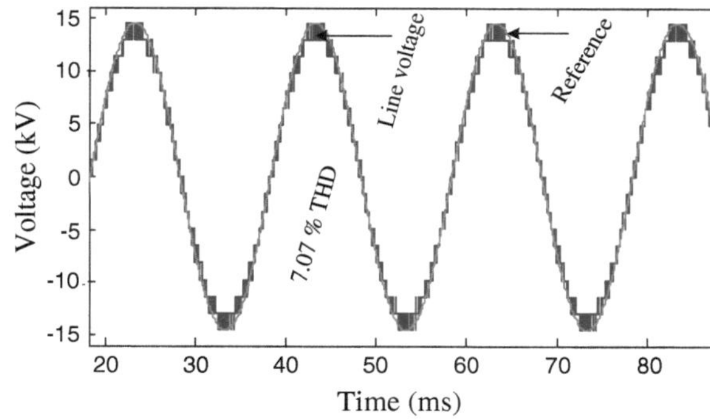

Fig. 3.66 Line voltage of 11-level neutral point clamped converter

the reverse recovery of the clamping diodes, and this is a major design challenge in high-voltage high-power systems. A list of the number of power components required for each converter topology is shown in Table 3.11. As already stated, the availability of IGBT and diode modules is also considered when designing the converter. For the 5-level NPC, FC and MMC converter topology, each IGBT switch is formed from the series connection of two 4.5 kV IGBTs so the number of IGBTs is 48. To enable a converter output phase current of 250 A, the simulation result is used to determine the current rating of the power semiconductors.

3.4.2 Detailed Design and Analysis of 5-Level and 11-Level FC Converters

A total of 6 and 45 clamping capacitors are required for each phase of the 5-level and 11-level FC converters, respectively. These large numbers increase the converter size and cost. In addition, they reduce the overall lifetime of the converter. The capacitor voltage balancing problem also becomes a challenging issue with such a high number of components. An 11 kV 5-level FC converter also requires 4 DC-link capacitors with a rated voltage of 4,044 V. Each voltage level contributes 4,044 V to the line- or phase-voltage waveforms. The line voltage waveform of a 5-level FC converter is shown in Fig. 3.68. Due to capacitor voltage imbalance, the voltage waveform of the FC converter is severely distorted compared with that of the NPC converter. Therefore, the harmonic performance is not as good as that of the NPC converter. Figure 3.69 illustrates the harmonic spectrum of the 5-level FC converter. In addition to 135 auxiliary capacitors, an 11 kV 11-level FC converter requires 10 DC-link capacitors with a rated voltage of 1,618 V.

Each voltage level contributes 1,618 V to the line- or phase-voltage waveforms. The line voltage waveform of an 11-level FC converter is shown in Fig. 3.70. Although, the 11-level converter output voltage waveform matches the reference sine signal, the performance is not as good as that of the NPC converter. The output line voltage frequency spectrum of an 11-level FC converter is shown in Fig. 3.71.

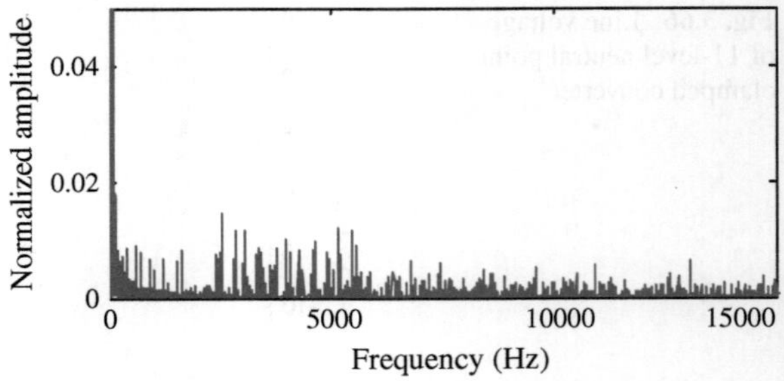

Fig. 3.67 Frequency spectrum of 11-level neutral point clamped converter

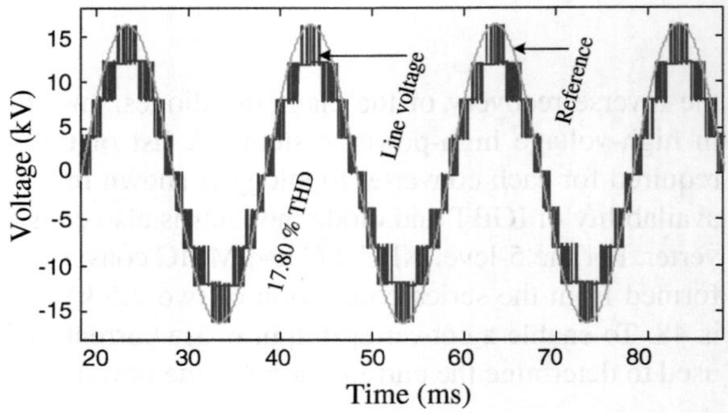

Fig. 3.68 Line voltage of 5-level flying capacitor converter

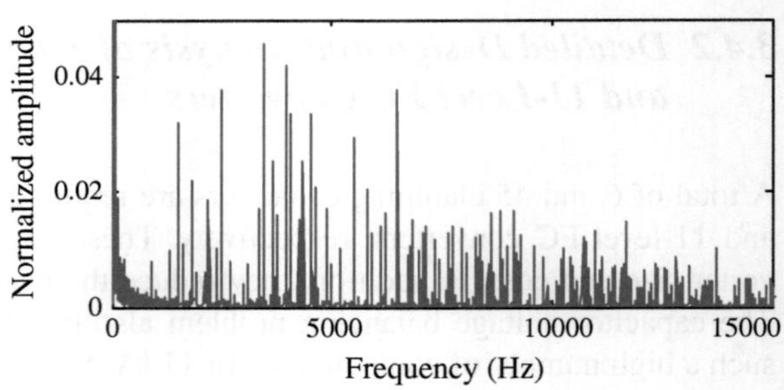

Fig. 3.69 Frequency spectrum of 5-level flying capacitor converter

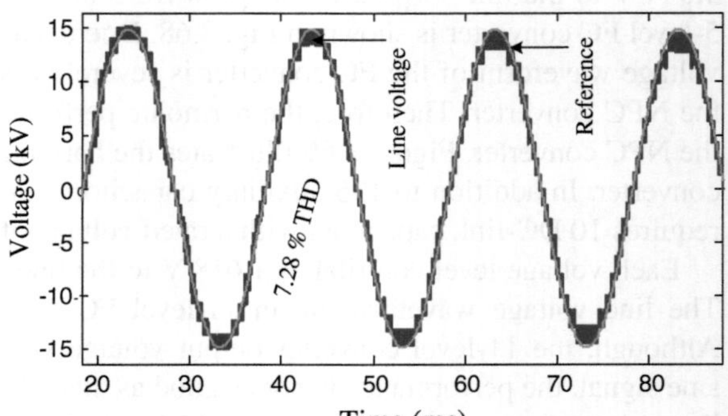

Fig. 3.70 Line voltage of 11-level flying capacitor converter

3.4.3 Detailed Design and Analysis of 5-Level and 11-Level MMC Converters

There are no blocking diodes or clamping capacitors in the MMC converter topology. The number of components scales linearly with the number of levels. Hence, the overall number of components is much lower than that of other multilevel converter topologies. The individual modules are similar and totally modular in construction, which makes it easy to implement for any number of levels. The higher number of attainable levels provides more scope for reducing harmonics. The high number of levels means that it is possible to connect the converter to the medium-voltage grid directly. However, unlike the NPC and FC converters, the MMC converter requires multiple isolated and balanced DC supplies. For example, a 3-phase 11-level MMC converter requires 15 isolated and balanced DC supplies. As the number of voltage levels increases, the voltage rating of these DC supplies decreases to a low level and this is favorable for some applications. The voltage rating of these DC supplies actually determines the voltage ratings of the active switching devices for each H-bridge inverter cell. The voltage balancing of multiple DC supplies is important for good output voltage waveforms. The output line voltage waveform of a 5-level MMC converter is shown in Figs. 3.72, and 3.73 shows its frequency spectrum. As calculated, an 11 kV 11-level MMC converter generates a 9-level line voltage waveform with 18.13 % THDs. The DC supplies voltage rating of an 11 kV 11-level MMC converter is 1,618 V. The available

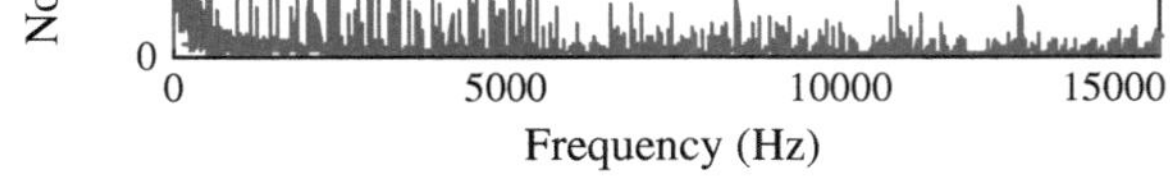

Fig. 3.71 Frequency spectrum of 11-level flying capacitor converter

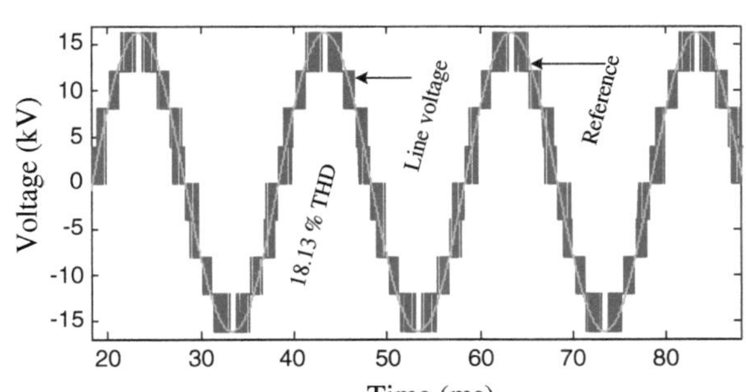

Fig. 3.72 Line voltage of 5-level modular multilevel cascaded converter

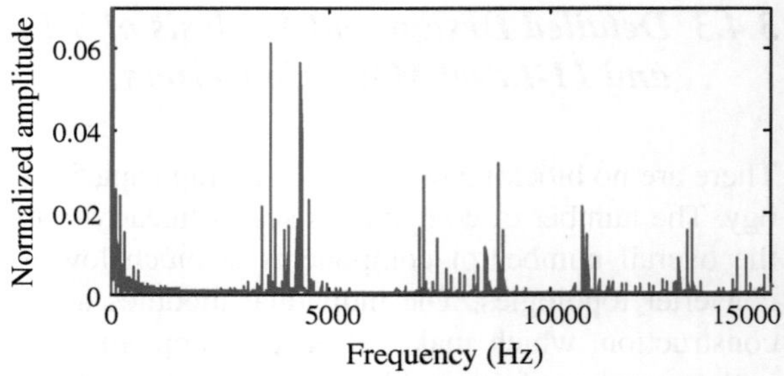

Fig. 3.73 Frequency spectrum of 5-level modular multilevel cascaded converter

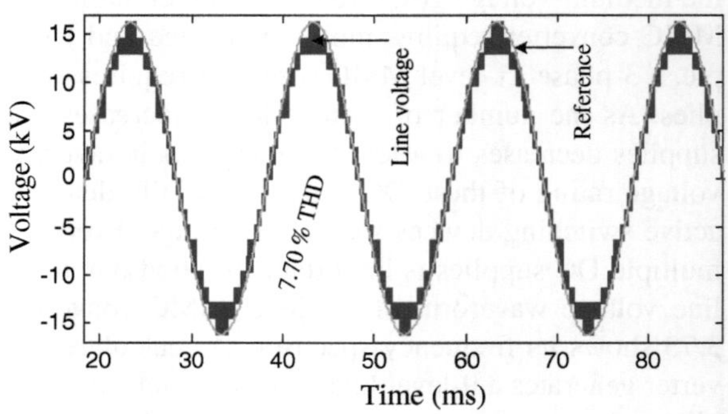

Fig. 3.74 Line voltage of 11-level modular multilevel cascaded converter

3.3 kV IGBT may be used to design the 11 kV 11-level converter because this IGBT is recommended for 1,800 V maximum applications.

With these IGBTs, 90 % DVUF can be ensured and this is important to confirm the high performance/cost of the power converter. Figure 3.74 plots the line voltage waveform of an 11-level MMC converter. There are 20 steps in the peak-to-peak line voltage waveform, and each step contributes 1,618 V to the line voltage of 32,360 V peak-to-peak. The 11 kV 11-level MMC converters give 7.70 % THDs, which is the highest among the three multilevel converter topologies. The frequency spectrum of the line voltage of an 11-level MMC converter is depicted in Fig. 3.75.

3.4.4 Performance Analysis and Topology Selection

The MMC converter does not require any auxiliary devices whereas NPC and FC converters require a huge number of auxiliary diodes and capacitors. Although at the 3-level, the number of auxiliary devices of the FC converter is comparable with that of the NPC converter, at a higher number of levels, it is a few times higher. For example, 3-phase 3-level FC and NPC converters require three and six auxiliary devices, respectively, and 3-phase 17-level FC and NPC converters

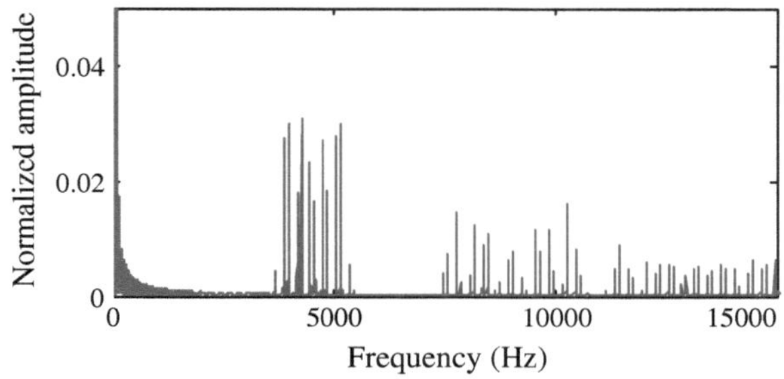

Fig. 3.75 Frequency spectrum of 11-level modular multilevel cascaded converter

Table 3.12 Cost and performance of different multilevel converters [16]

Number of levels	5			11		
Topology	NPC	FC	MMC	NPC	FC	MMC
Total comp.	66	66	48	114	195	60
THD (%)	17.26	17.80	18.13	7.07	7.28	7.70
Complexity	19	29	29	41	55	55
Total cost (AU$)	90,961	117,408	82,027	115,663	109,545	82,159

require 360 and 90 auxiliary devices, respectively. Figure 3.76 shows a comparison of the auxiliary device requirement in the FC and NPC converters. Table 3.12 summarizes the THDs for different multilevel converter topologies.

Among these three converter topologies, the NPC converter topology gives the best harmonic performance. The harmonic performance of the MMC converter topology is not as good as that of the NPC converter topology. The harmonic content decreases rapidly with the increase in the number of levels. This means that by increasing the levels of the converter, it is possible to keep the output voltage THDs to be less than, or equal to 5 % (according to the standards of IEEE1547 and IEC61727). Due to the modularity and the lower number of component requirements, the MMC converter would probably be the natural choice for high-level converter applications. Among these three converters, the 11-level MMC converter is the low-cost high-performance converter, and it is suitable for the direct connection to the medium-voltage grid. The price data quoted for the semiconductor devices, and capacitors were collected from the Galco Industrial Electronics and Farnell catalogs where devices were chosen from the same family so that it was possible to fit in with requirements.

The IGBTs chosen have integrated freewheel diodes, and hence, the diodes do not appear in the costing. The current rating of most devices is selected on the basis of simulation results. Table 3.12 also summarizes the estimated cost of different converter systems. The number of semiconductors increases with the number of levels but the change in terms of cost is small because the price of the lower-rated device is comparatively much lower. Because of the lower voltage and current requirements, the total semiconductor cost of the 11-level MMC converter is lower than that of all other topologies. The number of arithmetic

Table 3.13 Comparison of different multilevel converter topologies [16]

Number of levels	5			11		
Topology	NPC	FC	MMC	NPC	FC	MMC
Performance	0.92	0.97	1	0	0.02	0.06
Complexity	0	0.28	0.28	0.61	1	1
Cost	0.25	1	0	0.95	0.78	0.003
Total index	1.17	2.25	1.28	1.56	1.80	1.063

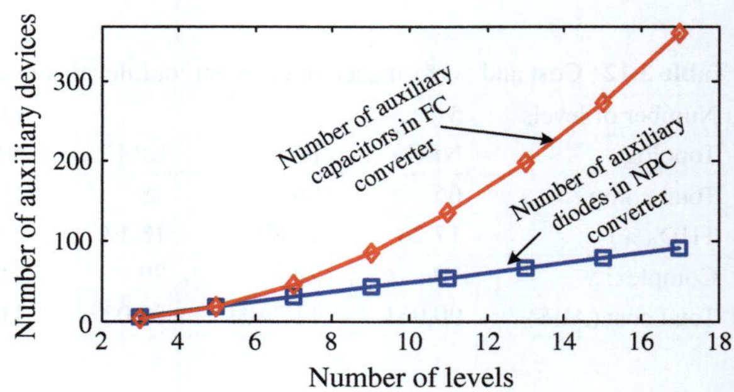

Fig. 3.76 Number of auxiliary devices in NPC and FC converters [32]

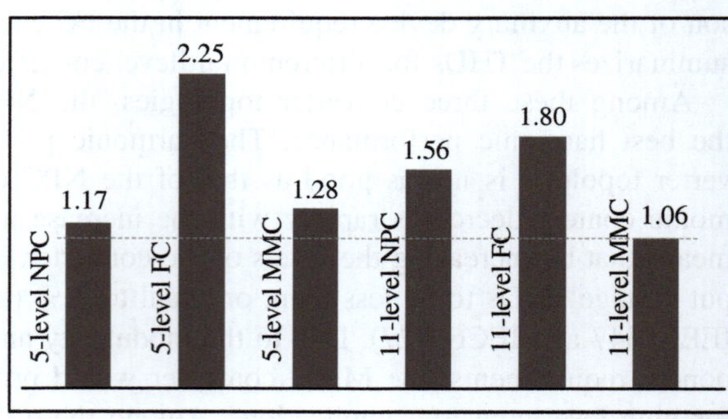

Fig. 3.77 Overall comparisons of multilevel converters [32]

and logic operations (ALOs) for the switching section is calculated through the MATLAB/Simulink environment. The number of ALOs is used to compare the complexity of the converters, as summarized in Table 3.12.

Based on the performance, control complexity, and semiconductor cost, an overall comparative study is carried out among the converter topologies mentioned. If y is the given value, $y_{\min}$ the minimum value, and $y_{\max}$ the maximum value in a group of data, the normalized index value can be calculated as

$$I_d = \frac{y - y_{\min}}{y_{\max} - y_{\min}}. \tag{3.25}$$

Table 3.13 summarizes the index values of Table 3.12. The lower the complexity, cost and THDs, the better the converter, i.e., a lower index value indicates a better converter performance/cost. The total index values of all converters are plotted in Fig. 3.77. For the lower-level number converter, the NPC converter topology may be the best choice. But for the higher-level number converter, the MMC converter topology may be the best choice. Moreover, in the MMC converter, the individual modules are identical and completely modular in construction, and thus, in case of a fault in one of these modules, it is possible to replace it quickly and easily. Therefore, the MMC is the proper converter topology to design a medium-voltage converter for the step-up transformer-less direct grid connection of renewable generation systems (Fig. 3.77).

3.5 Summary

Among the three basic multilevel converter topologies, the NPC converter topology gives the best harmonic performance. The number of auxiliary diodes in an NPC converter linearly increases with the number of converter level. This large number of diodes affects the reverse recovery of the clamping diodes which is a major design challenge in medium- or high-voltage applications. Therefore, the NPC converter topology is not suitable for medium- or high-voltage applications where a higher number of converter level is required.

The harmonic performance of the FC converter topology is slightly better than that of the MMC converter. In the FC converter, the number of required auxiliary capacitors is quadratically related to the number of converter levels. When the number of converter levels is high, the number of required capacitors will make the system impractical to implement. The FC converter requires an extra control circuit to pre-charge the clamping capacitors with the required voltage level, which decreases the system performance and increases the complexity and cost of the converter. These clamping capacitors increase the weight and volume of the converter and significantly reduce its lifetime. Therefore, the use of FC converter is still within the 5-level converter topology.

Unlike the NPC and FC converter, the MMC converter does not require any auxiliary diodes or capacitors. The MMC converter would be the natural choice to design medium-voltage power converters for step-up transformer-less grid integration of renewable generation units, because there is no requirement for any auxiliary devices, and individual modules are identical and completely modular in construction. The MMC converters can enable high-level number attainability as well as significant reduction in the semiconductor cost of the converter.

References

1. Islam MR, Guo YG, Zhu JG (2014) A high-frequency link multilevel cascaded medium-voltage converter for direct grid integration of renewable energy systems. IEEE Trans Power Electron 29(8):4167–4182

2. Islam MR, Guo YG, Zhu JG (2014) A review of offshore wind turbine nacelle: technical challenges, and research and developmental trends. Renew Sustain Energy Rev 33:161–176
3. Islam MR, Guo YG, Zhu JG (2014) A multilevel medium-voltage inverter for step-up-transformer-less grid connection of photovoltaic power plants. IEEE J Photovoltaics 4(3):881–889
4. Broeck HW, Skudelny HC, Stanke GV (1988) Analysis and realization of a pulse width modulator based on voltage space vectors. IEEE Transactions Ind Appl 24(1):142–150
5. Baker BK, Bannister LH (1975) Electric power converter. US Patent 3 867 643, Feb 1975
6. Peng FZ, Lai JS (1997) Multilevel cascade voltage-source inverter with separate dc source. US Patent 5 642 275, June 1997
7. Aiello MF, Hammond PW, Rastogi M (2001) Modular multi-level adjustable supply with series connected active inputs. US Patent 6 236 580, May 2001
8. Islam MR, Guo YG, Zhu JG (2011) Performance and cost comparison of NPC, FC and SCHB multilevel converter topologies for high-voltage applications. In: 2011 international conference on electrical machines and systems, Beijing, China, 20–23 Aug 2011, pp 1–6
9. Massoud MA, Finney SJ, Williams BW (2004) Multilevel converters and series connection of IGBT evaluation for high-power, high-voltage applications. In: Proceedings of the second international conference on power electronics, machines and drives, University of Edinburgh, UK, 31 Mar–2 Apr 2004, pp 1–5
10. Nabae A, Takahashi I, Akagi H (1981) A new neutral-point clamped PWM inverter. IEEE Trans Ind Appl 17(5):518–523
11. Manjrekar MD, Lipo TA (1998) A hybrid multilevel inverter topology for drive applications. In: Proceedings of the thirteenth annual applied power electronic conference exposition, Anaheim, California, 15–19 Feb 1998, pp 523–529
12. Corzine K, Familiant Y (2002) A new cascaded multilevel H-bridge drive. IEEE Trans Power Electron 17(1):125–131
13. Lai JS, Peng FZ (1996) Multilevel converters—a new breed of power converters. IEEE Trans Ind Appl 32(3):509–517
14. Nami A, Zare F, Ghosh A, Blaabjerg F (2010) Multi-output DC-DC converters based on diode clamped converters configuration: topology and control strategy. IET Trans Power Electron 3(2):197–208
15. Meynard TA, Foch H (1992) Multi-level conversion: high voltage choppers and voltage-source inverters. In: Proceedings of the 23rd annual IEEE power electronic specialised conference, Madrid, 29 June–3 July 1992, pp 397–403
16. Islam MR, Guo YG, Zhu JG (2014) Multilevel converters for step-up transformer-less direct grid integration of renewable generation units with medium voltage smart microgrids. In: Large scale renewable power generation: advances in technologies for generation, transmission and storage. Springer, Berlin Heidelberg, pp 127–149
17. Xu L, Agelidis VG (2002) Flying capacitor multilevel PWM converter based UPFC. IEE Proc Electric Power Appl 149(4):304–310
18. Xu L, Agelidis VG (2007) VSC transmission system using flying capacitor multilevel converters and hybrid PWM control. IEEE Trans Power Delivery 22(1):693–701
19. Islam MR, Guo YG, Lin ZW, Zhu JG (2014) An amorphous alloy core medium frequency magnetic-link for medium voltage photovoltaic inverters. J Appl Phys 115(17):17E710-1–17E710-3
20. Islam MR, Guo YG, Zhu JG (2013) A medium-frequency transformer with multiple secondary windings for medium-voltage converter based wind turbine generating systems. J Appl Phys 113(17):17A324-1–17A324-3
21. Marchesoni M, Mazzucchelli M, Tenconi S (1990) A nonconventional power converter for plasma stabilization. IEEE Trans Power Electron 5(2):212–219
22. Hammond PW (1997) A new approach to enhance power quality for medium voltage AC drives. IEEE Trans Ind Appl 33(1):202–208
23. Peng FZ, Lai JS, McKeever JW, VanCoevering J (1996) A multilevel voltage-source inverter with separate DC sources for static var generation. IEEE Trans Ind Appl 32(5):1130–1138

24. Tolbert LM, Peng FZ, Habetler TG (1999) Multilevel converters for large electric drives. IEEE Trans Ind Appl 35(1):36–44
25. Meynard TA, Foch H (1998) Electronic device for electrical energy conversion between a voltage source and a current source by means of controllable switching cells. US Patent 5737201 A, Apr 1998
26. Islam MR, Guo YG, Jafari M, Malekjamshidi Z, Zhu JG (2014) A 43-level 33 kV 3-phase modular multilevel cascaded converter for direct grid integration of renewable generation systems. In: Proceedings of IEEE innovative smart grid technologies conference (ISGT) Asia, Kuala Lumpur, Malaysia, 20–23 May 2014, pp 1–6
27. Wilkinson RW, Meynard TA, Mouton HT (2006) Natural balance of multi cell converters: the general case. IEEE Trans Power Electron 21(6):1658–1666
28. Meynard TA, Foch H, Thomas P, Cournault J, Jakob R, Nahrstaedt M (2002) Multicell converters: basic concepts and industry applications. IEEE Trans Ind Electron 49(5):955–964
29. Carrara G, Gardella S, Marchesoni M, Salutari R, Sciutto G (1992) A new multilevel PWM method: a theoretical analysis. IEEE Trans Power Electron 7(3):497–505
30. Rodriguez R, Franquelo LG, Kouro S, Leon JI, Portillo RC, Prats MAM, Perez MA (2009) Multilevel converters: an enabling technology for high-power applications. Proc IEEE 97(11):1786–1817
31. Krug D, Bernet S, Fazel SS, Jalili K, Malinowski M (2007) Comparison of 2.3 kV medium voltage multilevel converters for industrial medium voltage drives. IEEE Trans Ind Electron 54(6):2979–2992
32. Islam MR, Guo YG, Zhu JG (2013) Multiple-input-multiple-output medium frequency-link based medium voltage converter for direct grid connection of photovoltaic arrays In: Proceedings of international conference on electrical machines and systems (ICEMS2013), Busan, Korea, 26–29 Oct 2013, pp 202–207

Chapter 4
Design and Characterization of High-Frequency Magnetic Links Used in Power Electronic Converters

Abstract Nowadays, it is common to use high-frequency magnetic links in designing grid-connected power electronic converters, which can provide electrical isolation without increasing system volume and weight. The modular multilevel cascaded (MMC) converter topology has received significant attention in designing medium or high-voltage converters, but it requires balanced multiple isolated DC sources. To couple a renewable energy source to the MMC converter, a high-frequency magnetic link with multiple secondary windings can be an excellent option and is explored in this chapter. The high-frequency magnetic link is used to generate the isolated balanced multiple DC supplies for all of the H-bridge inverter cells of the MMC converter from a single, low-voltage power source. Compared with the conventional power frequency transformers operated at 50 or 60 Hz, the high-frequency magnetic links (in the range of a few kHz to MHz) have much smaller and lighter magnetic cores and windings, and thus much lower costs. The selection of core material and the design of magnetic link are multiphasic problems, which highly affects the efficiency of the magnetic link. This chapter presents the design and characterization of high-frequency magnetic links with advanced magnetic materials.

Keywords High-frequency · Magnetic links · Grid isolation · Voltage imbalance · Design optimization · Electromagnetic characteristics · Amorphous alloy · Nanocrystalline

4.1 Introduction

It is a common fashion to use high-frequency magnetic links in power electronic converters. According to the structure of the converter circuits, the high-frequency magnetic links can provide a few essential functions, such as step-up or step-down of the voltage levels, isolation of load from source, and generation of isolated multiple sources from single or multiple sources. The modular multilevel cascaded (MMC) converter requires multiple isolated DC sources that must be

M.R. Islam et al., *Power Converters for Medium Voltage Networks*,
Green Energy and Technology, DOI 10.1007/978-3-662-44529-7_4

balanced, and therefore, its application is not straightforward, especially in wind power generation systems [1]. For example, a 3-phase 13-level converter, as shown in Fig. 4.1, requires eighteen isolated and balanced DC supplies. In this context, high-frequency magnetic links with multiple secondary windings could be the possible option to generate isolated and balanced multiple DC supplies for all the H-bridge inverter cells of MMC converter from a single DC or AC source, e.g., a PV array or wind turbine [2, 3]. Increasing the operation frequency can greatly reduce the physical volume of the transformer that has already been widely used in the switched mode power supplies. Operated at 1.2 kHz, the weight and size of a 3-MW, 1.2-kHz transformer can be less that 8 % of an equivalent 50-Hz unit [4].

In 2011, Pereda and Dixon [5] proposed a high-frequency magnetic link to generate multiple DC supplies for all of the auxiliary H-bridge cells of asymmetric cascaded multilevel inverters. In the asymmetric cascaded multilevel converter,

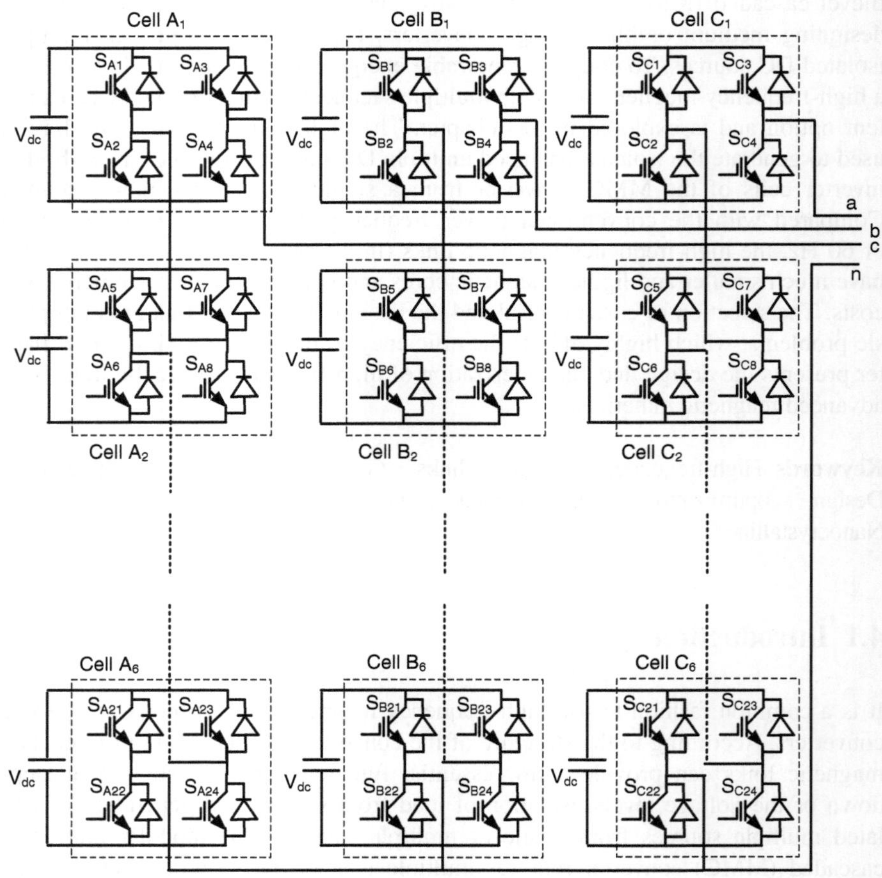

Fig. 4.1 Power circuit of a 3-phase 13-level MMC converter

all cascaded modules have different power ratings. This diminishes the modularity and thus limits the high-level attainability and increases the cost of the converter. In this converter system, the high-frequency magnetic link generates multiple sources from a single source with different voltages and power ratings. Moreover, in the asymmetric cascaded multilevel converter system, only the auxiliary H-bridges are connected through the high-frequency magnetic link. The main H-bridges are supplied directly from the source, which means that there is no electrical isolation between the grid and load. Therefore, the use of this converter is limited to the isolated winding motor applications [6]. For step-up-transformerless grid-connected renewable generation systems, the medium-voltage converter should have the capability to ensure electrical isolation and minimize the common mode issues. In this case, all H-bridge modules (auxiliary and main) require connecting through the high-frequency magnetic link.

In 2013, a high-frequency magnetic link with multiple identical secondary windings was proposed to interconnect wind turbine generator with MMC converter [7]. The identical secondary windings of the common magnetic link generate multiple isolated sources for the MMC converter. Figure 4.2 shows the basic block diagram of common high-frequency magnetic-link-based medium-voltage converter system. Although the common magnetic link may ensure equal voltages at the secondary terminals, the topology diminishes modularity of the power conversion system, which is very important for medium- or high-voltage high-power system. The implementation of high-power high-frequency inverter is the main challenging issue. Moreover, the leakage inductance may also limit the power capacity of the high-frequency magnetic link. In this regard, identical multiple four windings (single primary and three secondaries) high-frequency magnetic links can be connected in cascaded form to have same current passing through all primary windings of all magnetic links.

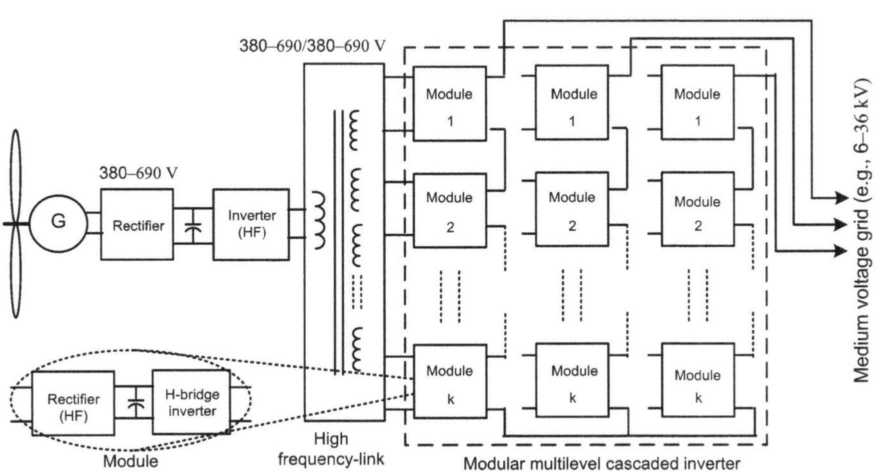

Fig. 4.2 Common magnetic-link-based medium-voltage converter for wind power generation systems [8]

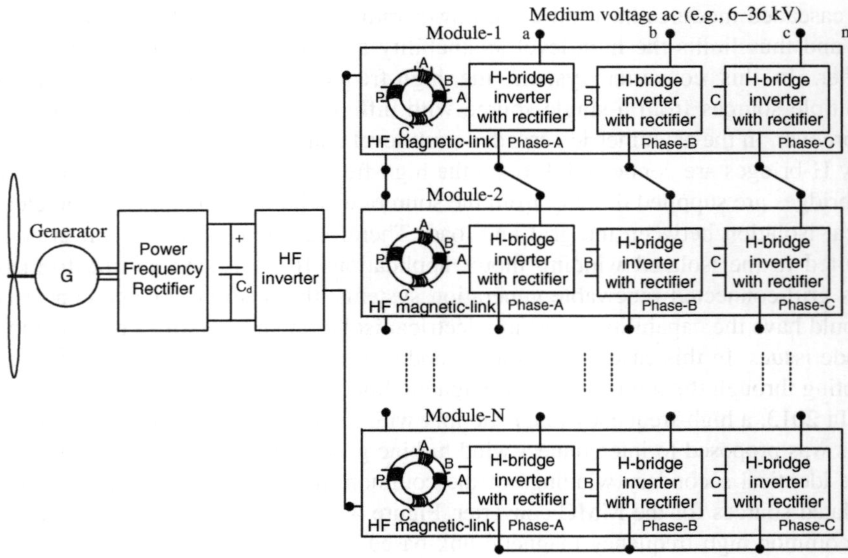

Fig. 4.3 Cascaded high-frequency magnetic-link-based medium-voltage converter for wind power generation systems

Figure 4.3 shows the basic block diagram of cascaded high-frequency magnetic-link-based MMC converter. The cascaded magnetic links give freedom to use multiple low-power magnetic cores other than a single high-power core, which helps to increase the modularity and reduce the leakage inductances. Still this topology uses single high-frequency inverter to energize all high-frequency magnetic links, whose implementation is really critical when the power level is high.

In order to keep the modularity of whole power conversion system, the high-frequency magnetic link as well as high-frequency inverter should also be modular. Each identical magnetic link can be excited by identical high-frequency inverter. All high-frequency inverters may be energized by the same power source, which helps generate equal voltage at the secondary terminals and ensure balanced DC supplies for the MMC converter. In this topology, each module consists of a high-frequency inverter, a four windings magnetic link, and three H-bridge inverters with high-frequency rectifiers. Each module may provide three-level line-to-line voltage. Figure 4.4 shows a modular converter for medium-voltage grid connection of wind turbine generators. With this topology, it is very simple to change the converter power level by addition or subtraction of modules.

In power frequency transformers, the most widely used magnetic materials are silicon-steel sheets, nickel-steel sheets, and nanocrystalline ribbons. With higher excitation frequencies, the skin and proximity effects, hysteresis losses, and dielectric losses will increase significantly. Therefore, various better magnetic materials with high saturation flux densities and low specific power losses have been developed in the last few decades. Among these materials, the amorphous

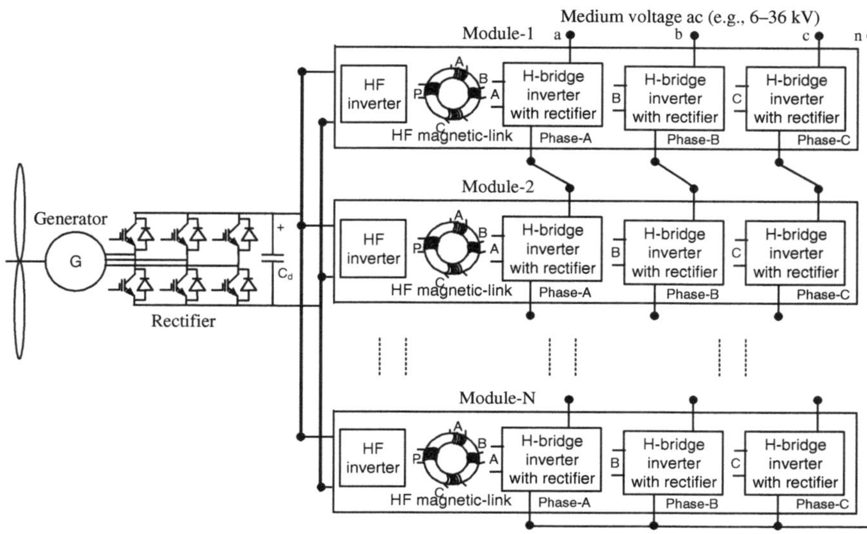

Fig. 4.4 Modular medium-voltage converter for wind power generation systems

and nanocrystalline are the most commonly used core materials in recent high-frequency applications. This reduced weight and volume medium-frequency transformer can be integrated in the MMC converter structure to overcome all the existing limitations, such as the requirement for multiple isolated DC supplies, balancing of DC-link voltages, grid isolation, and common mode issues [9].

This chapter presents the development and characterization of a high-frequency magnetic link with multiple secondary windings to couple the renewable energy source to the MMC converter. The high-frequency magnetic link is used to generate the isolated balanced multiple DC supplies for all of the H-bridge inverter cells of the MMC converter from a single low-voltage power source. Compared with the conventional power frequency transformers, the high-frequency magnetic links have much smaller and lighter magnetic cores and windings, and thus much lower costs. The high-frequency magnetic-link-based multilevel converter systems will have the following advantages: (i) capable of generating multiple isolated sources from a single or multiple sources, (ii) inherent DC-link voltage balancing, (iii) inherent grid isolation capability, and (iv) a wide range of MPPT operation.

4.2 Design of High-Frequency Magnetic Links

In the transformer design, the winding electromotive force (emf) is proportional to the number of turns, frequency, and the magnetic flux linked to the winding, or

$$e_{(\mathrm{emf})} \propto N f \varphi \tag{4.1}$$

where N is the number of turns, f the frequency of excitation current, and Φ the magnetic flux linking to the winding. For a given power capacity, as the operating frequency increases, the required cross-sectional area of the magnetic core and the number of turns of the primary and secondary windings can be dramatically reduced. The following sections present the design and analysis of two high-frequency magnetic links of single primary and six secondary windings. The magnetic links can be used to generate six isolated balanced DC supplies of a three-phase 5-level MMC converter. The concept can be used to model other magnetic links for other power converter topologies by only changing the number of primary and secondary windings according to power converter requirements.

4.2.1 Flowchart of High-Frequency Magnetic-Link Design

Taking into account the operating frequency of commercially available power semiconductor devices, one can find a frequency window in the range of 1–20 kHz for the proposed system. According to the power inverter rating, the magnetic-link specifications, such as the rated power, frequency, excitation current, and voltage, can be calculated. From the specifications of magnetic link and the data sheets of core materials, magnetic-link initial parameters can be calculated. These parameters can be used as the initial values of optimization process. The volume and weight of the transformer were optimized by selecting proper parameters. Different factors are considered during the design process, such as the winding dimensions, hole reserve for natural cooling, maximum temperature limits, maximum power loss, availability of core material stripe dimensions (standard available widths in the range of 2.5–50 mm and thickness of 20 μm), parasitic parameters, skin and proximity effects, and possibility to induce identical voltage in multiple secondary windings. Single-layer windings have low AC/DC resistance ratios, which would increase significantly the winding and core dimensions. For simplicity of the winding process, a toroidal core structure is adopted. Finally, the inner and outer diameters, and height of the core are chosen as 6.5, 10.5, 25 mm, respectively. The flowchart of the proposed methodology for the multiple secondary windings high-frequency magnetic-link design is depicted in Fig. 4.5. As shown, the design process involves multiphysics problems with some critical decision making tasks.

4.2.2 Core Material Selection

Along with the advent of new power semiconductor devices, different soft magnetic materials are conceived with a high magnetic saturation and low core loss to reduce the weight and volume of conventional power transformers. The grain-oriented silicon-steel sheet, which are commonly used as the core material for power frequency transformers, are not suitable for medium-frequency applications because of the

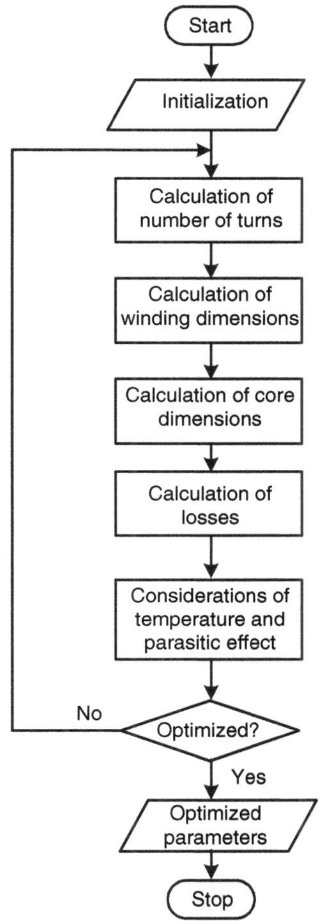

Fig. 4.5 Flowchart of high-frequency magnetic-link design process [10]

heavy eddy current loss [11]. The soft ferrites have been widely used in medium- and high-frequency inductors and transformers due to their low price and general availability. Because of the low saturation flux density [11] (only 0.3–0.5 T), which results in the transformer's large size, they are not suitable for large power applications. The amorphous alloy and nanocrystalline magnetic materials have excellent magnetic characteristics, such as high permeability, high saturation flux density, and relatively low core losses at medium-/high-frequency range [12–14]. Metglas is a commercially available amorphous material manufactured by Hitachi Metals, Japan. Figure 4.6 shows a photograph of Metglas stripe. The Metglas magnetic alloys 2705M and 2714A are cobalt-based materials with maximum flux density of 0.77 and 0.57 T, and the specific core losses are about 6 W/kg and 3 W/kg, respectively, at 10 kHz sinusoidal excitation of 0.3 T. The alloy 2,826 MB is iron–nickel-based material with maximum flux density of 0.88 T and specific core loss of 30 W/kg at

Fig. 4.6 Metglas amorphous alloy 2605S3A stripe of 20 μm thickness and 25 mm width

Table 4.1 Properties of Metglas alloys [15]

Parameters	2605S3A	2705M	2826MB	2714A	2605SA1
Maximum flux density (T)	1.41	0.77	0.88	0.57	1.56
Core loss at 10 kHz, 0.3 T (W/kg)	7	6	30	3	20
Basic material	Iron	Cobalt	Iron nickel	Cobalt	Iron
Resistivity (μΩcm)	138	136	138	142	130
Curie temperature (°C)	358	365	353	225	395
Magnetostriction (ppm)	20	<0.5	12	<0.5	27
Mass density (g/cm^3)	7.29	7.80	7.90	7.59	7.18
Ribbon thickness (μm)	18	22	29	15	25
Minimum ribbon width (mm)	2.5	2.5	2.5	2.5	5
Maximum ribbon width (mm)	50	50	12.7	50	213

10-kHz sinusoidal excitation of 0.3 T. The Metglas alloys 2605SA1 and 2605S3A are iron-based material with maximum flux density of 1.56 and 1.41 T, and the specific core losses are about 20 and 7 W/kg, respectively, at 10-kHz sinusoidal excitation of 0.3 T. Table 4.1 summarizes the properties of five different Metglas alloys, which are commercially available.

There are a few vendors providing commercially manufactured nanocrystalline magnetic materials, such as Hitachi Metals, Vacuumschmelze, MH&W International, and AMMET, etc. Commercially available nanocrystalline magnetic materials are Finemet, Vitroperm, and Nanoperm, which are manufactured by Hitachi Metals, Vacuumschmelze, and MH&W International, respectively. Figure 4.7 shows a photograph of Vitroperm nanocrystalline stripe. Table 4.2 summarizes the properties of nanocrystalline materials. The precursor material of Finemet is amorphous metal obtained by rapid quenching the molten metal,

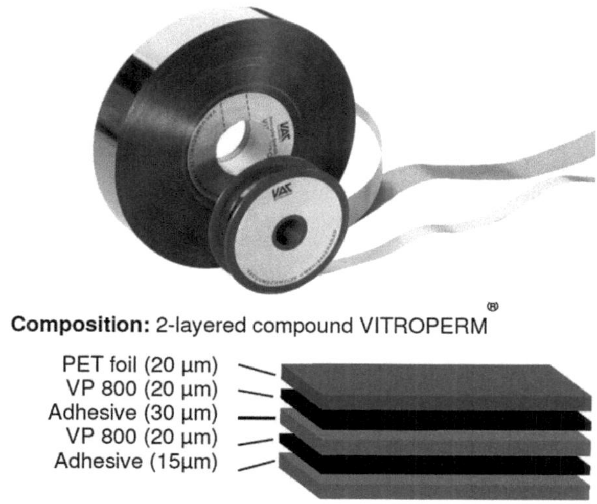

Fig. 4.7 Compound VITROPERM stripe [16]

Table 4.2 Properties of nanocrystalline magnetic materials

Parameters	Finemet	Vitroperm	Nanoperm
Maximum flux density (T)	1.23	1.20	1.20
Core loss at 100 kHz, 0.3 T (W/kg)	85	80	110
Basic material	Fe, Si, B	Fe, Cu, Nb, Si, B	Fe, Cu, Nb, Si, B
Resistivity (μΩcm)	120	115	115
Curie temperature (°C)	570	600	600
Magnetostriction (ppm)	<1	<0.5	<0.5
Mass density (g/cm^3)	7.30	7.35	7.35
Ribbon thickness (μm)	18	18	17

consisting of Fe, Si, B, and small amounts of Cu and Nb. By applying heat treatment to the alloy at higher temperature than its crystallization temperature, this alloy forms nanocrystalline structure (grain size of approximately 10 nm).

Although the nanocrystalline material has lower specific core loss than Metglas, its saturation flux density (about 1 T) is much lower than that of Metglas, which is up to 1.56 T. Taking into account the flux density, specific core loss, cost, and availability, the Metglas alloys 2605S3A and 2605SA1 stripe of 20 and 30 μm thickness, respectively, have been chosen as the core materials. In order to develop test cores, the core material Metglas alloys 2605S3A and 2605SA1 ribbons were collected from Metglas Inc.

The coefficients, k', m', and n' in Steinmetz Eq. (4.2) deduced from the datasheet for this material are experimental results under sinusoidal voltage excitation, where f is the frequency in kHz and B the magnitude of flux density in T. Table 4.3 summarizes the datasheet coefficients of Steinmetz equation. Usually, the high-frequency magnetic links are operated under non-sinusoidal voltage excitations, i.e., square-wave voltage [17]. Therefore, for this design, new coefficients

Table 4.3 Coefficients of Steinmetz equation with sinusoidal wave excitation of Metglas alloy 2605SA1

Excitation voltage type	Coefficients		
	k'	m'	n'
Sinusoidal (from datasheet)	6.50	1.51	1.74

Table 4.4 Coefficients of Steinmetz equation with square-wave high-frequency excitation

Core material	Coefficients		
	k'	m'	n'
2605S3A	6.571	1.422	1.999
2605SA1	6.256	1.580	1.619

are calculated by measurements under square-wave voltage excitations. The newly derived coefficients and datasheet coefficients are listed in Table 4.4.

$$P_{\text{core}} = k' f^{m'} B^{n'} \tag{4.2}$$

Using these two sets of coefficients, the specific core losses of Metglas alloy 2605SA1 have been compared. About 20–30 % extra loss is observed due to non-sinusoidal excitation waveform. Figure 4.8 shows the specific core loss of Metglas alloy 2605SA1 under square-wave voltage excitation. Compared with Metglas alloy 2605SA1, 2605S3A gives much lower core loss. Figure 4.9 plots the specific core loss of Metglas alloy 2605S3A.

4.2.3 Calculation of Number of Turns

Coils of high-frequency magnetic links are usually wound with Litz wires. High number of turns may increase the volume, weight, cost, and winding complexity of the high-frequency magnetic links. On the other hand, sufficient number of turns is

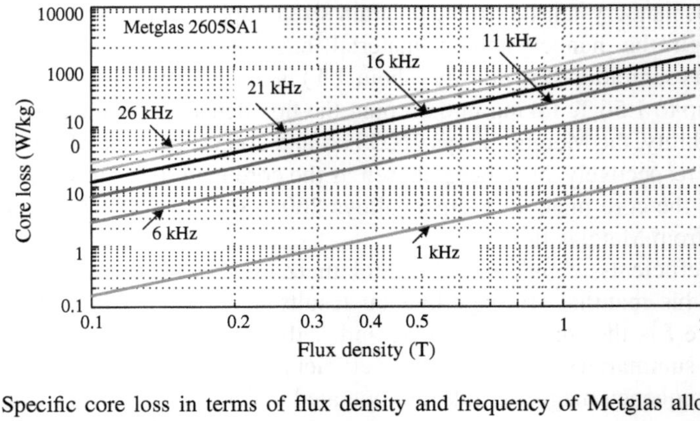

Fig. 4.8 Specific core loss in terms of flux density and frequency of Metglas alloy 2605SA1 under square-wave excitation

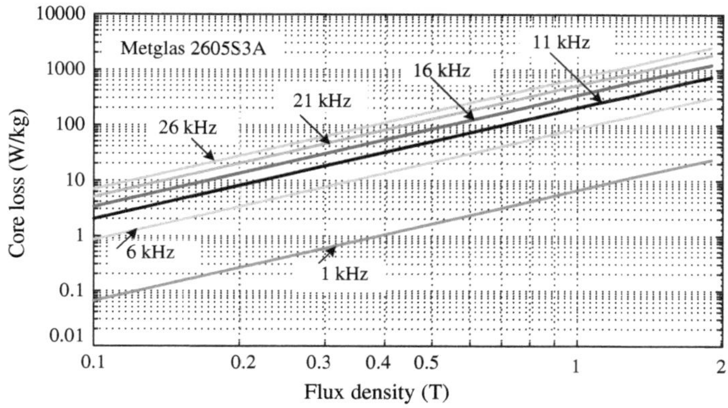

Fig. 4.9 Specific core loss in terms of flux density and frequency of Metglas alloy 2605S3A under square-wave excitation

required for proper electromagnetic inductions. Therefore, selection of an optimal number of turns is important for the design of high-frequency magnetic links.

By Faraday's law, the voltage, v, and flux, φ, of a transformer can be related by

$$v(t) = N\frac{\mathrm{d}\varphi}{\mathrm{d}t} \tag{4.3}$$

where N is the number of turns. When the transformer works with a square-wave voltage, according to (4.3), a triangular flux is required to generate the square-wave voltage as shown in Fig. 4.10, where T is the period of excitation voltage, $V_{\max}$ the maximum excitation voltage, and $\varphi_{\max}$ the maximum flux.

The triangular flux can be expressed mathematically as

$$\varphi(t) = \begin{cases} \frac{\varphi_{\max}}{T/4}\left(t - \frac{T}{4}\right) & 0 \le t \le \frac{T}{2} \\ \frac{-\varphi_{\max}}{T/4}\left(t - \frac{3T}{4}\right) & \frac{T}{2} \le t \le T \end{cases}. \tag{4.4}$$

The expression of the voltage can be deduced by substituting (4.4) into (4.3) as

$$v(t) = \begin{cases} N\frac{\varphi_{\max}}{T/4} & 0 \le t \le \frac{T}{2} \\ -N\frac{\varphi_{\max}}{T/4} & \frac{T}{2} \le t \le T \end{cases}. \tag{4.5}$$

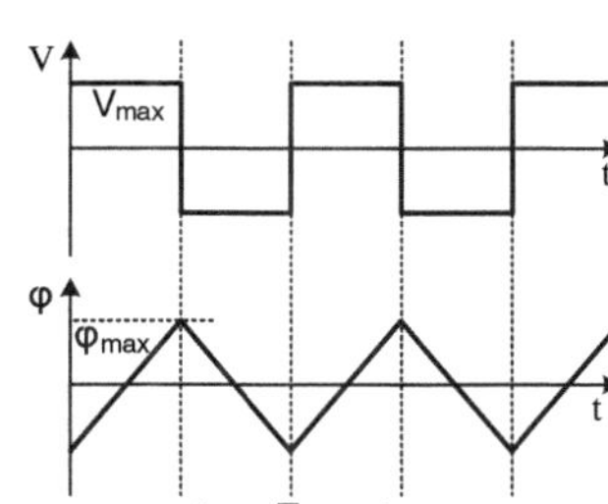

Fig. 4.10 Voltage and flux in a square-wave transformer

Let

$$V_{\max} = N\frac{\varphi_{\max}}{T/4} \tag{4.6}$$

and (4.5) can be rewritten as

$$v(t) = \begin{cases} V_{\max} & 0 \le t \le \frac{T}{2} \\ -V_{\max} & \frac{T}{2} \le t \le T \end{cases}. \tag{4.7}$$

The rms value of the excitation voltage can be calculated as

$$V_{\mathrm{rms}} = \sqrt{\frac{1}{T}\left\{V_{\max}^2\left(\frac{T}{2}\right) + (-V_{\max})^2\frac{T}{2}\right\}} \tag{4.8}$$

$$V_{\mathrm{rms}} = \sqrt{\frac{2}{T}\left\{V_{\max}^2\left(\frac{T}{2}\right)\right\}} = V_{\max} \tag{4.9}$$

$$V_{\mathrm{rms}} = V_{\max} \tag{4.10}$$

If f is the frequency of the excitation voltage, $B_{\max}$ the maximum flux density, and A the cross-sectional area of the transformer core, from (4.6) and (4.10), the expression of the number of turns can be deduced as

$$V_{\mathrm{rms}} = V_{\max} = N\frac{\varphi_{\max}}{T/4} \tag{4.11}$$

$$V_{\mathrm{rms}} = 4fN\varphi_{\max} \tag{4.12}$$

$$V_{\mathrm{rms}} = 4fNAB_{\max} \tag{4.13}$$

Since $\varphi_{\max} = B_{\max}A$, (4.12) can be rewritten as

$$\therefore N = \frac{V_{\mathrm{rms}}}{4fAB_{\max}}. \tag{4.14}$$

If the transformer excitation frequency is 10 kHz, core size is 5 cm^2 (2.5 × 2 cm), and the flux density is 1 T, the minimum number of turns required by the primary winding of the transformer can be calculated as

$$N_{\mathrm{p}} = \frac{V_{\mathrm{rms(pri)}}}{4fAB_{\max}} = \frac{210}{4 \times 10 \times 10^3 \times 5 \times 10^{-4} \times 1} = 10.50$$

and the required minimum number of turns of each secondary winding as

$$N_{\mathrm{s}} = \frac{V_{\mathrm{rms(sen)}}}{4fAB_{\max}} = \frac{374}{4 \times 10 \times 10^3 \times 5 \times 10^{-4} \times 1} = 18.70.$$

Finally, the designs are considered with 14 turns for primary windings and 25 turns for each secondary winding.

4.2.4 Winding Wire Selection

The current density within the conductors becomes highly uneven as the excitation frequency increases, due to mainly the skin and proximity effects. With the increase of excitation frequency, the current density will reduce inside the conductor and increase near the surface. This is called the skin effect. As a result, although the total amount of current within the conductor is not affected, the current density distribution across the conductor becomes non-uniform, and the AC resistance significantly higher than the DC resistance. The skin effect can be evaluated by the skin depth, which is defined as the radial distance from the surface of the conductor to where the value of current density is 37 % smaller than its value at the surface [18]. The relationship between the skin depth and the operating frequency can be deduced as

$$\delta = \sqrt{\frac{\rho}{\pi \mu f}}. \tag{4.15}$$

where μ is the permeability of the material, ρ the resistivity, and f the frequency. Figure 4.11 depicts the skin depth versus frequency of a copper conductor.

If R_{dc} is the DC resistance of a winding carrying a current of I, the skin effect loss in a Litz winding can be calculated by [18]

$$P_{\mathrm{skin}} = R_{\mathrm{dc}} \frac{\gamma}{4} I^2 \tau_1(\gamma). \tag{4.16}$$

where

$$\gamma = \frac{d}{\delta\sqrt{2}} \tag{4.17}$$

and

$$\tau_1(\gamma) = \frac{\mathrm{ber}(\gamma)\mathrm{bei}'(\gamma) - \mathrm{bei}(\gamma)\mathrm{ber}'(\gamma)}{\mathrm{ber}'(\gamma)^2 + \mathrm{bei}'(\gamma)^2} \tag{4.18}$$

where

$$\mathrm{ber}(\gamma) = \mathrm{Re}\left[J(\gamma e^{j3\pi/4})\right], \tag{4.19}$$

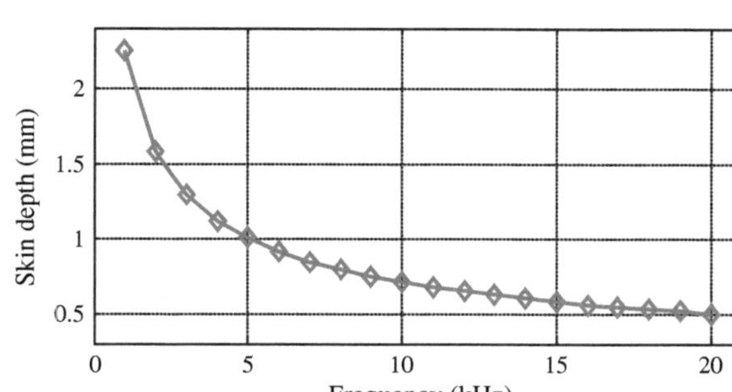

Fig. 4.11 Skin depth versus excitation frequency of a copper conductor

$$\mathrm{bei}(\gamma) = \mathrm{Im}\left[J(\gamma e^{j3\pi/4})\right], \tag{4.20}$$

$$\mathrm{ber}'(\gamma) = \frac{\mathrm{ber}(\gamma) + \mathrm{bei}(\gamma)}{\sqrt{2}} \tag{4.21}$$

and

$$\mathrm{bei}(\gamma) = \frac{-\mathrm{ber}(\gamma) + \mathrm{bei}(\gamma)}{\sqrt{2}} \tag{4.22}$$

Figure 4.12 shows the skin effect loss in a round conductor for different penetration ratios, d/δ.

On the other hand, the AC current in a wire also generates a magnetic field that enters the adjacent conductors and induces emfs inside these conductors, resulting in non-uniform current distribution in the conductors. This is called the proximity effect, which highly depends on the excitation frequency [18]. Although the total current of the conductor does not change; the current density in the conductor will be reduced near the adjacent wire and reinforced on the opposite side through a redistribution of current density. Proximity effects can be classified into two types, internal proximity effects—the effects on each of the strands due to the field generated by the strands, and external proximity effects—the effects on an isolated round conductor within an external field. If l_s is the total length of a single strand, n_s the total number of strands, r_b the radius of the bundle, and σ the conductivity, the power loss due to the internal proximity effects can be calculated by [18]

$$P_{\mathrm{p\_int}} = -\frac{\gamma}{\sigma}\frac{n_s I^2 l_s}{4\pi r_b^2}\tau_2(\gamma). \tag{4.23}$$

where

$$\tau_2(\gamma) = \frac{\mathrm{ber}_2(\gamma)\mathrm{ber}'(\gamma) + \mathrm{bei}_2(\gamma)\mathrm{bei}'(\gamma)}{\mathrm{ber}'(\gamma)^2 + \mathrm{bei}'(\gamma)^2}. \tag{4.24}$$

The power losses due to an external magnetic field, H_e, can be calculated by [18]

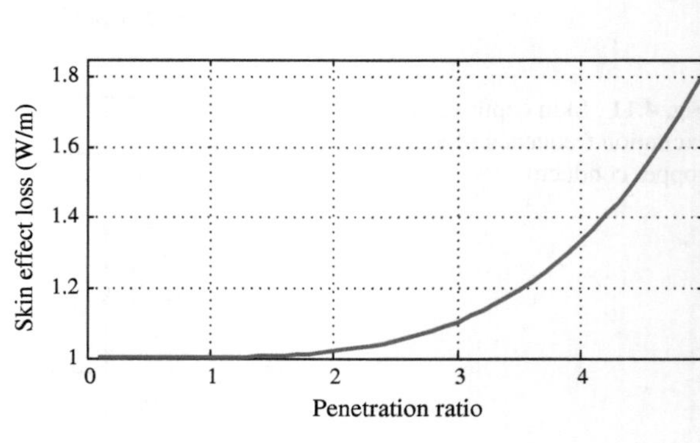

Fig. 4.12 Power losses due to skin effect in a round conductor for different penetration ratios

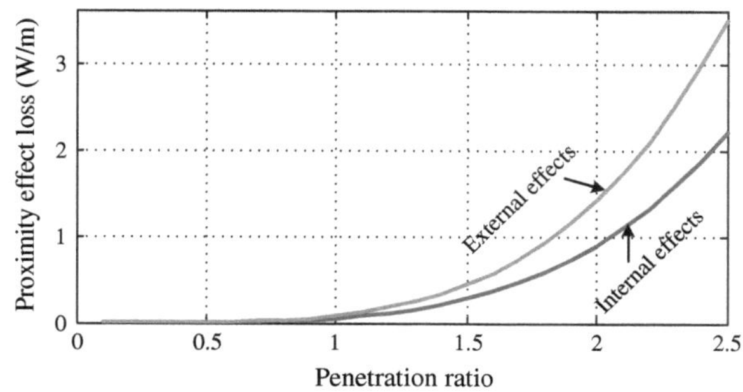

Fig. 4.13 Power losses due to proximity effects in a round conductor for different penetration ratios

$$P_{\text{p\_ext}} = -\frac{2\pi\gamma}{\sigma} n_s H_e^2 \tau_2(\gamma). \tag{4.25}$$

Figure 4.13 shows the power loss in a round conductor due to proximity effects.

The skin and proximity effects will increase the AC losses in high-frequency windings. A special type of wire has been conceived named the Litz wire, a conductor consisting of insulated strands twisted or braided together. Such a design equalizes the flux linkages of individual strands causing the current to spread uniformly throughout the conductor, and the AC to DC resistance ratio tends to approach unity. In a conductor, the AC/DC resistance ratios, K_r, depends strongly on the number of layers, U, the conductor diameter, d, and the operating frequency. The AC resistance of the U-th layer for solid round wires can be represented as [19]

$$K_r = \frac{R_{ac}}{R_{dc}} = \frac{\xi}{2}\left[M(\xi) + (2U-1)^2 D(\xi)\right]. \tag{4.26}$$

where

$$\xi = \frac{0.886d}{\delta} \tag{4.27}$$

$$M(\xi) = \frac{\sinh(\xi) + \sin(\xi)}{\cosh(\xi) - \cos(\xi)} \tag{4.28}$$

$$D(\xi) = \frac{\sinh(\xi) - \sin(\xi)}{\cosh(\xi) + \cos(\xi)} \tag{4.29}$$

and

$$\delta = \frac{0.071}{\sqrt{f}} \tag{4.30}$$

For a particular core shape and defined fill factor, there should be an optimal selection of the Litz wire strand number and diameter to realize the minimum loss. Figure 4.14 shows the AC/DC resistance ratio of a U-layer winding at a 10-kHz

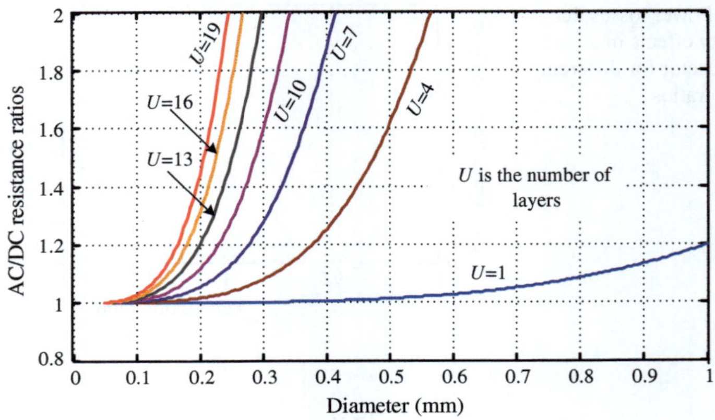

Fig. 4.14 Graphical representation of the AC/DC resistance ratio K_r, at 10 kHz, showing the effect of conductor diameter and number of layers on AC resistance

excitation current. It shows that at medium or high frequencies, associated with a small skin depth and proximity effect, the number of layers as well as the conductor diameter should be kept as small as possible. Moreover, the insulated strands should be twisted or braided together to equalize the flux linkages throughout the conductors. To achieve this and so as to reduce the winding loss, a winding of Litz wire with small number of layers should be always used in a medium-/high-frequency link.

The size of transformer-winding copper wire depends on the primary and secondary side currents, i_p and i_s, e.g., 6 and 1 A, respectively. The current density, J, is chosen as 4 A/mm^2, appropriate for a frequency of 10 kHz. The number of strands of the Litz wires for the primary and secondary windings can be chosen as 13 and 3, respectively. The required minimum cross-sectional area of the primary winding single strand can be calculated as

$$a_p = \frac{\frac{i_p}{13}}{J} = 0.115\ \text{mm}^2 \tag{4.31}$$

and single-strand diameter can be calculated as

$$d_p = \sqrt{\frac{4a_p}{\pi}} = 0.383\ \text{mm}. \tag{4.32}$$

Finally, for the primary windings, the Litz wires with a diameter of 0.4 mm are considered. Theoretically, the overall area of 13 strand wire is $13 \times 0.115 = 1.495$ mm^2. In practical, when 13 insulated strands are twisted or braided together, the overall diameter of the Litz wire is 2 mm and the cross-sectional area of the primary winding wire 3.2 mm^2. Also, the required minimum cross-sectional area of the secondary winding single strand can be calculated as

$$a_s = \frac{\frac{i_s}{3}}{J} = 0.083\ \text{mm}^2 \tag{4.33}$$

and the single-strand diameter can be calculated as

$$d_s = \sqrt{\frac{4a_s}{\pi}} = 0.325\ \text{mm}. \tag{4.34}$$

Finally, for the secondary windings, the Litz wires with a diameter of 0.4 mm are considered. Theoretically, the overall area of 3 strand wire is $3 \times 0.083 = 0.249\ \text{mm}^2$. In practical, when 3 insulated strands are twisted or braided together, the overall diameter of the Litz wire is 1 mm and the cross-sectional area of the primary winding wire $0.79\ \text{mm}^2$. Hence, the design can be considered with the wire cross-sectional areas of 4 and $1.2\ \text{mm}^2$ for the primary and secondary winding, respectively. The area required by the primary and secondary windings is calculated as

$$A_W = N_p d_p + N_s d_s \tag{4.35}$$

Considering a toroid hole reserve factor of 8 for all the windings, the minimum required toroid hole area can be calculated by

$$A_{TH} = 8(N_p d_p + N_s d_s) \tag{4.36}$$

4.2.5 Parasitic Calculation

The leakage flux and stray capacitance may affect the performance of a high-frequency transformer as well as the operation of the PWM converter. These two components, leakage inductances and stray capacitances, are usually called parasitic parameters.

4.2.5.1 Stray Capacitances

The stray capacitances of a medium-frequency transformer due to the electrical coupling between any two conducting elements are highly structure dependent and distributed in nature. It is hard to model the effects of stray capacitances accurately because of their distributive nature, i.e., they can be formed between two turns in the same winding or two different windings, two layers in the same winding or two different windings, windings and core, and windings and ground, etc. Since the stray capacitances can significantly affect the performance of the transformer as well as the power electronic converters, the consideration of stray capacitances is important for the design of a medium-frequency transformer.

The proposed multiwinding transformer can be approximately represented by the two-winding transformer model. For medium-frequency applications, the lumped models of stray capacitances are sufficient and have been generally employed. The

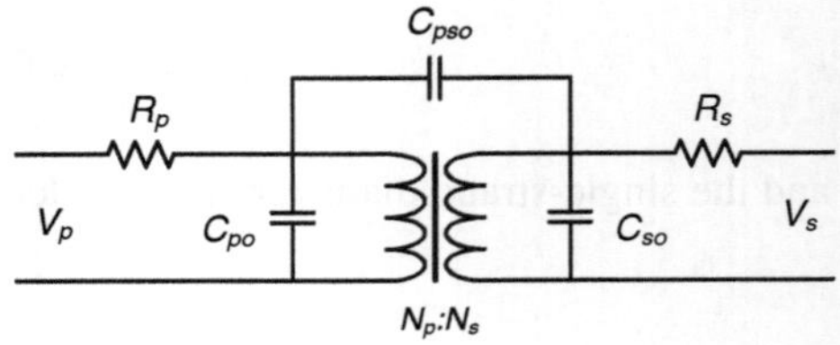

Fig. 4.15 A two-winding transformer with stray capacitances

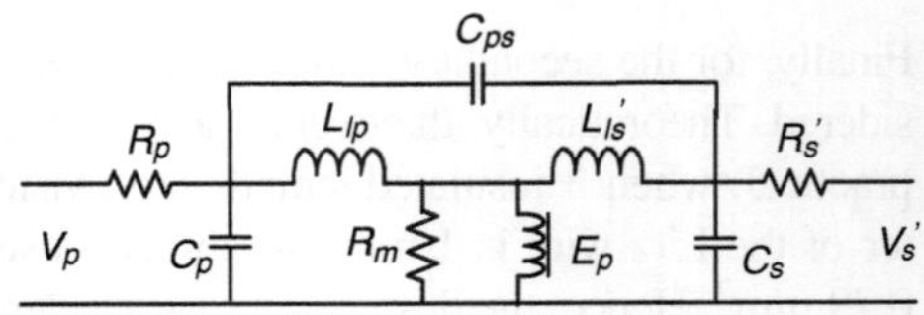

Fig. 4.16 Equivalent circuit of a two-winding transformer with stray capacitances; all parameters referred to the primary side [22]

π-shaped network of three lumped capacitances approach [20, 21] gives a clear view of the coupling capacitances using the concepts of self- and mutual capacitances. This approach represents the overall effects of all stray capacitances by three lumped capacitors, such as the self-capacitances of the primary and secondary windings, and the mutual capacitance between the primary and secondary windings [22]. Figure 4.15 illustrates the circuit of a two-winding transformer with stray capacitance effects, where R_{p} and R_{s} are the resistances of the primary and secondary windings and C_{po}, C_{so}, and C_{pso} are the self-capacitances of the primary and secondary windings, and the mutual capacitance between two windings, respectively.

If L_{lp} is the leakage inductance of the primary winding, L'_{ls} the equivalent leakage inductance of secondary winding referred to the primary side, E_{p} the magnetization back emf, and R_{m} and L_{m} are the equivalent core loss resistance and magnetization inductance, respectively, the equivalent circuit of a two-winding transformer referred to the primary side with stray capacitance effects can be represented by a network as shown in Fig. 4.16 [22].

The equivalent capacitances, C_{p}, C_{s}, and C_{ps}, can be calculated from

$$C_{\mathrm{p}} = C_{\mathrm{po}} + \left(1 - \frac{N_{\mathrm{s}}}{N_{\mathrm{p}}}\right) C_{\mathrm{pso}} \tag{4.37}$$

$$C_{\mathrm{s}} = \left(\frac{N_{\mathrm{s}}}{N_{\mathrm{p}}}\right)^2 C_{\mathrm{so}} + \left(\frac{N_{\mathrm{s}}}{N_{\mathrm{p}}}\right)\left(\frac{N_{\mathrm{s}}}{N_{\mathrm{p}}} - 1\right) C_{\mathrm{pso}} \tag{4.38}$$

and

$$C_{\mathrm{ps}} = \frac{N_{\mathrm{s}}}{N_{\mathrm{p}}} C_{\mathrm{pso}}. \tag{4.39}$$

If R_{p} and R_{s} can be ignored and the primary and secondary windings are short circuited, it is possible to measure C_{pso} approximately as the capacitance between the primary and secondary. In order to calculate the C_{pso}, an external inductor with known inductance value, L_{ex} can be used to make a resonance circuit and then from two resonance frequencies, f_1 and f_2, the C_{pso} can be calculated from

$$C_{pso} = \frac{1}{(2\pi f_2)^2 L_{ex}} - \frac{1}{(2\pi f_1)^2 L_{ex}}. \quad (4.40)$$

Using (4.39) and (4.40), C_{ps} can be calculated.The approximate values of capacitances C_p and C_s can be calculated from the short-circuit test of Fig. 4.15 with the following relations

$$C_p = C_{eqp} - C_{ps} \quad (4.41)$$

and

$$C_s = C_{eqs} - C_{ps}. \quad (4.42)$$

where the capacitances C_{eqp} and C_{eqs} are the approximated equivalent capacitances seen from the primary and secondary sides with the other side short circuited, which makes the circuit just like a single-coil inductor. The equivalent capacitances C_{eqp} and C_{eqs} can be calculated from the step response of the circuit under the condition of one-side short circuited. If ΔV_c is the capacitor-charging voltage and I_c is the average capacitor current over a time period, Δt, the stray capacitance of the coil inductor can be calculated from

$$C_{eq} = \frac{\Delta t I_c}{\Delta V_c}. \quad (4.43)$$

4.2.5.2 Leakage Inductances

Usually, the leakage inductance is defined as the ratio of the leakage flux linking a winding, λ_l, and the current passing through the winding, i, which is appropriate in the case of linear magnetic cores. In order to develop the dynamic circuit model of a high-frequency transformer, differential leakage inductance may be used due to its non-linear magnetic properties, which can be defined as [22]

$$L_{dl} = \frac{d\lambda_l}{d} i. \quad (4.44)$$

The differential leakage inductances of the primary and secondary windings can be defined by

$$L_{dlp} = \frac{d\lambda_{lp}(i_p)}{di_p}. \quad (4.45)$$

and

$$L_{dls} = \frac{d\lambda_{ls}(i_s)}{di_s}. \quad (4.46)$$

If i_{mp} and i_{ms} are the magnetization currents seen from the primary and secondary sides, respectively, the mutual inductances between the primary and secondary windings can be defined by

$$L_{\mathrm{dmp}} = \frac{d\lambda_{\mathrm{mp}}}{\mathrm{d}i_{\mathrm{mp}}} = N_{\mathrm{p}} \frac{\mathrm{d}\Phi_{\mathrm{m}}}{\mathrm{d}i_{\mathrm{mp}}}. \tag{4.47}$$

and

$$L_{\mathrm{dms}} = \frac{\mathrm{d}\lambda_{\mathrm{ms}}}{\mathrm{d}i_{\mathrm{ms}}} = N_{\mathrm{s}} \frac{\mathrm{d}\Phi_{\mathrm{m}}}{\mathrm{d}i_{\mathrm{ms}}}. \tag{4.48}$$

where Φ_{m} is the mutual flux.

An equivalent circuit of a two-winding medium-frequency transformer is shown in Fig. 4.17 [22].

If $\lambda_{\mathrm{p}}(i_{\mathrm{m}}) = N_{\mathrm{p}}\Phi_{\mathrm{m}}(i_{\mathrm{m}}) + N_{\mathrm{p}}\Phi_{\mathrm{lp}}(i_{\mathrm{p}})$is the total flux linkage of the primary winding and $\lambda_{\mathrm{s}}(i_{\mathrm{m}}) = N_{\mathrm{s}}\Phi_{\mathrm{m}}(i_{\mathrm{m}}) - N_{\mathrm{s}}\Phi_{\mathrm{ls}}(i_{\mathrm{s}})$ the total flux linkage of the secondary winding, and Φ_{lp} and Φ_{ls} are the leakage flux of the primary and secondary windings, respectively, the differential self-inductances of the primary and secondary windings can be calculated as

$$L_{\mathrm{dp}} = \frac{\partial \lambda_{\mathrm{p}}}{\partial i_{\mathrm{p}}}. \tag{4.49}$$

and

$$L_{\mathrm{ds}} = \frac{\partial \lambda_{\mathrm{s}}}{\partial i_{\mathrm{s}}}. \tag{4.50}$$

For self- and mutual inductances the leakage inductances of the primary and secondary windings can be calculated as

$$L_{\mathrm{dlp}} = L_{\mathrm{dp}} - L_{\mathrm{dmp}} \tag{4.51}$$

and

$$L_{\mathrm{dls}} = L_{\mathrm{ds}} - L_{\mathrm{dms}}. \tag{4.52}$$

4.2.6 Design Optimization

A frequency window in the range of 1–20 kHz is considered in the optimization. The differential evolution algorithm (DEA) can be used as the optimization algorithm. DEA was first presented in 1997 as a population-based stochastic global optimization method [23]. As demonstrated by many studies, it converges fast and is robust and simple to implement, requiring only a few control parameters.

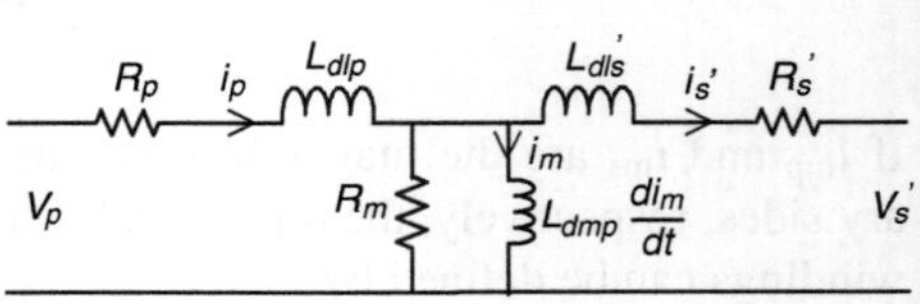

Fig. 4.17 Equivalent circuit of a two-winding transformer with leakage inductances; all parameters referred to the primary side [22]

It has been widely employed for optimization of electromagnetic devices and has achieved many improvements [24, 25].

The procedure of DEA, very similar to that of the genetic algorithm, consists of three processes, namely mutation, crossover, and selection. It starts by initializing the population randomly in the design space. The individuals in the population are then perturbed with others through mutation and crossover operators, and a new population consisting of the most promising solution can be generated by applying a selection criterion [23, 24]. Different factors are considered during the optimization, such as the winding dimensions, hole reserve for natural cooling, maximum temperature limits, maximum power loss, availability of core material stripe dimensions (available sizes of alloys 2605S3A and 2605SA1: widths of 2.5–50 and 5–213 mm, respectively, and thickness of 20 μm), parasitic parameters, skin and proximity effects, and possibility to induce identical voltage in multiple secondary windings. Power loss, size, and weight of the high-frequency magnetic link have been considered the main objective functions of optimization process.

In the implementation, the algorithm parameters of DEA are mutation factor of 0.8, crossover factor of 0.8, the maximum number of iteration of 1,000, and the maximum stall generation of 100 (as the stop criterion) [25]. The optimal parameters of 2.5-kVA magnetic links are summarized in Table 4.5. The designs show that magnetic alloy 2605S3A has a slightly larger volume and weight but much lower specific core loss than that of alloy 2605SA1 and is suitable for high-frequency magnetic links.

Ansoft Maxwell 3D solves the electromagnetic field problems for a given model with appropriate materials, boundaries, and source conditions, applying Maxwell's equations over a finite region of space. There are two types of solutions available in Maxwell 3D, such as magnetic fields and electric fields. Category magnetic fields cover three solvers, such as magnetostatic to calculate static magnetic fields, forces, torques, and inductances caused by DC currents, eddy current to calculate sinusoidally varying magnetic fields, forces, torques, and impedances caused by AC currents and oscillating external magnetic fields, and transient magnetic to calculate transient magnetic fields caused by time-varying or moving electrical sources and permanent magnets. Electric fields also cover three solvers, such as electrostatic to calculate static electric fields, forces, torques, and capacitances caused by voltage distributions and charges, DC conduction to calculate voltage, electric field, and the current density calculated from the potential, and transient electric to calculate transient electric fields caused by time-varying voltages, charge distributions, or current excitations in inhomogeneous materials.

In the Ansoft Maxwell 3D environment, two magnetic cores are modeled using the parameters obtained from optimization and non-sinusoidal

Table 4.5 Optimal parameters for 2.5-kVA magnetic links [26]

Core material	Parameters						
	WD (cm)	HT (cm)	B (T)	f (kHz)	Vol. (cm^3)	Wt. (kg)	P_{core} (W/kg)
2605S3A	2.22	2.78	1	6.05	169	1.23	71
2605SA1	2.29	2.61	1	6.25	165	1.18	96

high-frequency characteristics of Metglas alloys 2605S3A and 2605SA1. When the Maxwell starts a new project, it is automatically added to the project tree. By selecting Project > Insert Maxwell Design, a new Maxwell design can be added to the project. The Maxwell desktop consists of a numbers of functional icons, and windows, toolbars, etc. Figure 4.18 shows a part of Maxwell desktop. There are two geometry models available in Maxwell 3D, i.e., Cartesian (XY) model and Axisymmetric (RZ) model to model project in the 3D modeler window (graphics area).

Using the Modeler Materials toolbar, default materials can be selected for the project. Electromagnetic properties of new materials can be added through View/Edit Material window as shown in Fig. 4.19. Figure 4.20 shows the Maxwell desktop windows to create coils and assign excitation current. By selecting the menu item Show Conduction Path, it is possible to visualize the conduction path in 3D model. The menu item Mesh Operations has the options to select maximum length of elements and maximum number of elements for refinement. Figure 4.21 shows the Maxwell desktop to select element size and numbers. An analysis setup can be created by selecting menu item Add Solution Setup, which provides options to set adaptive and transient setups, such as number of passes, percent error, stop time, and time step. Figure 4.22 shows the Maxwell desktop to create an analysis setup. After saving the whole project and checking the validity of the model, it is possible to start the solution process by selecting menu item Analyze All. By selecting the menu item Mag_B, one can create a field plot. Figure 4.23 shows the magnetic field distribution of Metglas alloy 2605S3A with square-wave voltage excitation at 6 kHz. Figure 4.24 shows the magnetic field distribution of Metglas alloy 2605SA1.

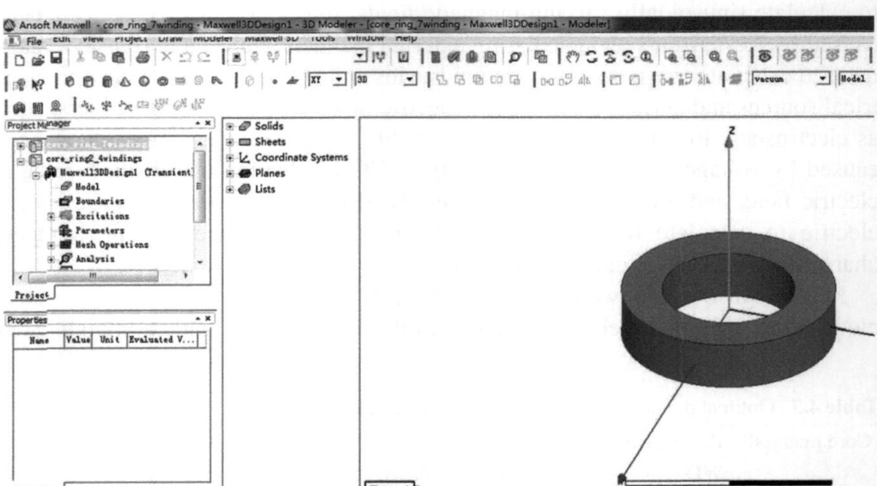

Fig. 4.18 Maxwell desktop with model

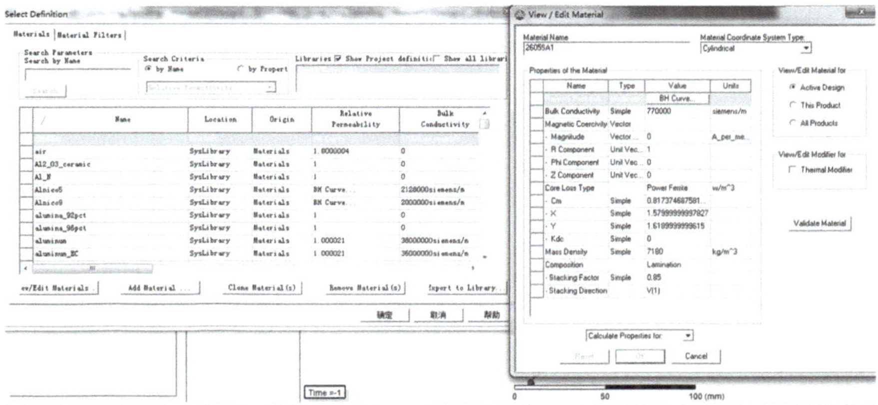

Fig. 4.19 Maxwell desktop windows to select material or add a new material

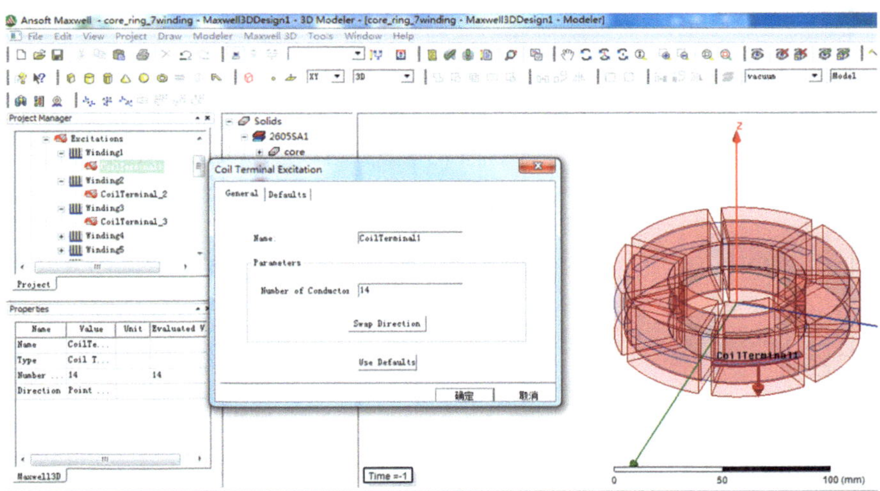

Fig. 4.20 Maxwell desktop windows to create excitation coil

4.2.7 Transformer Core Development

Based on the optimization results, the optimal core dimensions are chosen as 6.5-cm inner diameter (ID), 10.5 cm outer diameter (OD), and 2.5 cm height (HT) for the design, as shown in Fig. 4.25.

The Metglas sheet was glued with Araldite 2011 on the surface of each layer, providing both the electrical insulation and mechanical bonding. During the wrapping process, equal and opposite forces were applied to make a uniform distribution of Araldite. Figure 4.26 shows a photograph of the wrapping process.

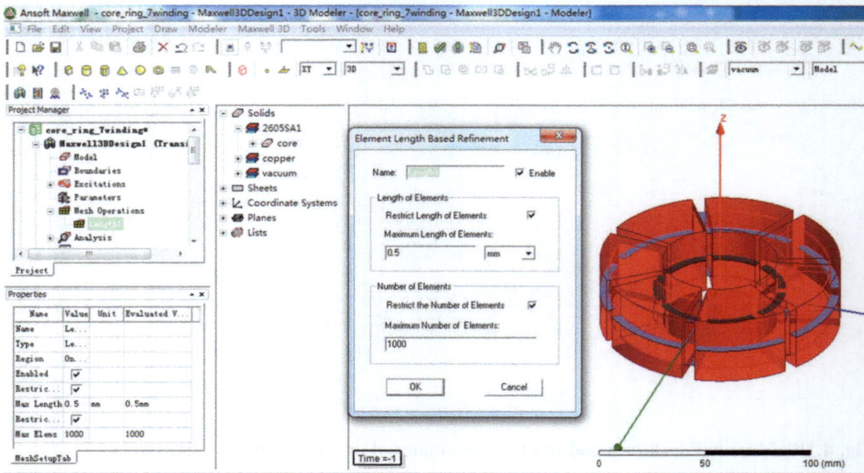

Fig. 4.21 Maxwell desktop windows to select element size and numbers

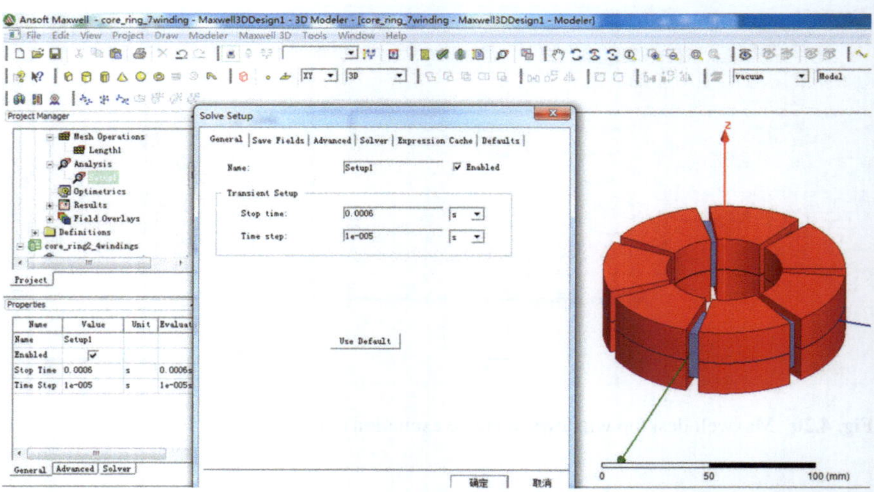

Fig. 4.22 Maxwell desktop with solve setup window

After wrapping, the frame was quickly removed before the Araldite dried up. The core volume and mass are 133.52 cm^3 and 0.96 kg, respectively. Figure 4.27 shows a photograph of the Metglas core with frame; just after wrapping process.

The dried core is shown in Fig. 4.28. The core was covered by transformer tape. To minimize the proximity effect, Litz wires are used for windings with a

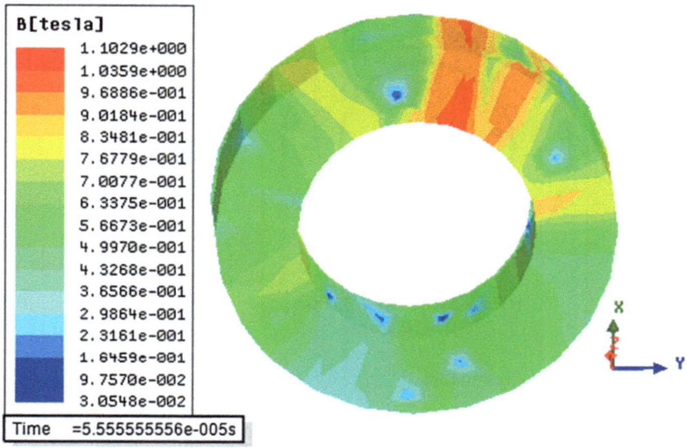

Fig. 4.23 Magnetic field illustration for Metglas alloy 2605S3A-based high-frequency magnetic link in the Ansoft Maxwell 3D [26]

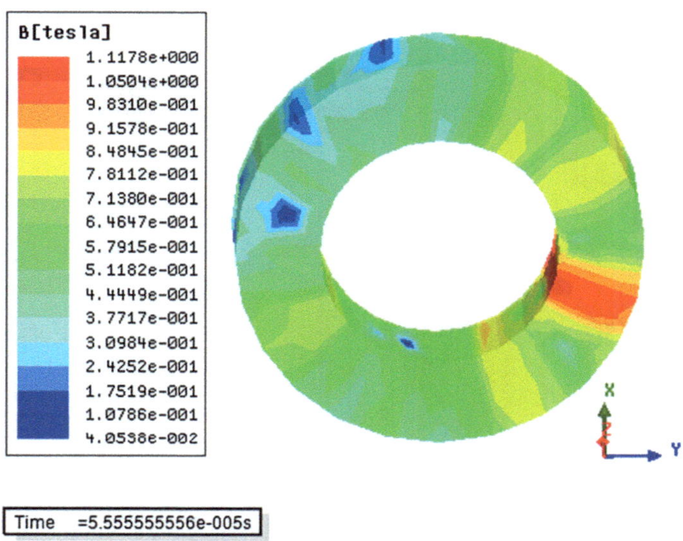

Fig. 4.24 Magnetic field illustration for Metglas alloy 2605SA1-based high-frequency magnetic link in the Ansoft Maxwell 3D [26]

single-layer placement. The electromagnetic performances were analyzed and compared with the experimental results in Sect. 4.3. Figures 4.29 and 4.30 show the photograph of developed transformers with Metglas alloys 2605SA1 and 2605S3A, respectively.

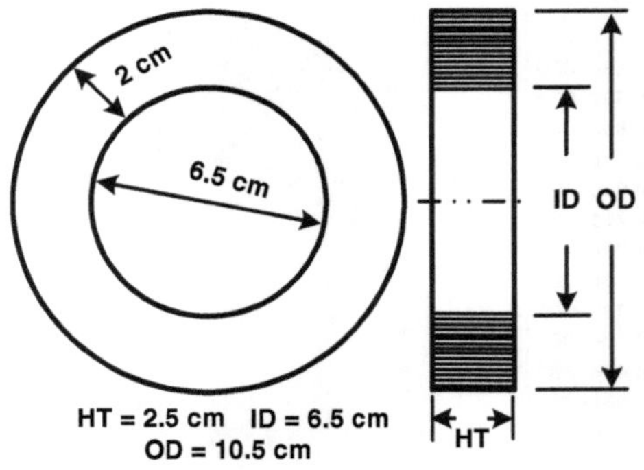

Fig. 4.25 Transformer core dimensions

Fig. 4.26 Transformer core development setup (raping of Metglas sheet)

Fig. 4.27 A photograph of the transformer core with frame; just after wrapping

Fig. 4.28 Photograph of the transformer core with Metglas alloy 2605SA1 ribbon of 30 μm thickness and 25 mm of width

Fig. 4.29 Photograph of high-frequency magnetic link with Metglas alloy 2605SA1 core

Fig. 4.30 Photograph of high-frequency magnetic link with Metglas alloy 2605S3A core

4.3 Characterization of High-Frequency Magnetic Links

4.3.1 Windings DC Resistance Calculation

The electrical resistance, *R*, of an element is defined as the ratio of the voltage, *V*, across the element and the current, *I*, through that element, i.e.,

$$R = \frac{V}{I}. \tag{4.53}$$

For given element, R depends on the resistivity, ρ, of the material (temperature dependant), length, L, and cross-sectional area, A, i.e.,

$$R = \rho \frac{L}{A}. \tag{4.54}$$

At 20 °C, the resistivity of copper is 1.678×10^{-8} Ωm. The average length of each turn is 0.13 m. Since the primary winding has 14 turns and each secondary winding has 25 turns, the total winding wire lengths are 1.82 and 3.25 m, respectively. About 0.14 and 0.08 m extra wires are required for the end connection of primary and secondary windings, respectively. When the strands are twisted together, there is an increase in the length of the windings. About 1.14 and 1.09 % increases are observed in the length of primary and secondary windings, respectively. Therefore, the overall length $(1.82 + 0.14) \times 1.14 = 2.24$ m and $(3.25 + 0.08) \times 1.09 = 3.62$ m are considered for the primary and secondary windings, respectively. The diameters of primary and secondary winding copper wires are the same (0.4 mm), and also the cross-sectional areas (12.56×10^{-8} m^2). Therefore, the resistance of a single strand at 20 °C can be calculated by

$$\begin{aligned} R_{\mathrm{P}} &= 1.678 \times 10^{-8} \frac{2.24}{12.56 \times 10^{-8}} \\ &= 0.30\ \Omega \quad \text{(for single strand of primary winding)} \\ &= 0.0230\ \Omega \quad \text{(for primary winding, 13 strands) and} \end{aligned} \tag{4.55}$$

$$\begin{aligned} R_{\mathrm{S}} &= 1.678 \times 10^{-8} \frac{3.62}{12.56 \times 10^{-8}} \\ &= 0.483\ \Omega \quad \text{(for single strand of secondary winding)} \\ &= 0.161\ \Omega \quad \text{(for each secondary winding, 3 strands)} \end{aligned} \tag{4.56}$$

At the room temperature of 23 °C (during the test), the resistance can be calculated by

$$R_{\mathrm{t}_2} = R_{\mathrm{t}_1} \frac{234.5 + t_2}{234.5 + t_1} \tag{4.57}$$

$$R_{\mathrm{t}_2} = R_{\mathrm{t}_1} \frac{234.5 + 23}{234.5 + 20} = 1.012 \times R_{\mathrm{t}_1} \tag{4.58}$$

Therefore, 0.0233 and 0.162Ω resistances are considered theoretically for the primary and secondary windings, respectively.

A digital multimeter (DMM) was used to get an idea about the range of resistances. The DMM readings are tabulated in Table 4.6.

The Wheatstone bridge is also suitable for measuring a small unknown electrical resistance by balancing two legs of a bridge circuit. Figure 4.31 shows the Wheatstone bridge circuit diagram, and Fig. 4.32 describes the measurement technique.

Table 4.6 DC resistance measurement using DMM

Primary side P (Ω)	Secondary sides (Ω)					
	A	B	C	D	E	F
0.20	0.50	0.50	0.50	0.50	0.50	0.50

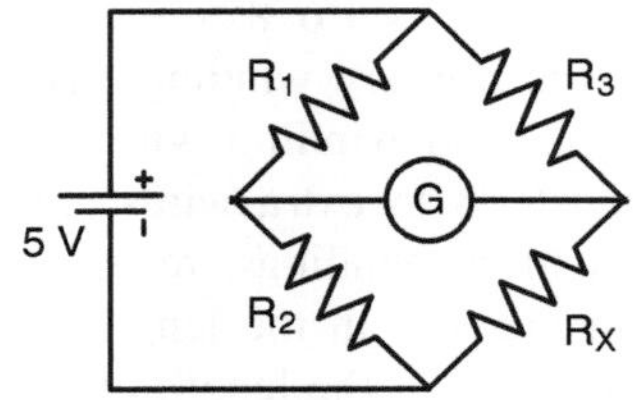

Fig. 4.31 Wheatstone bridge circuit diagram

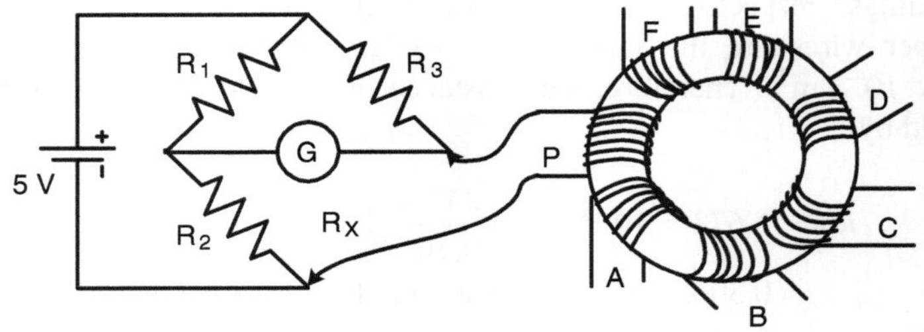

Fig. 4.32 DC resistance measurement technique

A photograph of the experimental setup is shown in Fig. 4.33. When the galvanometer deflection is zero (i.e., the bridge circuit is balanced), we have the ratio of

$$\frac{R_2}{R_1} = \frac{R_X}{R_3}. \tag{4.59}$$

$$\text{Therefore, } R_X = \frac{R_2}{R_1} \times R_3. \tag{4.60}$$

The Wheatstone bridge is used to measure the resistance of the coils since the coil resistance is very small. The lead wire resistance is also considered and subtracted from the measured values, as tabulated in Table 4.7. The measured winding resistances are compared with theoretical values and the variation in percentage is also tabulated in Table 4.7.

4.3.2 Voltage Transformation Ratio Calculation

If the number of turns of primary coil (P) is N_P, the numbers of turns of secondary coils (A, B, C, D, E, and F) are N_A, N_B, N_C, N_D, N_E and N_F, and the corresponding induced voltages in the primary and secondary coils are v_P, v_A, v_B, v_C, v_D, v_E, and

Fig. 4.33 Photograph of DC resistance measurement experimental setup

Table 4.7 DC resistance measurement using Wheatstone bridge

	Primary side	Secondary sides					
		A	B	C	D	E	F
Total (Ω)	0.043	0.191	0.190	0.205	0.185	0.182	0.182
Leading wires	0.019	0.029	0.029	0.029	0.029	0.029	0.029
Winding (Ω)	0.024	0.163	0.161	0.176	0.156	0.153	0.153
Error (%)	+3	+0.6	−0.6	+8.6	−3.7	−5.5	−5.5

v_F, respectively, the voltage transformation ratio can be expressed by (the same number of turns in all secondaries)

$$\frac{N_A}{N_P} = \frac{N_B}{N_P} = \frac{N_C}{N_P} = \frac{N_D}{N_P} = \frac{N_E}{N_P} = \frac{N_F}{N_P} \tag{4.61}$$

$$= \frac{v_A}{v_P} = \frac{v_B}{v_P} = \frac{v_C}{v_P} = \frac{v_D}{v_P} = \frac{v_E}{v_P} = \frac{v_F}{v_P}. \tag{4.62}$$

The theoretical turn ratio can then be calculated as

$$= \frac{374}{210} = 1.781.$$

The coil P is excited by a 50-Hz alternating current, i_P, and the corresponding secondary coil-induced voltages, v_P, v_A, v_B, v_C, v_D, v_E, and v_F, obtained from the respective coils, P, A, B, C, D, E and F, are listed in Tables 4.8 and 4.9. The transformation ratios of all secondary coils, TR_A, TR_B, TR_C, TR_D, TR_E, and TR_F, are calculated with respect to the primary coil. The measured ratios compared with the calculated values

Table 4.8 Transformation ratio calculations at 50 Hz (A–C)

P-coil		A-coil		B-coil		C-coil	
i_P	v_P	v_A	TR_A	v_B	TR_B	v_C	TR_C
1.00	1.16	2.070	1.784	2.068	1.782	2.066	1.781
1.60	1.25	2.226	1.781	2.225	1.780	2.226	1.781
2.25	1.38	2.460	1.782	2.46	1.782	2.45	1.775
3.29	1.63	2.890	1.773	2.890	1.773	2.887	1.771
Avg. ratio		1.780		1.779		1.777	
Error (%)		−0.05		−0.11		−0.22	

Table 4.9 Transformation ratio calculations at 50 Hz (D–F)

P-coil		D-coil		E-coil		F-coil	
i_P	v_P	v_D	TR_D	v_E	TR_E	v_F	TR_F
1.00	1.16	2.067	1.781	2.068	1.782	2.068	1.782
1.60	1.25	2.229	1.783	2.229	1.783	2.227	1.781
2.25	1.38	2.46	1.782	2.45	1.775	2.46	1.782
3.29	1.63	2.889	1.772	2.886	1.770	2.890	1.773
Avg. ratio		1.779		1.777		1.779	
Error (%)		−0.11		−0.22		−0.11	

are also listed in Tables 4.8 and 4.9. The measured ratios fairly coincide with the calculated values, considering the very small change in voltage with the large change in excitation current since at this frequency, the winding resistances are quite small. Utility supply with single-phase auto-transformer can be used to calculate the voltage transformation ratios at 50-Hz sinusoidal excitation. A Fluke clip-on wattmeter is used to measure the voltage and current of the coils. Figure 4.34 shows the experimental setup to calculate voltage transformation ratios at 50-Hz sinusoidal excitation.

Voltage transformation ratios are also calculated with a 1-kHz excitation current. The calculated transformation ratios are compared with the theoretical values as listed in Tables 4.10 and 4.11. The ratios are highly coincident with the theoretical values and better than power frequency transformation ratios.

The voltage transformation ratios have also been measured at 10 kHz and compared to the calculated value (1.781). The ratios and their percentage variation from the calculated value are summarized in Tables 4.12 and 4.13. The high-frequency transformer generates 6 almost equal and isolated sources (i.e., all secondary windings show similar voltage transformation characteristics). Such a similarity of characteristics is obligatory to generate balanced multiple sources for the MMC system.

4.3.3 Measurement of Losses

High-frequency performance has been measured and appraised in the laboratory. The high-frequency (e.g., 1–12 kHz) square-wave signal is generated by an IGBT-based H-bridge inverter supplied by a controlled 210-V DC voltage source. The

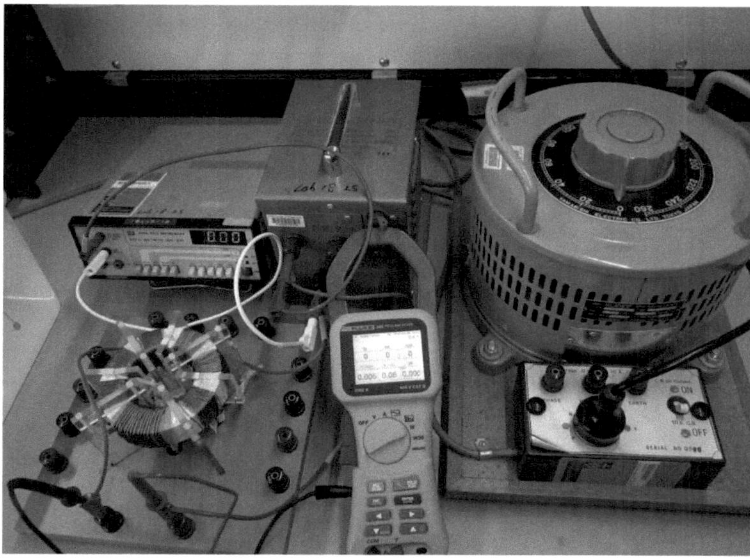

Fig. 4.34 Photograph of experimental setup to calculate voltage transformation ratios at 50-Hz sinusoidal excitation

Table 4.10 Transformation ratio calculations at 1 kHz (A–C)

P-coil	A-coil		B-coil		C-coil	
v_P	v_A	TR_A	v_B	TR_B	v_C	TR_C
14.20	25.31	1.782	25.30	1.781	25.28	1.780
16.00	28.51	1.781	28.50	1.781	28.48	1.780
16.82	29.97	1.781	29.95	1.780	29.93	1.779
Avg. ratio	1.7813		1.7806		1.7797	
Error (%)	+0.016		−0.022		−0.073	

Table 4.11 Transformation ratio calculations at 1 kHz (D–F)

P-coil	D-coil		E-coil		F-coil	
v_P	v_D	TR_D	v_E	TR_E	v_F	TR_F
14.20	25.29	1.781	25.30	1.781	25.30	1.781
16.00	28.48	1.780	28.50	1.781	28.52	1.782
16.82	29.93	1.779	29.95	1.780	29.96	1.781
Avg. ratio	1.7800		1.7806		1.7813	
Error (%)	−0.056		−0.022		+0.016	

Tektronix DPO 2024 Digital Phosphor Oscilloscope with a P5200 high-voltage differential probe and the Tektronix TCPA300 current probe are used to observe the waveforms. The total loss (core loss plus copper loss) is measured by a Voltech

Table 4.12 Transformation ratio calculations at 10 kHz (A–C)

P-coil	A-coil		B-coil		C-coil	
v_P	v_A	TR_A	v_B	TR_B	v_C	TR_C
68.2	121	1.774	121	1.774	122	1.788
104	186	1.788	187	1.798	187	1.798
207	367	1.773	366	1.768	365	1.763
Avg. ratio	1.778		1.780		1.783	
Variation (%)	−0.16		−0.05		+0.11	

Table 4.13 Transformation ratio calculations at 10 kHz (D–F)

P-coil	D-coil		E-coil		F-coil	
v_P	v_D	TR_D	v_E	TR_E	v_F	TR_F
68.2	121	1.774	121	1.774	122	1.788
104	186	1.788	187	1.798	186	1.788
207	366	1.768	367	1.773	368	1.777
Avg. ratio	1.776		1.781		1.784	
Variation (%)	−0.28		0.00		+0.16	

PW3000A universal power analyzer. Figure 4.35 shows a photograph of the experimental setup for high-frequency non-sinusoidal excitation. Figure 4.36 shows the measured primary and secondary side voltage waveforms. The secondary winding current waveform is shown in Fig. 4.37.

All secondary windings are excited one by one with a square-wave excitation current at different frequencies ranging from 4 to 12 kHz. Table 4.14 summarizes the measured losses of all secondary windings. From the oscilloscope data, the

Fig. 4.35 A photograph of the experimental setup for characterization of high-frequency magnetic links

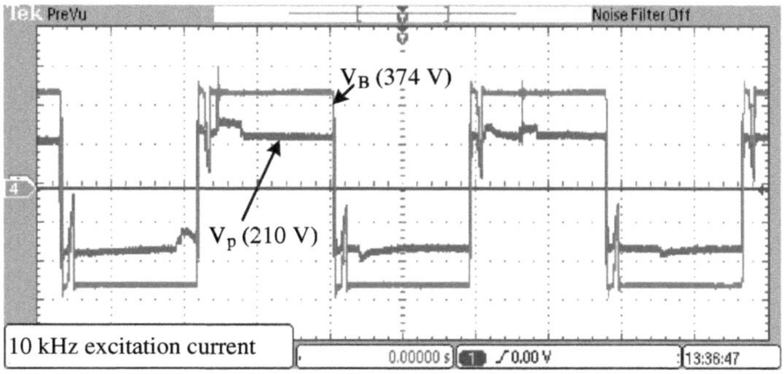

Fig. 4.36 Measured primary and secondary side voltage waveforms

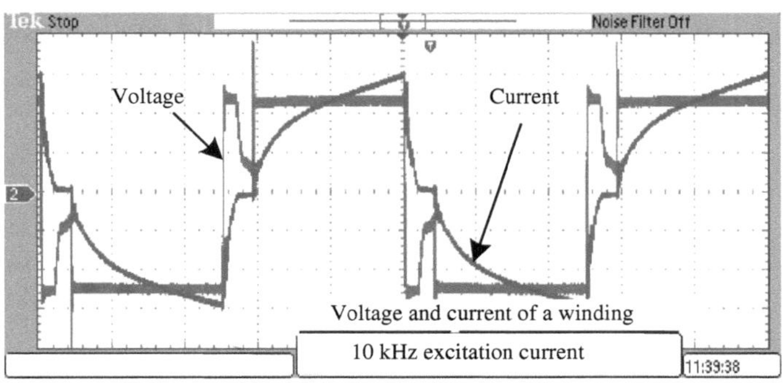

Fig. 4.37 Measured voltage and current waveforms

copper loss of each winding at a different frequency ranging from 4 to 12 kHz was calculated by using the DC resistances (0.024 Ω for the primary and 0.16 Ω for the secondary), since the AC/DC resistance ratios, *Kr*, in this design are almost unity due to the use of the Litz wires. Figure 4.38 shows that the total losses of all secondary windings measured at different excitation frequencies ranging from 4 to 12 kHz are almost the same. Measurements were also conducted by exciting the secondary windings one by one with different excitation currents (0.1–0.6 A) and a fixed frequency of 10 kHz. Table 4.15 summarizes the losses of all six secondary windings with different values of excitation current. As shown in Fig. 4.39, at 10 kHz, all secondary windings also have very similar total losses. Such a similarity of characteristics is also obligatory to generate balanced multiple sources for the MMC system. The proper balancing of all sources means the elimination of the extra control circuit with the switching algorithm.

The core loss against flux density was measured within the frequency ranging from 6 to 12 kHz. The specific core loss of Metglas alloy 2605SA1 measured was 157 W/kg under 10-kHz square-wave excitation of magnitude 0.8 T as shown in Fig. 4.40, while the specific core loss of 125 W/kg is reported by the material

Table 4.14 Total loss calculations with respect to each secondary coil with 0.35-A excitation current

Coil A excited			Coil B excited		
f (kHz)	*v* (V)	*P* (W)	*f* (kHz)	*v* (V)	*P* (W)
4.015	111.08	30.24	4.015	109.81	29.91
6.045	155.23	46.39	6.045	155.10	46.61
8.068	197.82	62.34	8.068	200.10	62.53
10.035	230.40	74.20	10.035	230.00	74.47
12.095	247.81	81.42	12.095	246.70	80.60
Coil C excited			Coil D excited		
f (kHz)	*v* (V)	*P* (W)	*f* (kHz)	*v* (V)	*P* (W)
4.015	110.60	30.45	4.015	109.83	29.91
6.045	156.38	46.67	6.045	156.87	46.67
8.068	199.77	62.75	8.068	200.30	63.08
10.035	229.90	75.14	10.035	228.30	73.50
12.095	248.80	81.70	12.095	246.90	81.06
Coil E excited			Coil F excited		
f (kHz)	*v* (V)	*P* (W)	*f* (kHz)	*v* (V)	*P* (W)
4.015	110.03	30.11	4.015	110.56	30.38
6.045	156.30	46.20	6.045	156.26	46.19
8.068	199.14	62.29	8.068	198.25	61.74
10.035	230.18	75.64	10.035	230.80	74.96
12.095	248.40	81.87	12.095	247.40	81.52

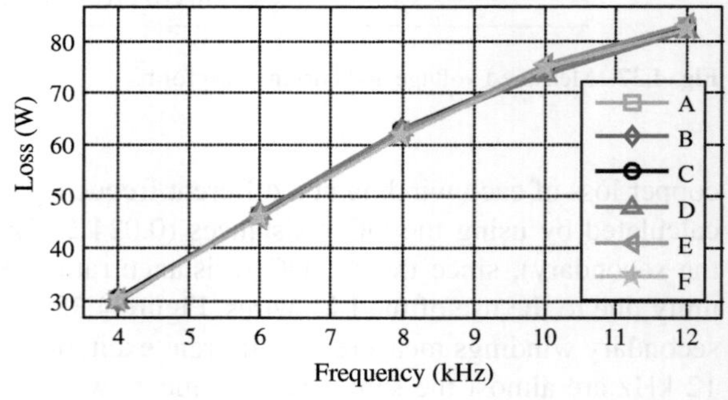

Fig. 4.38 Measured total losses of Metglas alloy 2605SA1 against excitation frequencies at excitation current of 0.35 A; only one coil is energized at a time while the others are open circuited

manufacturer under 10-kHz sinusoidal excitation of magnitude 0.8 T. In comparison with the material manufacturer's data, about 20–30 % extra loss is observed due to the non-sinusoidal excitation waveform.

4.3.4 Measurement of B–H Characteristics

The magnetic field intensity and magnetic flux density are calculated by measuring the P-coil excitation voltage and current and the open-circuit terminal voltage of the secondary side D-coil. As the coils are uniformly wound on the toroidal core, the

Table 4.15 Total loss calculations with respect to each secondary coil at 10-kHz Metglas alloy 2605SA1-based magnetic link)

Coil A excited			Coil B excited		
i (A)	*v* (V)	*W* (W)	*i* (A)	*v* (V)	*W* (W)
0.105	31.01	2.56	0.100	27.81	2.12
0.201	75.05	13.18	0.201	72.30	12.44
0.299	171.19	48.06	0.298	169.74	47.58
0.396	259.9	92.06	0.4	261.50	92.81
0.501	289.4	113.33	0.505	280.91	108.21
0.599	302.3	122.99	0.602	291.30	116.45
Coil C excited			**Coil D excited**		
i (A)	*v* (V)	*W* (W)	*i* (A)	*v* (V)	*W* (W)
0.101	29.09	2.28	0.100	27.57	2.13
0.202	75.85	13.34	0.201	74.76	13.10
0.300	173.12	48.98	0.296	168.02	46.89
0.402	261.50	93.01	0.396	258.70	91.42
0.496	286.50	111.45	0.495	287.80	112.05
0.596	299.00	120.97	0.601	303.10	122.57
Coil E excited			**Coil F excited**		
i (A)	*v* (V)	*W* (W)	*i* (A)	*v* (V)	*W* (W)
0.101	28.20	2.19	0.101	29.24	2.31
0.201	74.04	12.90	0.201	76.42	13.42
0.302	174.3	49.60	0.298	171.52	48.22
0.399	260.00	92.08	0.398	259.50	91.94
0.505	289.60	113.61	0.499	288.10	112.17
0.601	302.7	122.10	0.602	302.30	122.79

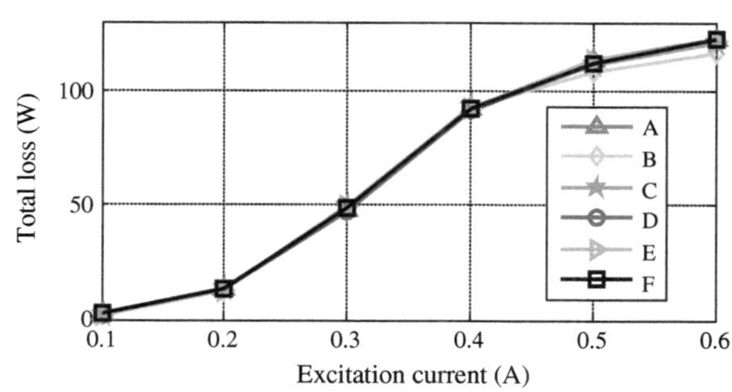

Fig. 4.39 Measured total losses of Metglas alloy 2605SA1 against excitation currents of 10 kHz; only one coil is energized at a time while the others are open circuited

magnetic field intensity H and magnetic flux density B within the core can be considered as uniform. By the Ampere's law, the field intensity can be calculated as

$$H = \frac{N_1 i(t)}{l_e}. \tag{4.63}$$

where N_1 is the number of turns in the primary coil, $i(t)$ the excitation current, and l_e the mean length of the core.

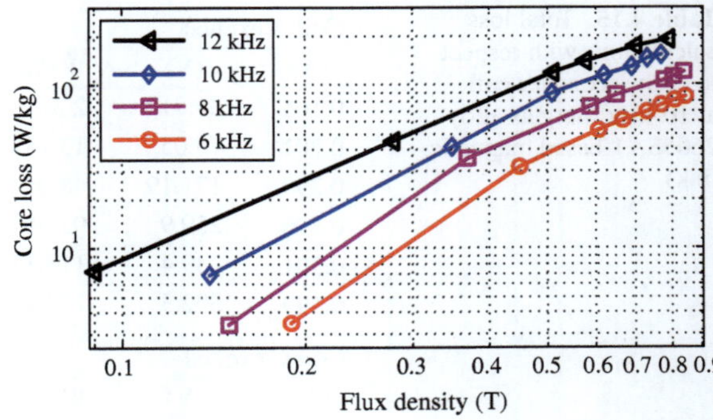

Fig. 4.40 Measured core losses of Metglas alloy 2605SA1 at different frequency ranging from 6 to 12 kHz; experiments were carried out at 40 °C

By the Faraday's law, the magnetic flux density in the core can be calculated as

$$B = \frac{1}{N_2 A_e} \int V_L \mathrm{d}t. \tag{4.64}$$

where N_2 is number of turns in the pickup coil (D-coil) and A_e the cross-sectional area of the core, and V_L the pickup coil voltage.

Different magnitude of 10-kHz excitation currents (e.g., 1–3 A) are applied to the primary windings. The *B*–*H* loops are plotted in Fig. 4.41 based on experimental data. Up to 0.5 T, the flux density rises sharply with a constant field intensity of about 50 A/m. At 3 A, the maximum magnetic flux density is about 0.8 T with the field intensity of about 600 A/m, which satisfies the design requirement. The *B*–*H* loops are also measured at different temperatures ranging from 40 to 100 °C as depicted in Fig. 4.42. The maximum flux density remains approximately constant for this temperature range, as illustrated in Fig. 4.43. The FLUKE infrared temperature probe 80T-IR and FLIR infrared thermal imaging camera i7 were used to measure the temperature of the high-frequency magnetic links. Figure 4.44 shows the thermal image of the high-frequency magnetic link when the probe was pointed at the surface of the core. Quite lower temperature is observed at the center of the toroid hole. Figure 4.45 shows the thermal image of the high-frequency magnetic link when the probe was pointed at the center of the toroid hole.

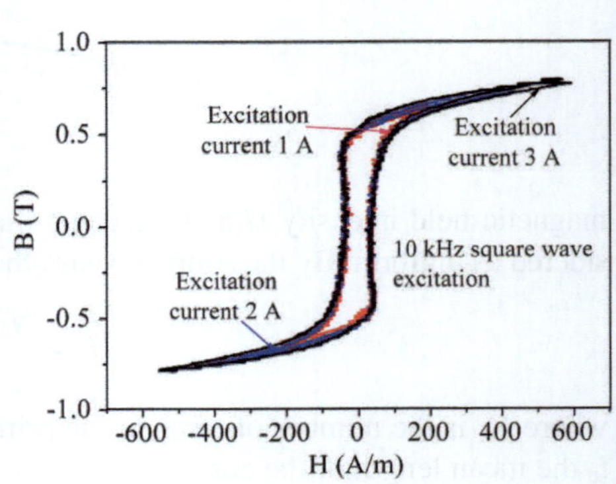

Fig. 4.41 Measured *B*–*H* loops of Metglas alloy 2605SA1 at 40 °C; maximum magnetic flux density at 1, 2, and 3 A is 0.65, 0.74, and 0.80 T, respectively

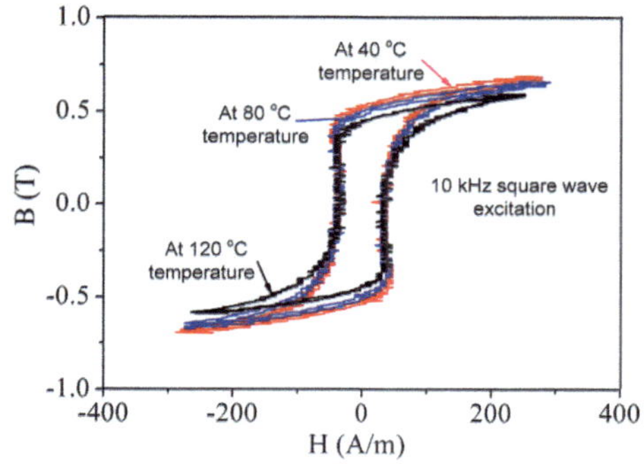

Fig. 4.42 Measured *B–H* loops of Metglas alloy 2605SA1 at different temperature ranging from 40 to 100 °C

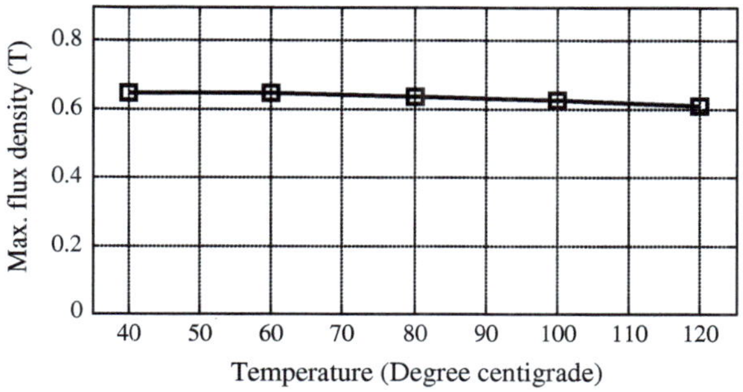

Fig. 4.43 Measured maximum magnetic flux density of Metglas alloy 2605SA1 at different temperature ranging from 40 to 120 °C: maximum magnetic flux density at 40, 60, 80, 100, and 100 °C is 0.65, 0.65, 0.64, 0.63, and 0.62 T, respectively

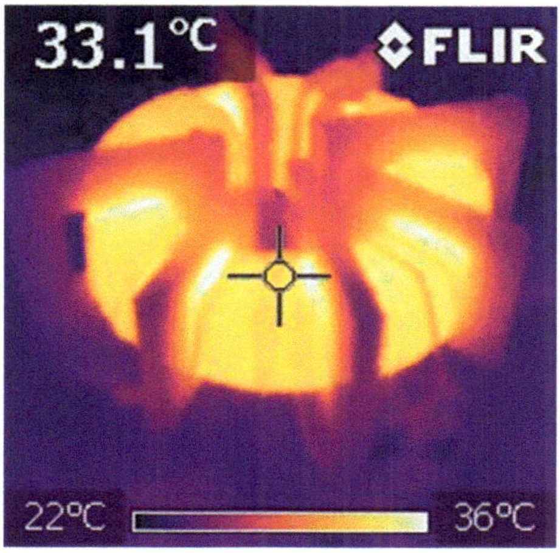

Fig. 4.44 Thermal image of high-frequency magnetic link when the probe was pointed at the *surface* of the core

Figure 4.46 shows the *B–H* loops of Metglas alloy 2605S3A based on experimental data with excitation currents of 3 and 5 A. The plotted *B–H* loops have been compared with the material manufacturer's data and found to be highly consistent. The maximum flux density was also measured under different temperature

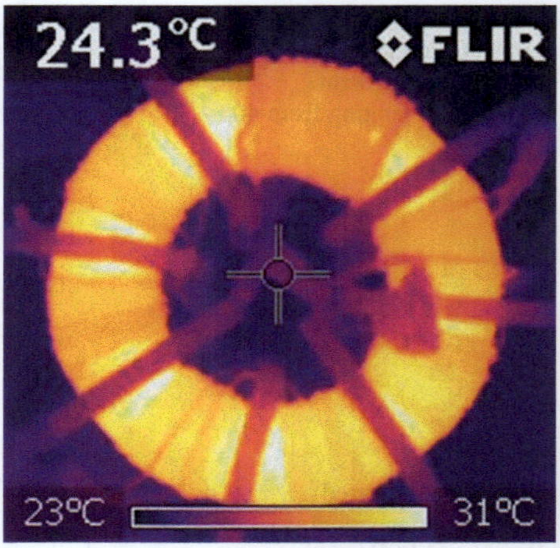

Fig. 4.45 Thermal image of high-frequency magnetic link when the probe was pointed at the *center* of the core

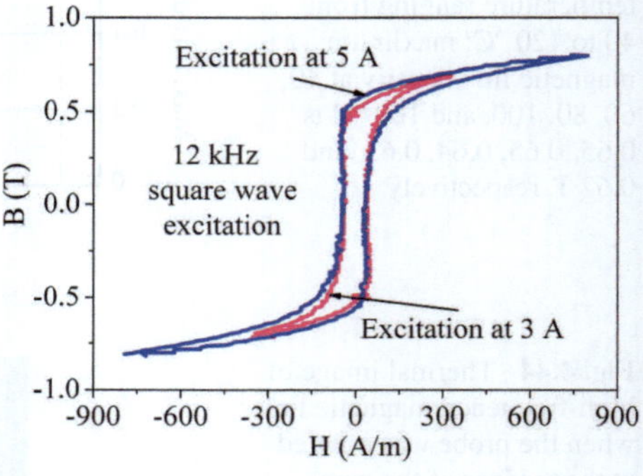

Fig. 4.46 Measured *B–H* loops of Metglas alloy 2605S3A at excitation current of 3 and 5 A. The link is excited by 12-kHz *square-wave* primary voltages

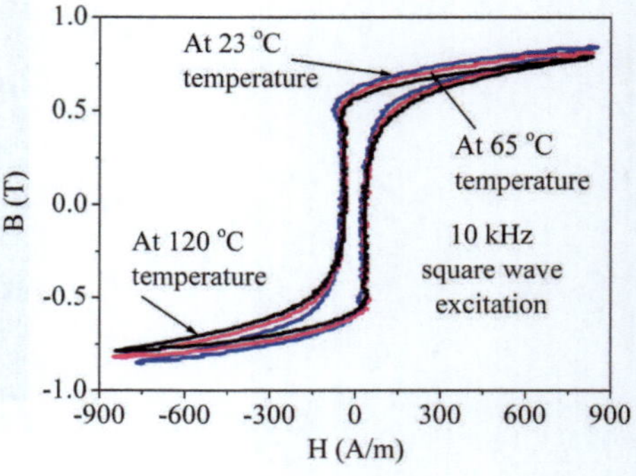

Fig. 4.47 Measured *B–H* loops of Metglas alloy 2605S3A at various temperatures. The link is excited by 12-kHz *square-wave* primary voltages

conditions, which is almost stable for the temperature range of 23–120 °C. Figure 4.47 shows the *B–H* loops of Metglas alloy 2605S3A at various temperatures. Figure 4.48 plots the *B–H* loops at different frequencies. The *B–H* loops

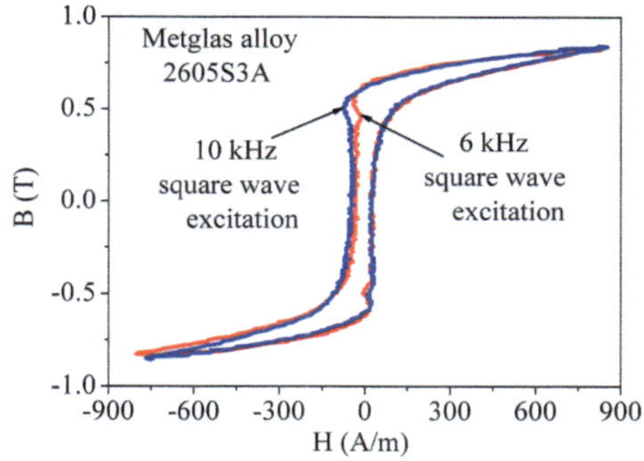

Fig. 4.48 Measured *B–H* loops of Metglas alloy 2605S3A at different frequencies

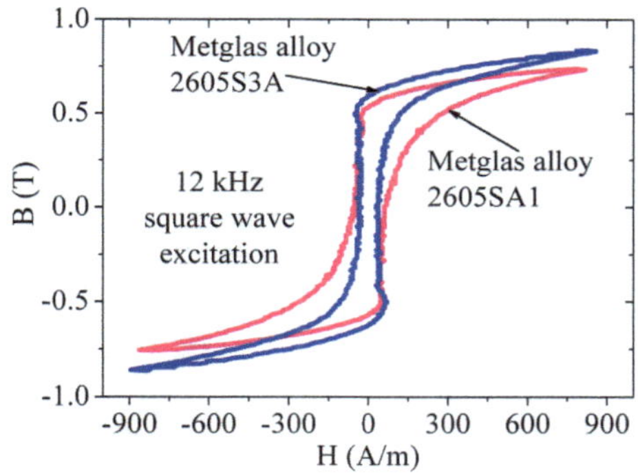

Fig. 4.49 Measured *B–H* loops of Metglas alloys 2605SA1 and 2605S3A; the links are excited by 12-kHz *square-wave* primary voltages

of Metglas alloy 2605S3A are compared with that of Metglas alloy 2605SA1, as shown in Fig. 4.49. In comparison with 2605SA1, the 2605S3A shows a slightly narrower *B–H* loop, i.e., lower core loss.

4.3.5 Measurement of DC-Link Voltages

The output of each secondary winding is connected to a fast recovery diode-based rectifier with a low-pass RC filter circuit. An Agilent Technologies DS06034A oscilloscope with a P5200 high-voltage differential probe is used to observe the voltages. The DC-link voltages were found approximately equal at about 370 V, which can serve satisfactorily as the isolated and balanced DC sources for the proposed MMC converter. The voltage waveforms of one secondary winding and corresponding rectified DC voltage are shown in Fig. 4.50. Without using any special control algorithm regarding the capacitor voltage unbalancing, the prototype converter generates a satisfactory output-voltage waveform as the transformer can provide balanced sources for all of the H-bridge inverter cells of the MMC converter. Figure 4.51 shows the output phase voltage of the prototype 5-level MMC converter.

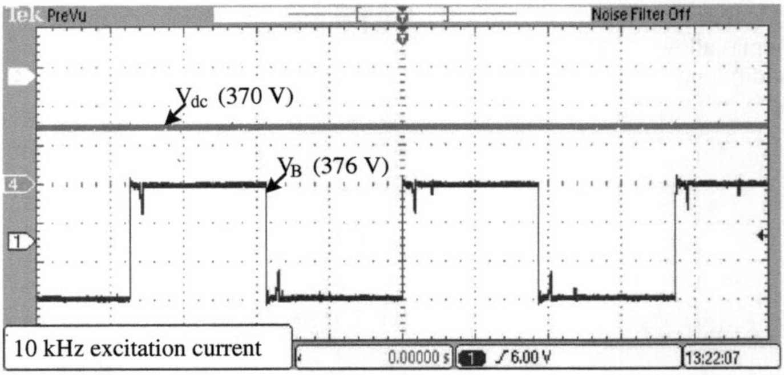

Fig. 4.50 DC-link voltage of an H-bridge inverter cell

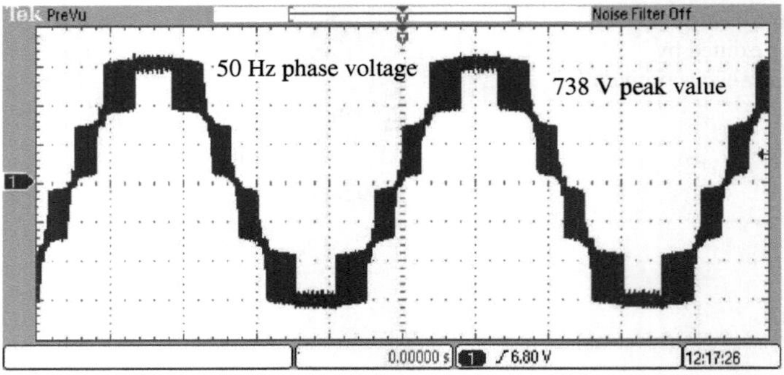

Fig. 4.51 Measured phase voltage of the prototype 5-level MMC converter before the LC line filter circuit

4.4 Summary

The amorphous alloy or nanocrystalline magnetic material-based high-frequency magnetic link can be a good solution to provide multiple isolated and balanced DC supplies for the MMC converter to step-up the low voltage of a photovoltaic array or wind generator to a medium 3-phase AC voltage suitable for power transmission from remote (e.g., offshore) wind farms to the main grid transmission lines. It can also overcome the common mode and voltage imbalance problems. It is found that magnetic alloy 2605S3A, which has a narrower *B*–*H* loops and lower specific core loss than those of 2605SA1 at the high-frequency range, can be an excellent choice for the development of high-frequency magnetic links for grid-connected power converters.

References

1. Islam MR, Guo YG, Zhu JG (2014) A review of offshore wind turbine nacelle: technical challenges, and research and developmental trends. Renew Sustain Energy Rev 33:161–176
2. Islam MR, Guo YG, Zhu JG (2011) H-bridge multilevel voltage source converter for direct grid connection of renewable energy systems. In: Proceedings of IEEE Power and Energy Society innovative smart grid technology conference, 13–16 Nov 2011, Perth, Australia, pp 1–7
3. Islam MR, Guo YG, Zhu JG (2012) A Transformer-less compact and light wind turbine generating system for offshore wind farms. In: Proceedings of 2012 IEEE international conference on power and energy (PECon2012), 2–5 Dec 2012, Kota Kinabalu, Malaysia, pp 605–610
4. Prasai A, Yim JS, Divan D, Bendre A, Sul SK (2008) A new architecture for offshore wind farms. IEEE Trans Power Electron 23(3):1198–1204
5. Pereda J, Dixon J (2011) High-frequency link: a solution for using only one DC source in asymmetric cascaded multilevel inverters. IEEE Trans Ind Electron 58(9):3884–3892
6. Islam MR, Guo YG, Zhu JG (2014) A multilevel medium-voltage inverter for step-up-transformer-less grid connection of photovoltaic power plants. IEEE J Photovoltaics 4(3):881–889
7. Islam MR, Guo YG, Zhu JG (2013) A medium-frequency transformer with multiple secondary windings for medium-voltage converter based wind turbine generating systems. J Appl Phys 113(17):17A324-1–17A324-3
8. Islam MR, Guo YG, Zhu JG (2014) A high-frequency link multilevel cascaded medium-voltage converter for direct grid integration of renewable energy systems. IEEE Trans Power Electron 29(8):4167–4182
9. Islam MR, Guo YG, Lin ZW, Zhu JG (2014) An amorphous alloy core medium frequency magnetic-link for medium voltage photovoltaic inverters. J Appl Phys 115(17):17E710-1–17E710-3
10. Islam MR, Guo YG, Zhu JG (2013) Multiple-input-multiple-output medium frequency-link based medium voltage converter for direct grid connection of photovoltaic arrays. In: Proceedings of international conference on electrical machines and systems (ICEMS2013), 26–29 Oct 2013, Busan, Korea, pp 202–207
11. Agheb E, Bahmani MA, Hoidalen HK, Thiringer (2012) Core loss behavior in high frequency high power transformers-II: arbitrary excitation. J Renew Sustain Energy 4(3):033113–033113-11
12. Burdt R, Curry RC, McDonald KF, Melcher P, Ness R, Huang C (2006) Evaluation of nanocrystalline materials, amorphous metal alloy, and ferrites for magnetic pulse compression applications. J Appl Phys 99(8):08D911–08D911-3
13. Zhang Z, Sharma P, Makino A (2012) Role of Si in high Bs and low core-loss $Fe_{85.2}B_{10-x}P_4Cu_{0.8}Si_x$ nano-crystalline alloy. J Appl Phys 112(10):103902–103902-8
14. Makino A, Kubota T, Yubuta K, Inoue A, Urata A, Matsumoto H, Yoshida S (2011) Low core losses and magnetic properties of Fe85-86Si1-2B8P4Cu1 nanocrystalline alloy with high B for power applications. J Appl Phys 109(7):07A302–07A302-5
15. Metglas Inc. Technical bulletin. Available at: http://metglas.com/products/magnetic_materials/. Accessed on 20 Jan 2014
16. Vacuumschmelze. Compound Viteroperm. Available at http://vacuumschmelze.com. Accessed on 03 Jul 2013
17. Pleite J, Prieto R, Asensi R, Cobos JA, Olias E (1999) Obtaining a frequency-dependent and distributed-effects model of magnetic components from actual measurements. IEEE Trans Magn 35(6):4490–4502
18. Villar I (2010) Multiphysical characterization of medium-frequency power transformers. PhD dissertation, EPFL, 2010
19. Sippola M, Sepponen RE (2002) Accurate prediction of high-frequency power-transformer losses and temperature rise. IEEE Trans Power Electron 17(5):835–847
20. Baccigalupi A, Daponte P, Grimaldi D (1994) On a circuit theory approach to evaluate the stray capacitances of two coupled inductors. IEEE Trans Instrum Measur 43(5):774–776

21. Corsonello P, Daponte P, DeGrazia F, Grimaldi D (1992) On two-coupled inductors stray capacitances evaluation for CAD-oriented modeling. In: Proceedings of the 35th midwest symposium on circuits and systems, vol 2, 9–12 Aug 1992, Washington, DC, pp 1252–1255
22. Lu HY, Zhu JG, Hui SYR (2003) Experimental determination of stray capacitances in high frequency transformers. IEEE Trans Power Electron 18(5):1105–1112
23. Storn R, Price K (1997) Differential evolution—a simple and efficient heuristic for global optimization over continuous spaces. J Global Optim 11(4):341–359
24. Baatar N, Zhang DH, Koh CS (2013) An improved differential evolution algorithm adopting λ-best mutation strategy for global optimization of electromagnetic devices. IEEE Trans Magn 49(5):2097–2100
25. Lei G, Shao KR, Guo YG, Zhu JG, Lavers JD (2008) Sequential optimization method for the design of electromagnetic device. IEEE Trans Magn 44(11):3217–3220
26. Islam MR, Lei G, Guo YG, Zhu JG (2014) Optimal design of high-frequency magnetic-links for power converters used in grid connected renewable energy systems. IEEE Trans Magn, In Press, doi:10.1109/TMAG.2014.2329939

Chapter 5
FPGA-Based Digital Switching Controller for Multilevel Converters

Abstract The available microcontrollers at present can only provide about six pairs of PWM channels, which are clearly insufficient for multilevel converter systems. On the other hand, the field programmable gate array (FPGA) may provide multiple PWM generators according to the converter requirements. Unlike the digital signal processor (DSP) which runs a sequential program in its microprocessor, an FPGA may run all the operations in parallel with the clock signal. The capability of parallel processing of the FPGA affords the opportunity to the switching controller to update all gate signals simultaneously. Various design techniques and software environments are available for the modeling of switching control schemes with the FPGA technology. Most of the techniques require special software such as HDL coder, System Generator, PSIM and ModelSim, which increases the development time and cost. In this chapter, the most common software such as the MATLAB/Simulink- and Xilinx ISE-based alternative design technique is proposed, which may reduce the developmental time and cost of the switching controller.

Keywords Multilevel converters · Digital switching controller · FPGA technology · DSP technology · Design · Experimental verification

5.1 Introduction

Three basic technologies, the application-specific integrated circuit (ASIC), the digital signal processor (DSP), and the field programmable gate array (FPGA), have commonly been used for embedded applications. Due to some special features such as fast processing speed, capability of parallel processing, and ability to provide multiple PWM generators according to the converter requirements, the FPGA technology may be an excellent choice for medium-voltage multilevel converters [1]. Although several design techniques are available for the development of FPGA-based switching controller, most of them require special software, which

M.R. Islam et al., *Power Converters for Medium Voltage Networks*,
Green Energy and Technology, DOI 10.1007/978-3-662-44529-7_5

Fig. 5.1 Photograph of an ASIC by AMI

increases not only the developmental time but also the cost of the controller. In this chapter, a fully digital switching controller is developed for a 3-phase 5-level converter. The SK 30 GH 123 IGBTs are used to develop a prototype multilevel converter, and an XC3S500E FPGA is used to develop the switching controller. The most common software such as the MATLAB/Simulink- and Xilinx ISE-based alternative design technique is used, which may reduce the developmental time and cost of the switching controller. The simulation results serve as a preliminary validation of the proposed design technique, which will be finally verified by the experimental results. The developed switching scheme can be used for any multilevel converter configurations with very little change in the software environment. Moreover, the proposed design and implementation techniques may be useful for designing any other modern power converter's switching controller.

5.1.1 ASIC Technology

The ASIC technology is custom-designed for a particular application. ASICs provide the highest performance, because their design can be optimally configured with respect to the application's requirements. However, the costs of ASIC-based controllers are also the highest, mainly due to the nonrecurring engineering (NRE) costs such as mask manufacturing and the cost of hardware and software development. Figure 5.1 shows an ASIC by AMI.

5.1.2 DSP Technology

Compared to an ASIC, the DSP and FPGA technologies have much lower NRE costs. The DSP has a fixed hardware configuration, which is the hardware of the

Fig. 5.2 Photograph of a DSP by TI

basic functions of many signal processing algorithms. The DSP technology is a fast microprocessor, which has much flexibility with different features. The DSP technology is typically programmed in C/C++ or with assembly language for performance. This technology is well suited to extremely complex mathematical tasks, with conditional processing. However, the performance of DSP technology is limited by the clock rate, and the number of useful operations per clock. For example, the TMS320C6201 can achieve 400 millions multiplications per second with two multipliers and 200-MHz clock [2]. Figure 5.2 shows a photograph of DSP by TI.

5.1.3 FPGA Technology

An FPGA is a programmable logic device that consists of a large number of simple logic elements (LEs). The FPGA is programmed using a very-high-speed integrated hardware description language (VHDL), which programs the connections between the individual LEs to create logical functions (i.e., digital hardware). Figure 5.3 shows a photograph of Xilinx FPGA. The main limitations of FPGA technology are the number of LEs and signal propagation delay. Advancements in FPGA technologies have increased the number of the available LEs and reduced the propagation delay. Recent FPGAs have more multipliers. For example, Xilinx Spartan-6 XC6SLX150T has 180 multipliers that can operate at 100 MHz, which gives 18,000 millions multiplications per second [3]. In comparison with the DSP technology, the advanced FPGA technology gives 10–100 times faster processing speed. The yearly growth of FPGA capabilities compared to the traditional DSP is depicted in Fig. 5.4 [4]. The most important performance criteria of modern controllers are use of high performance control algorithms, easy to modify the control strategy and parameters, fully integrated single device controllers, low cost as well

Fig. 5.3 Photograph of an FPGA by Xilinx

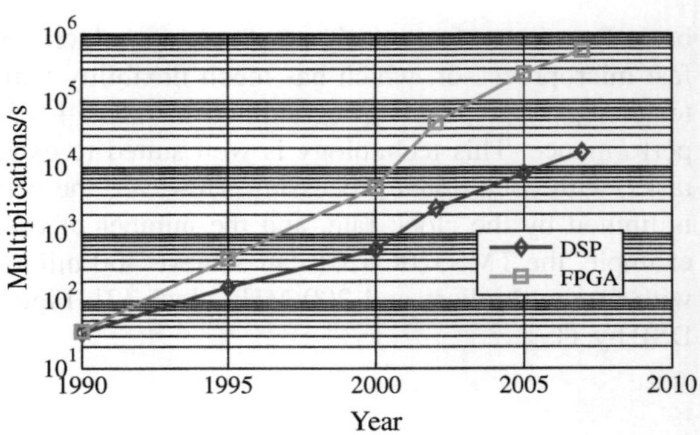

Fig. 5.4 Multiplications per second of texas instruments DSP and Altera FPGA

as implementation time, and high reliability and accuracy. In this regard, recent studies have shown that FPGA could be an appropriate alternative over DSP for many applications.

5.1.3.1 FPGA for Multilevel Converters

The PWM generator requirements of multilevel converters scale quadratically with the number of levels. For example, a 3-phase voltage source N-level modular multilevel cascaded (MMC) converter needs $3 \times (N - 1)$ pair PWM generators [5]. The power circuit of a 3-phase 5-level converter system is illustrated in Fig. 5.5. The available microcontrollers at present can only provide about six pairs of PWM channels, which are clearly insufficient for multilevel converter systems [6, 7]. On the other hand, the FPGA may provide multiple PWM generators according to the converter requirements. Unlike the DSP which runs a sequential program in its microprocessor, an FPGA may run all the operations in parallel with the clock

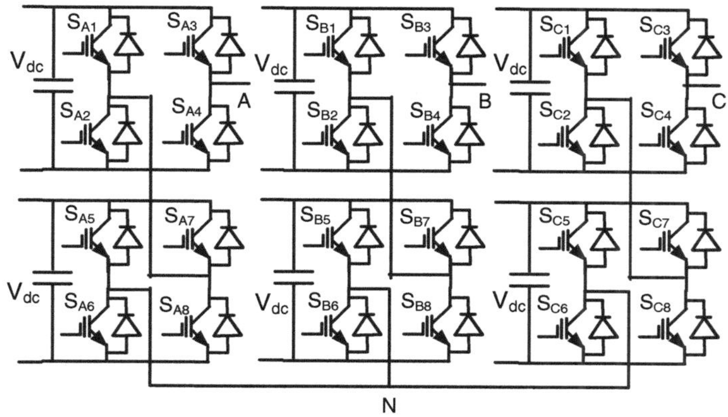

Fig. 5.5 Power *circuit* of a 3-phase 5-level converter

Table 5.1 A comparison between the DSP and the FPGA technology

Performance indicators	Options	
	DSP	FPGA
More than a few MHz sample rate	Hard to handle	Easy to handle
Programming language	Standard C/C++	VHDL
More than 30 Mb/s data rate	Hard to handle	Easy to handle
Conditional operations	Easy to implement	Hard to implement
Use of floating point	Easy to implement	Hard to implement
PWM output channel	Limited (about six pairs)	Can be configured as required
Parallel processing	Serial implementation	Serial/parallel implementation

signal. The capability of parallel processing of the FPGA provides the opportunity to the switching controller to update all gate signals simultaneously [8]. In most cases, the processing time is independent of the number of converter levels. Therefore, the FPGA technology is a natural choice for the control of multilevel converters. A comparison between the FPGA and the DSP technology is summarized in Table 5.1.

5.1.3.2 FPGA Developmental Boards and Kits

Xilinx and Altera are the current market leaders in the FPGA technology. These two vendors have developed different series of FPGA boards and kits. FPGA developmental boards and kits are specially designed and suitable for the design of prototype test platforms. Before selecting the board, the software environment also needs to be considered because different vendors may use different programming software. The selection of FPGA boards depends on many factors, such as clock

Table 5.2 FPGA developmental boards and kits

Brand	FPGA boards and kits	Clock MHz	I/O pins	LEs K	Memory kB	Multipliers	Price AUD
Altera	Arria II GX dev. kit	125	1,152	124	8,121	576	1,495
	Altera DE2-115 D&E board	50	528	114	486	266	595
	Altera DE3 dev. system	50	736	254	1,836 K	768	2,695
	Stratix III dev. kit	125	744	142	128 K	384	2,495
	Stratix IV E dev. kit	125	744	531	2,000 K	1,024	5,495
	Stratix V GX dev. kit	156	696	622	1,152 K	512	6,995
	EZ1CUSB-altera dev. board	50	128	12	29	2	219
	Cyclone III dev. kit	125	531	119	256 K	288	995
Xilinx	Spartan-3 starter board	50	173	4	27	12	139
	Spartan-3E 1600 dev. board	50	250	33	8	36	225
	Spartan-3E starter board	50	232	10	8	20	189
	Spartan-6 LX150T dev. kit	100	540	147	1,355	180	995

speed, number of LEs, number of I/O pins, number of multipliers, area of applications, market availability, and commercial price. A few commercially available developmental boards and kits are summarized in Table 5.2 [9, 10].

5.2 FPGA-Based Switching Controller Design Techniques

Various design techniques and software environments are available for the modeling of switching control schemes with the FPGA technology. The most commonly used design techniques are (i) modeling the switching circuit and target system in the MATLAB/Simulink environment and generation of the programming file with hardware description language (HDL) coder, as shown in Fig. 5.6 and (ii) modeling the switching circuit and target system in the MATLAB/Simulink environment and generation of the programming file with the System Generator, as shown in Fig. 5.7. In order to verify the performance of generated VHDL code in the simulation environment, the ModelSim and PSIM with ModCoupler represent the possible options. These techniques require special software such as HDL coder, System Generator, PSIM, and ModelSim, which increases the developmental time and cost. Figure 5.8 shows the interfacing with target system in simulation environment. In this chapter, the most common

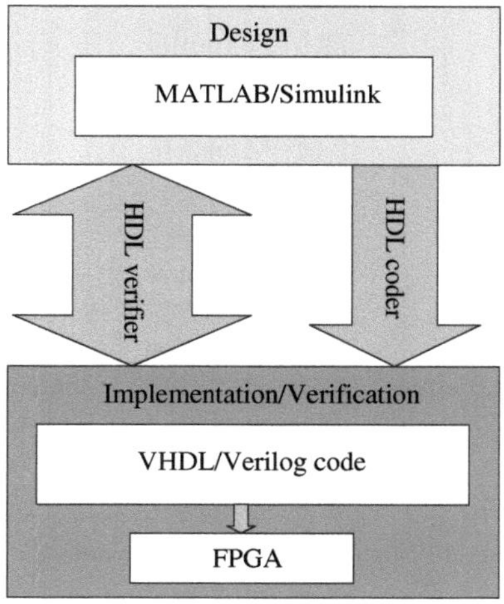

Fig. 5.6 MATLAB/Simulink and HDL coder-based FPGA programming

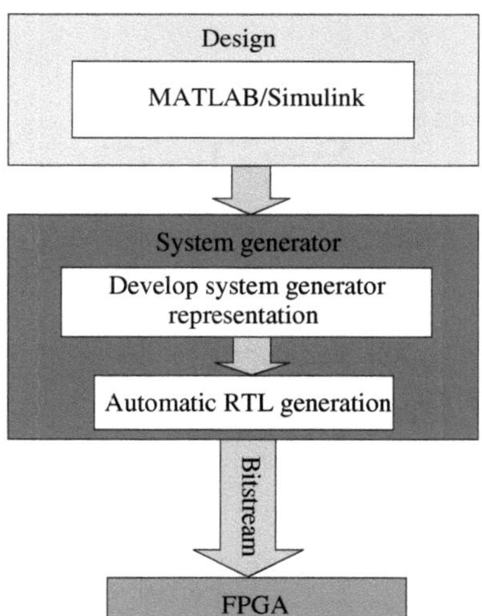

Fig. 5.7 MATLAB/Simulink- and system generator-based FPGA programming

software such as the MATLAB/Simulink- and Xilinx ISE-based alternative design technique is proposed which may reduce the developmental time and cost of the switching controller [1].

In a 5-level converter, two H-bridge cells on each phase are shown in Fig. 5.5. For each H-bridge cell, one reference signal has to be compared with two triangular function carrier signals (one for the left half-bridge and its inverted signal for

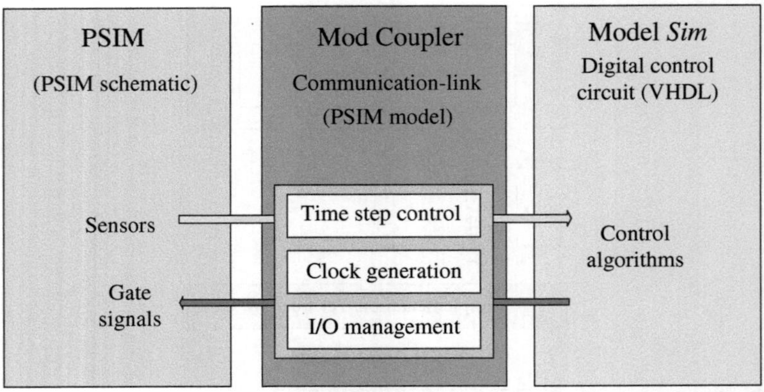

Fig. 5.8 Interfacing with target system in simulation environment

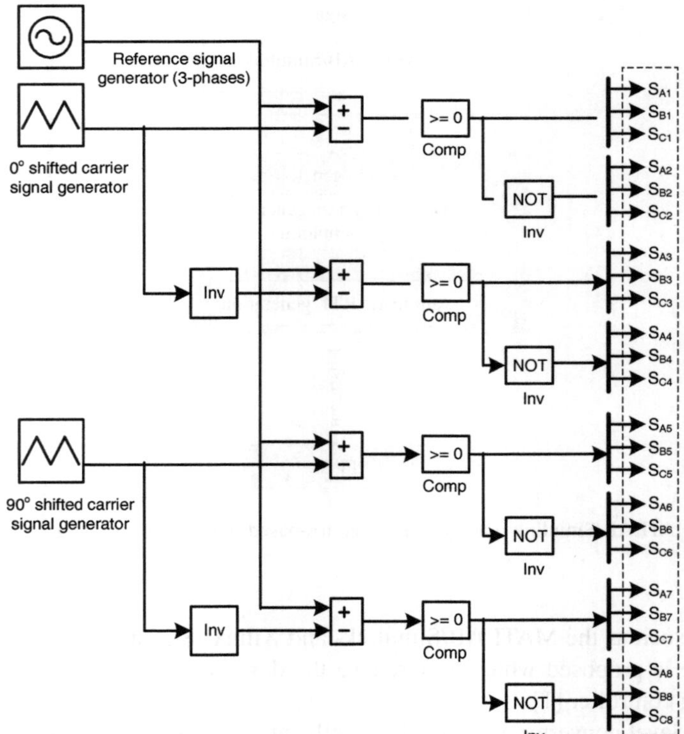

Fig. 5.9 Switching scheme for a 5-level MMC converter

the right half-bridge). The carrier for each H-bridge cell on a particular phase leg is phase-shifted by 360°/(m − 1) [11, 12]. Figure 5.9 shows the switching scheme for a 5-level converter [13].

On each phase leg, in total four half-bridge cells, only two carrier signals are required and the other two can be generated just by inverting them. In total, three reference signals are required, one for each phase leg. The frequency of the reference signal determines the frequency of the converter output voltage. The comparator module compares the carrier signal with the respective reference signal and generates control pulses including reasonable deadtime as required by switching devices. An onboard crystal (e.g., 50 MHz) is used for a clock source, and a clock divider reduces the clock frequency. A lookup table (LUT) is used to generate the reference signals, which makes the control circuit totally digital and integrated. The basic architecture of the FPGA-based switching controller is illustrated in Fig. 5.10.

In total, there are eight switching devices in a phase, requiring eight gate pulses to drive them. Including the inverted carrier signals, a total of four carriers are able to generate four gate pulses when comparing them with a reference signal. The other four gate pulses can be generated by just inverting these four gate pulses with a consideration of deadtime.

The available libraries of the Xilinx FPGA development software Xilinx ISE Design Suite 13.2 only have basic logic symbols, which are clearly insufficient for

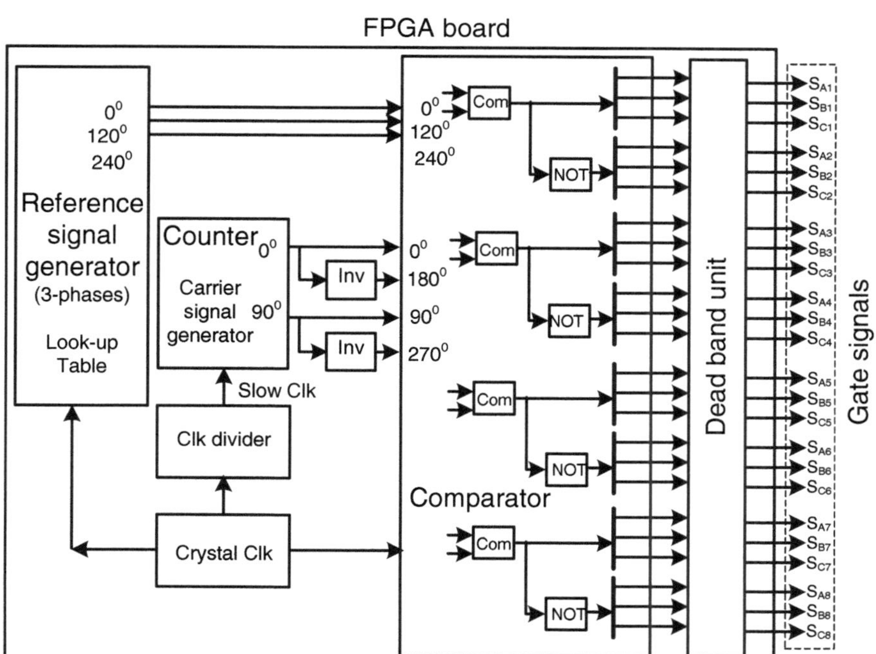

Fig. 5.10 Switching controller architecture

designing a sophisticated switching controller. According to the system requirement, new logic symbols need to be created by writing the VHDL program. In the Xilinx ISE software environment, it is difficult to model a target system (e.g., switching devices and loads). Therefore, the switching controller performance cannot be verified before implementation, which may be unsafe especially in power circuits. On the other hand, in the MATLAB/Simulink environment, the switching controller with target system can be modeled and the performance can be readily verified. Hence, during the modeling of switching controller in the Xilinx ISE environment, the MATLAB/Simulink results are considered as a reference. The Simulink and Xilinx ISE 13.2 Design Suite software-based design technique is illustrated in Fig. 5.11.

In this technique, the switching control scheme with the target system is modeled in the MATLAB/Simulink environment first. After getting satisfactory performance from Simulink, the updated model can be used for behavioral modeling of the switching controller in the Xilinx ISE environment. The behavioral simulation results can be compared with the Simulink results. After getting satisfactory simulation results, the design can be implemented with verification of timing simulation. Before connecting with the target system, the gate signals could be measured and compared with the Simulink results as well as theoretical values.

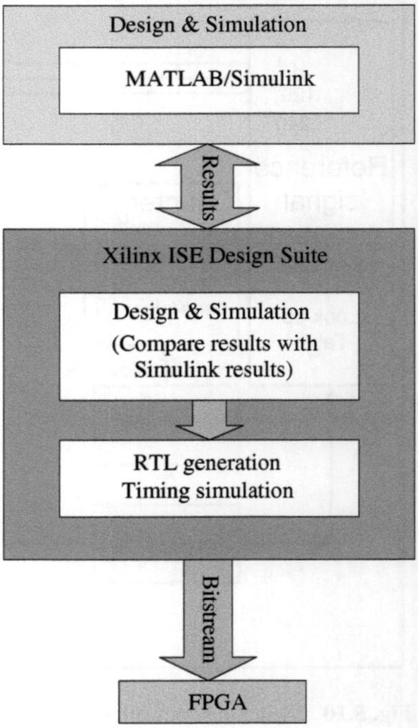

Fig. 5.11 Simulink and Xilinx ISE 13.2 Design Suite software-based design technique

5.3 Modeling and Schematic Symbol Creation Using VHDL

5.3.1 Carrier Signal Generation Unit

The MATLAB/Simulink model used to generate the carrier signals is depicted in Fig. 5.12. The second signal is phase-shifted by 90°. The carrier signals from Simulink are shown in Fig. 5.13.

The triangular carrier waves as shown in Fig. 5.13 can be generated by an up–down counter as depicted in Fig. 5.14. To obtain the correct PWM frequency, a clock divider is used to reduce the crystal frequency before it is used to clock the

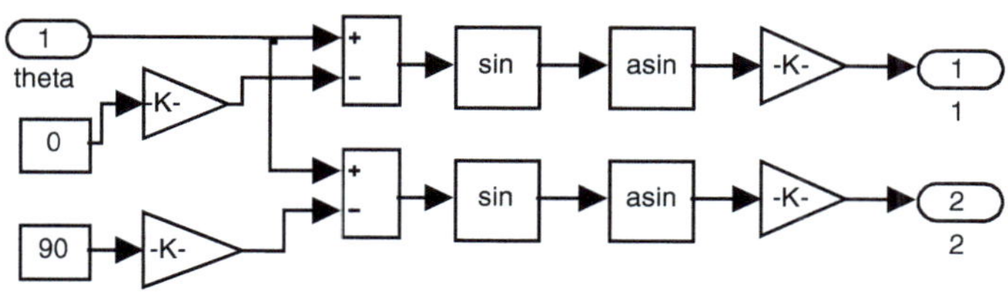

Fig. 5.12 MATLAB/Simulink model for two carrier signals, where $\alpha = \omega t$

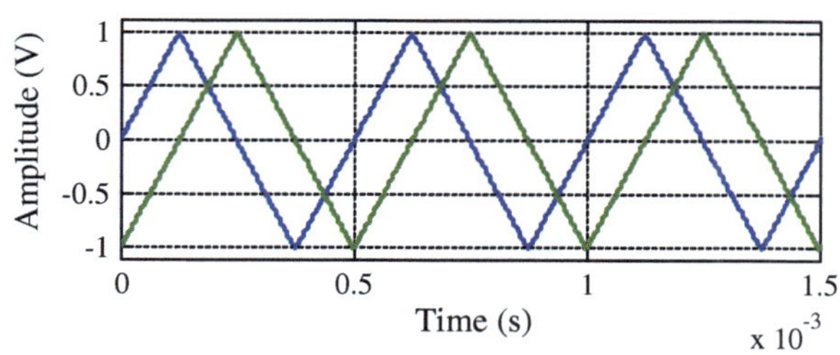

Fig. 5.13 Two carrier signals; one for the *top* H-bridge and 90° shifted carrier for the *bottom* H-bridge

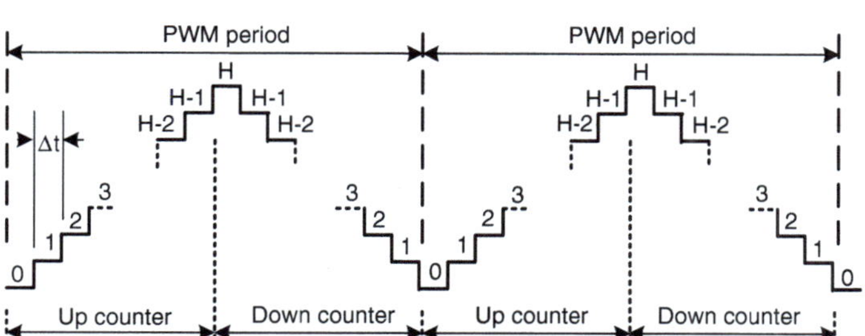

Fig. 5.14 *Up–down* counter-based carrier signal generation, where H is the highest count, and Δt the step size

counter unit. The frequency of the carrier signal should be many times higher than the reference signal. In this work, 1.5-kHz and 50-Hz frequencies are considered for the carrier and reference signals, respectively (i.e., carrier is set 30 times fast as the reference signal). For the 9-bit up–down counter, the total number of steps in a PWM period is $2 \times 2^9 = 1{,}024$. If switching frequency 1.5 kHz is considered, then the required counter clock frequency is $1{,}024 \times 1.5 \times 10^3$ Hz.

Therefore, the clock divider value, k_1, for a system clock frequency of 50 MHz is given by

$$k_1 = \frac{50 \times 10^6\ \text{Hz}}{1024 \times 1.5 \times 10^3\ \text{Hz}} = 32.552. \tag{5.1}$$

In this work, the value of k_1 has been considered to be 32, and with this value, the switching frequency is calculated as 1.525 kHz. The up–down counter and comparator unit combination generates a pulse center aligned within the PWM period. The center-aligned PWM has advantages over the edge-aligned PWM in that the outputs that control the inverter are not all switching on at the same time, at the beginning of every period, as they would do with edge-aligned PWM. This can help reduce noise on the converter power lines, thus increasing conversion efficiency.

The VHDL-code-based behavioral model of the up–down counter is designed and synthesized using the Project Navigator tool of the Xilinx ISE Design Suite 13.2 software. The syntax of the behavioral model is verified using the behavioral check syntax tool. After having analyzed the performance of the behavioral model using the ISim simulator, a schematic symbol is created as shown in Fig. 5.15a. Similarly, the 90° phase-shifted counter is created as depicted in Fig. 5.15b. An inverter or 180° phase shifter is synthesized to generate the inverted carrier signals from the output of the T_Cou_V_3 and B_Cou_V_3 counters. The schematic symbol of the inverter is illustrated in Fig. 5.16.

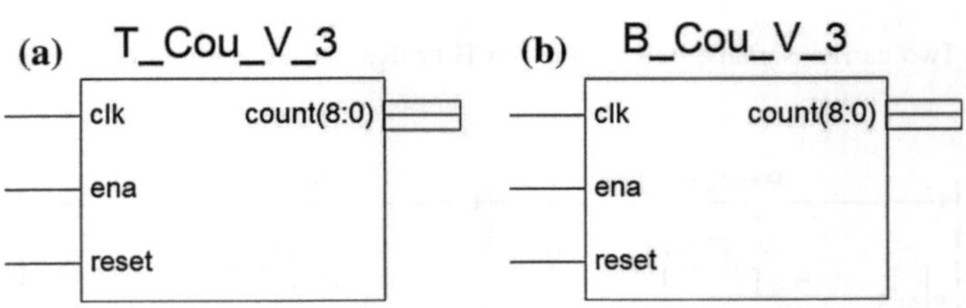

Fig. 5.15 Schematic symbols of *up–down* counter: **a** without phase shift and **b** with 90° phase shift

Fig. 5.16 Schematic symbols of inverter, 180° phase shifter

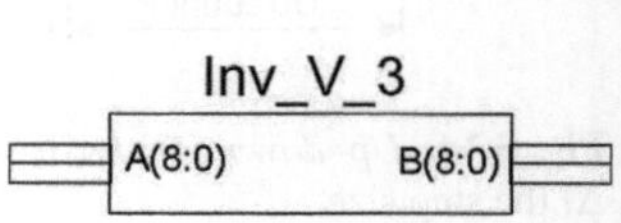

5.3.2 Reference Signal Generation Unit

In order to generate the 50-Hz 3-phase voltage output waveform, three reference signals with 120″ phase-shifted from each other are needed. The MATLAB/Simulink model for the generation of these reference signals is illustrated in Fig. 5.17, and the generated reference signals are shown in Fig. 5.18. The microcontrollers and DSPs are commonly used to generate reference signals. This additional external component increases the system size, weight, and cost and requires proper synchronization. In this instance, logic-based reference signal generation (programming the FPGA logic cells to generate reference signals) using direct digital synthesis (DDS) has attracted significant attention recently [14–16]. The DDS is an LUT-based technique to generate analog signal using stored digital samples. An LUT is actually a set of memory locations which contain the sampled values of a desired analog signal.

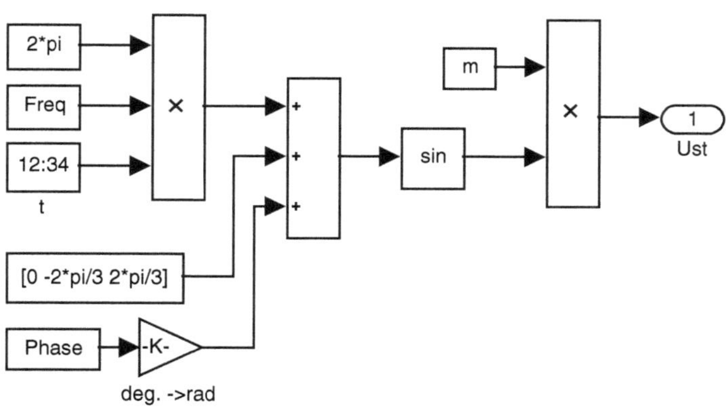

Fig. 5.17 Reference signals generation model

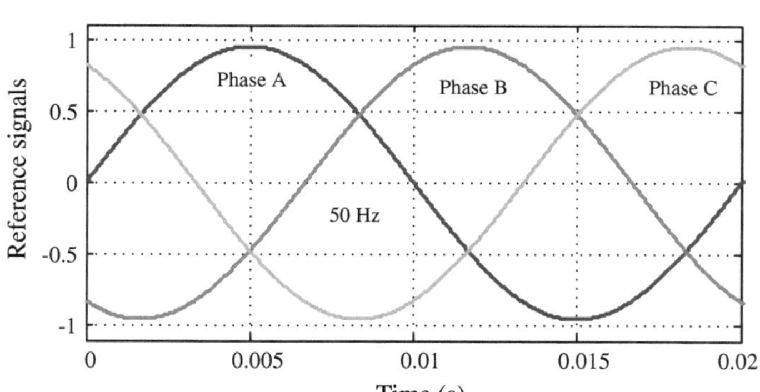

Fig. 5.18 Three-phase 50-Hz reference signals (in MATLAB/Simulink environment)

The frequency of the sine wave can be increased by skipping entries in the sine lookup table as the angular resolution $\Delta\theta$ depends on the number of entries, L_n for a cycle. The angular resolution can be calculated from

$$\Delta\theta = \theta(i+1) - \theta(i) = \frac{360^\circ}{L_n}. \tag{5.2}$$

where i varies from 0 to $L_n - 1$.

If A_{out} is the amplitude of the output signal, then the sample value can be calculated from [1]

$$S_R(i+1) = A_{\text{out}} \times \sin\left(\frac{360}{L_n} i + \Psi\right). \tag{5.3}$$

where Ψ is the phase angle. In order to keep the output span positive between 0 and A_{out}, the sample value stored in the LUT can be expressed as [1]

$$S_{\text{LUT}}(i+1) = \frac{A_{\text{out}}}{2} + \frac{A_{\text{out}}}{2} \times \sin\left(\frac{360}{L_n} i + \Psi\right) \tag{5.4}$$

The clock divider may determine the frequency of the reference signals. For example, if there are 256 entries in the LUT, to generate a 50-Hz reference signal from 256 entries using the 50-MHz clock, the clock divider value k_2 is given by

$$k_2 = \frac{50 \times 10^6 \text{ Hz}}{256 \times 50 \text{ Hz}} = 3{,}906.25. \tag{5.5}$$

The sample values of the reference signals can be calculated from (5.4). The reference signals for phases A, B, and C can be generated by considering the values of Ψ as 0°, 120°, and −120°, respectively. If the peak value, A_{out}, is 255, and L_n

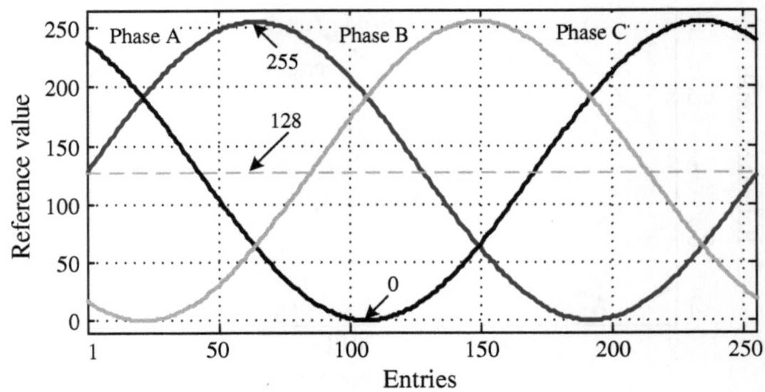

Fig. 5.19 Three-phase reference signals (in ISim Simulator environment)

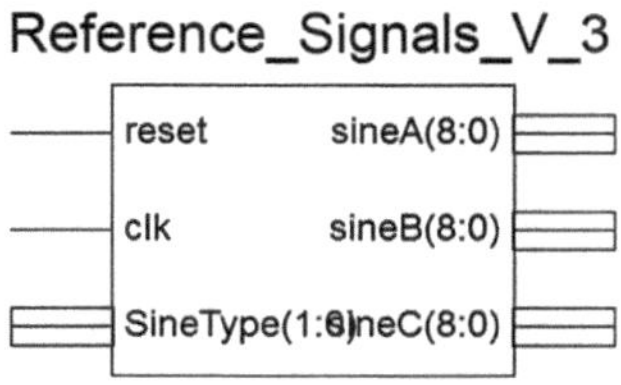

Fig. 5.20 Schematic symbol of reference signal block

256, the decimal sample values of three sine reference signals can be presented as shown in Fig. 5.19. In order to store the sample values, the decimal number needs to be converted to hexadecimal number.

A VHDL-code-based program is used to create the LUT, which contains the sine reference. Different carrier-based modulation techniques used in the traditional converters can also be used in the MMC converter system (e.g., SPWM and THPWM) [17]. Each technique has its unique advantages and disadvantages. In order to analyze the converter performance against different types of reference signals, sine and third-harmonic injected sine reference signals are stored in the LUT. Only one reference signal from each type is stored in the LUT. Based on the stored reference signal, three sine waves (each 120° out of phase from the others) are created by using three separate indexes into the LUT. Only one type reference is used at a time, which can be controlled by an onboard switch without modifying the program. After verifying the syntax of the behavioral model and analyzing the performance through the ISim Simulator environment, a schematic symbol is created as shown in Fig. 5.20.

5.3.3 Comparator Unit

One comparator unit is used for each of the half-bridge cells. The comparator units compare the carrier signal with a reference signal. Each comparator unit generates one switching signal for the top switching device of the half-bridge cell. The inverted form of this switching signal drives the bottom switching device. For the left half-bridge cell, one is asserted when the reference signal value is greater than or equal to the carrier signal value, and the other is asserted when the reference signal value is less than the carrier signal value. For the right half-bridge cell, one is asserted when the inverted carrier signal value is greater than or equal to the reference signal value, and the other is asserted when the inverted carrier signal value is less than the reference signal value. The MATLAB/Simulink model of the comparator unit for a 5-level converter is illustrated in Fig. 5.21. The four carrier signals and the three reference signals as required by the 3-phase 5-level converter are depicted in Fig. 5.22.

The overall switching system is implemented in the MATLAB/Simulink environment to verify the performance roughly. Each H-bridge cell requires four gate

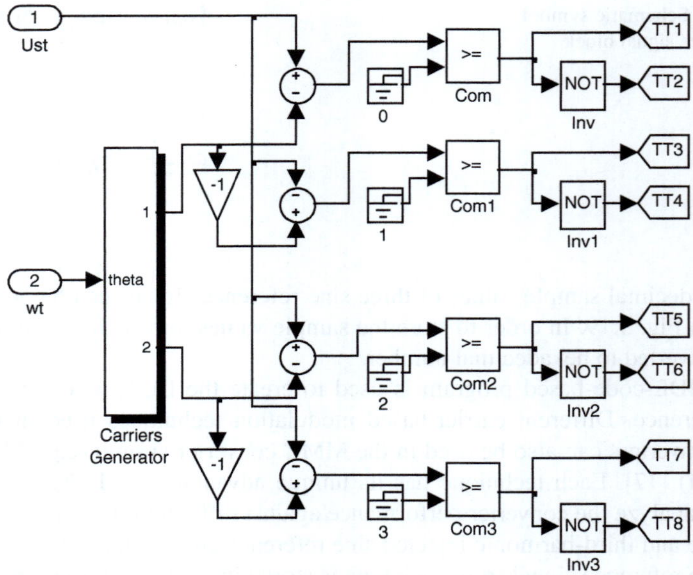

Fig. 5.21 Simulink block diagram of comparator module

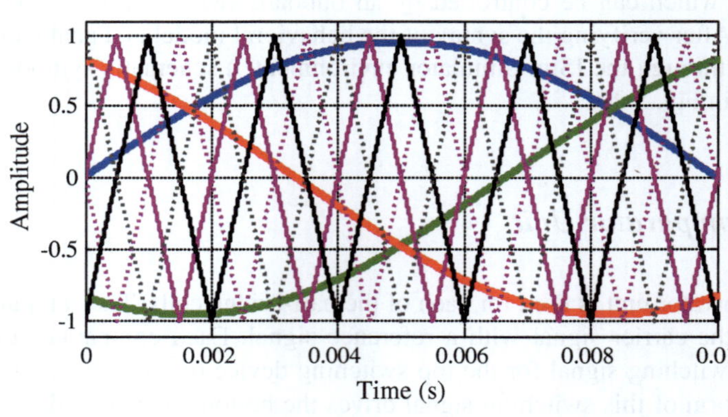

Fig. 5.22 Carrier and reference signals in Simulink environment: *thick waves* represent sine reference signals and *dotted triangular lines* represent the inverted form of carrier signals as presented by *triangular solid lines*

pulses as presented in separate figures. In a 3-phase 5-level converter, there are totally six H-bridge cells and two on each phase. In total, 24 gate pulses are illustrated in Figs. 5.23, 5.24, 5.25, 5.26, 5.27, and 5.28. In the Simulink environment, the deadband is not considered since the switching devices are ideally operated.

The concept as used in Simulink is implemented in the Xilinx ISE Design Suite 13.2 software environment using VHDL coding. Since practical power semiconductor devices do not switch-on or switch-off immediately, i.e., certain time is

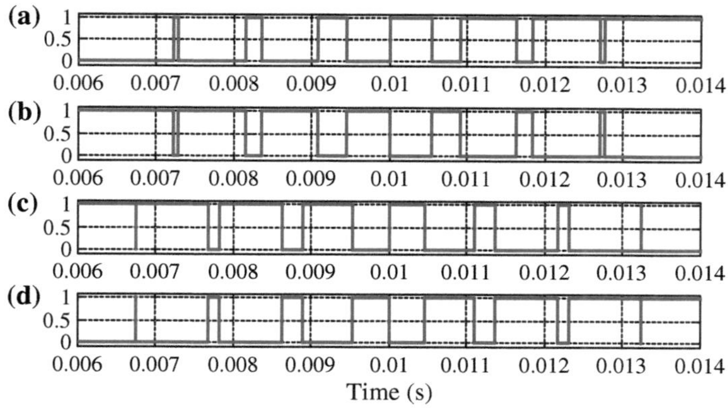

Fig. 5.23 Gate pulses for the *top* H-bridge cell in phase A: **a** *left top*, **b** *left bottom*, **c** *right top* and **d** *right bottom*

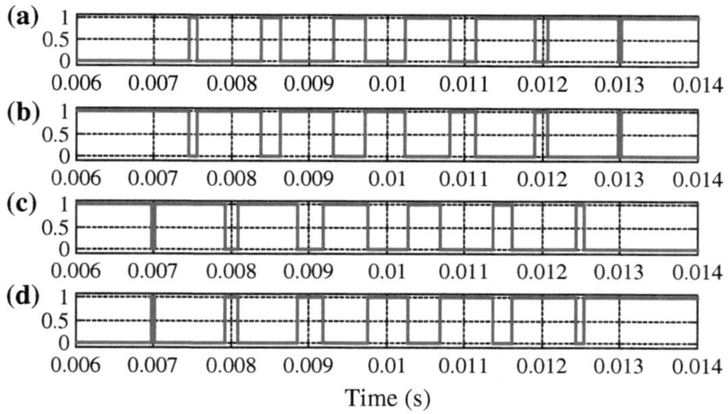

Fig. 5.24 Gate pulses for the *bottom* H-bridge cell in phase A: **a** *left top*, **b** *left bottom*, **c** *right top* and **d** *right bottom*

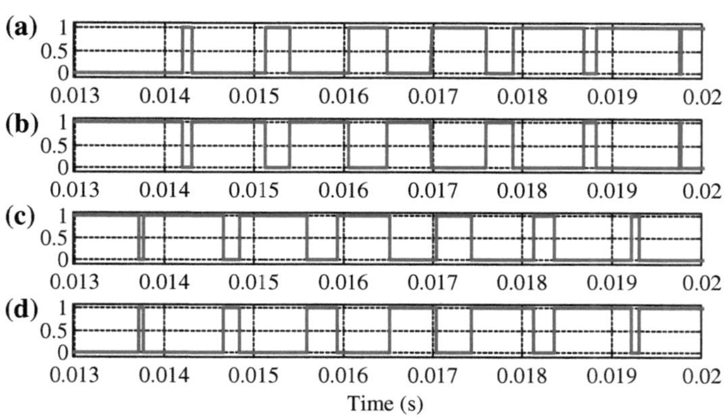

Fig. 5.25 Gate pulses for the *top* H-bridge cell in phase B: **a** *left top*, **b** *left bottom*, **c** *right top* and **d** *right bottom*

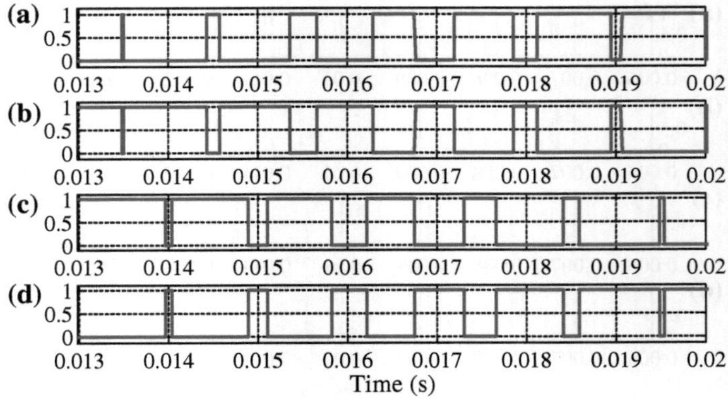

Fig. 5.26 Gate pulses for the *bottom* H-bridge cell in phase B: **a** *left top*, **b** *left bottom*, **c** *right top* and **d** *right bottom*

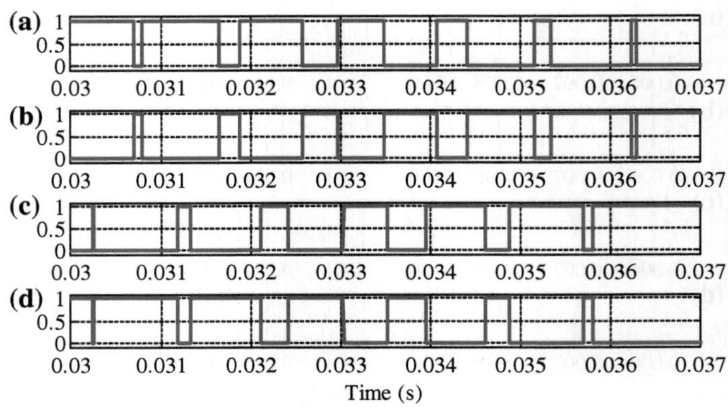

Fig. 5.27 Gate pulses for the *top* H-bridge cell in phase C: **a** *left top*, **b** *left bottom*, **c** *right top* and **d** *right bottom*

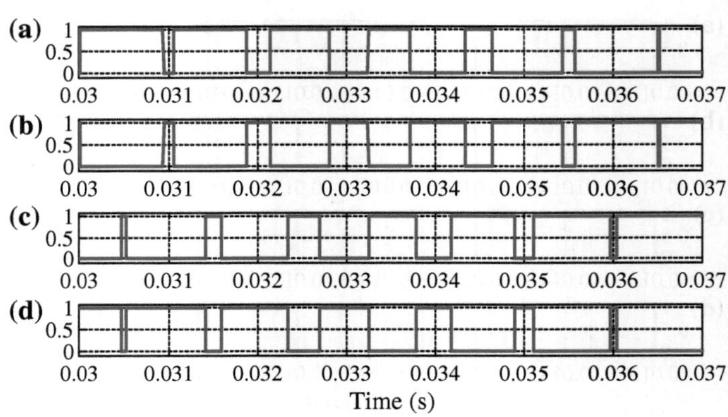

Fig. 5.28 Gate pulses for the *bottom* H-bridge cell in phase C: **a** *left top*, **b** *left bottom*, **c** *right top* and **d** *right bottom*

required to switch them on or off with the switching technique as illustrated in Fig. 5.29, there could be a period of time when both the switching devices of a half-bridge cell are on at the same time, causing a direct short circuit of the terminals of the respective source. In order to avoid this crucial problem, a deadband period is inserted between the turning off of one switching device and the turning on of the other switching device on a half-bridge leg. Thus, both switching devices remain off before one turns on, which ensures the safe operation of both switching devices. The switching technique with deadband is shown in Fig. 5.30, where one PWM output is asserted when the count value is greater than or equal

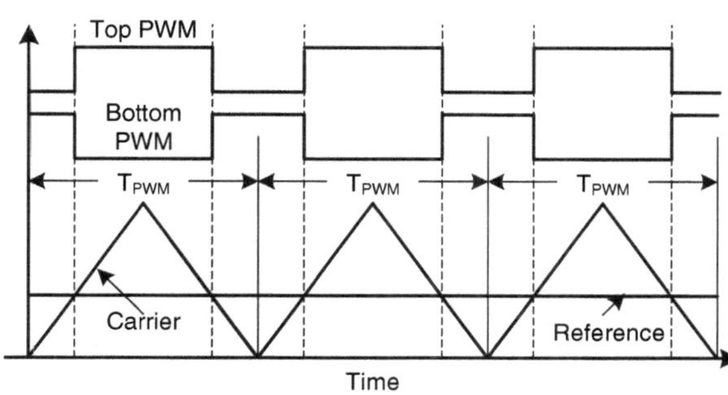

Fig. 5.29 Switching technique without deadband

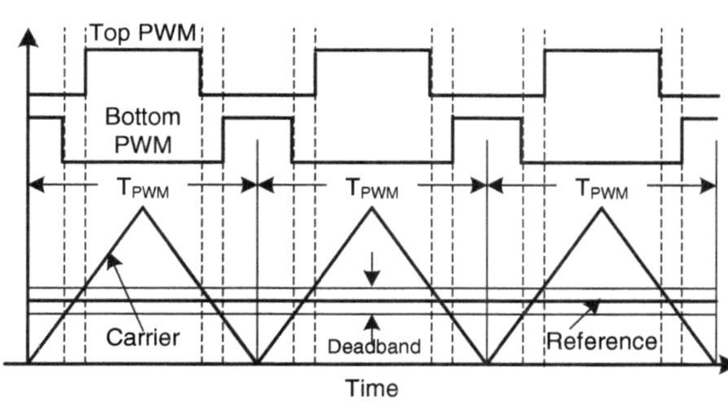

Fig. 5.30 Switching technique with deadband

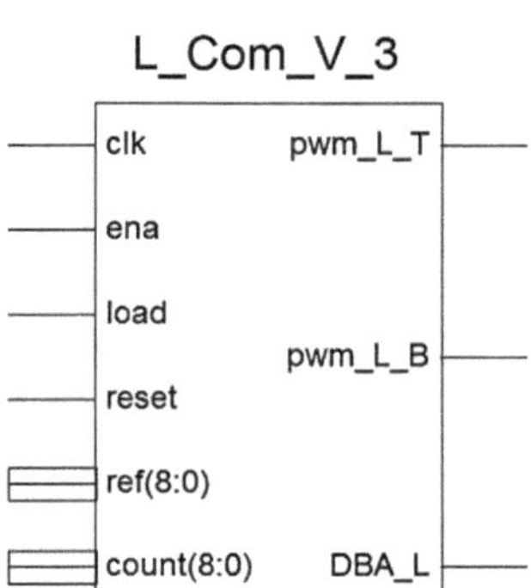

Fig. 5.31 Schematic symbol of the comparator and PWM pulse generator unit, which also has an output terminal called DBA_L to access the deadband wave against PWM waves at pwm_L_T and pwm_L_B

to the reference value plus half of the deadband value and the other PWM output is asserted when the carrier value is less than the reference value minus half of the deadband value.

The designed comparator as well as the PWM pulse generator is simulated by the ISim Simulator, and a schematic symbol is created from the VHDL program-based behavioral model. The schematic symbol for the comparator and PWM pulse generator unit is depicted in Fig. 5.31. The overall algorithm to synthesize a single fully digital integrated circuit (IC) to control a 5-level converter is illustrated in Fig. 5.32 [9].

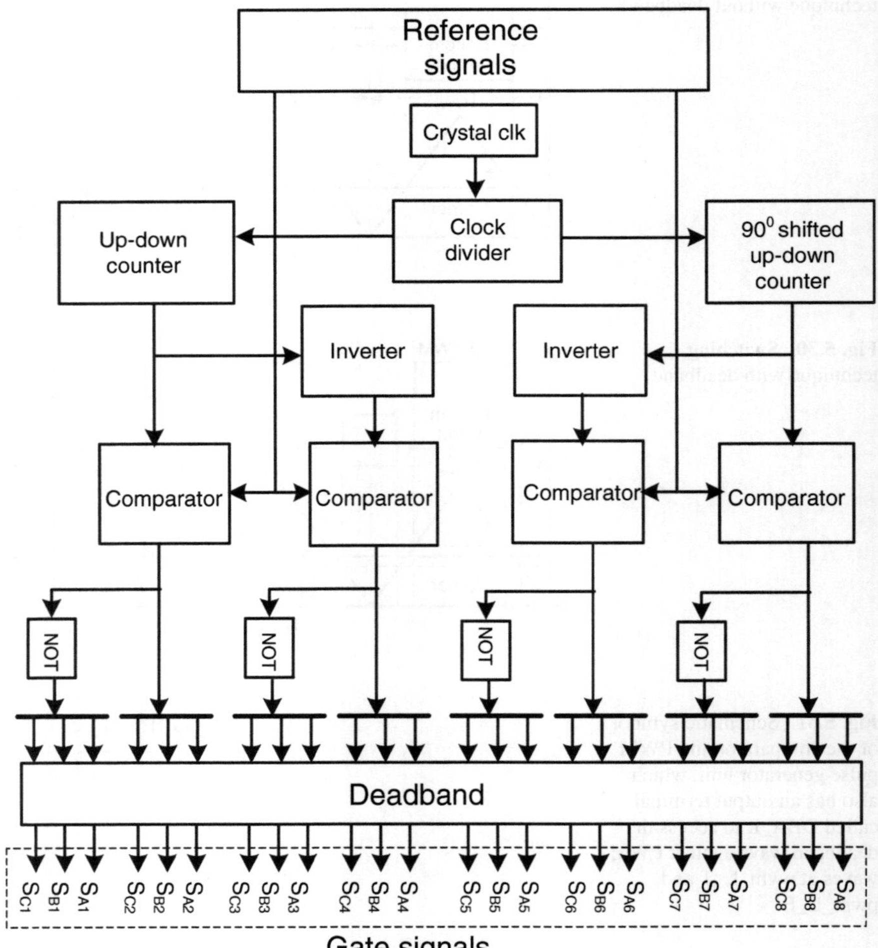

Fig. 5.32 Basic block diagram of FPGA-based 5-level converter control algorithm

5.4 Model of Complete Switching Controller

In the Project Navigator environment, a schematic-based complete switching controller for the 3-phase 5-level converter is developed using schematic symbols created by VHDL programs. The complete switching controller is depicted in Fig. 5.33, where totally there are 12 comparators in a three-column arrangement. Each column is related to a phase of the converter and each comparator unit corresponds to a half-bridge cell of the converter. The top and bottom counters are named T_Cou_V_3 and B_Cou_V_3, respectively. The comparator circuits corresponding to the left and right half-bridge cells of an H-bridge inverter cell are named L_Com_V_3 and R_Com_V_3, respectively. The deadtime is inserted within each comparator unit as a variable, which can be adjusted according to the requirements of the semiconductor devices. All units, such as carrier generators, reference generators and comparators, are connected in parallel with the same clock source, which ensures the parallel operation of FPGA. The complete switching controller is simulated using the behavioral simulation process. Then, the model is synthesized and followed by the implementation process. Before synthesizing the model, all I/Os are assigned to available I/O ports of the FPGA board. During the implementation process, the model verification is carried out. In the verification process, timing simulation is carried out to check the real-time performance.

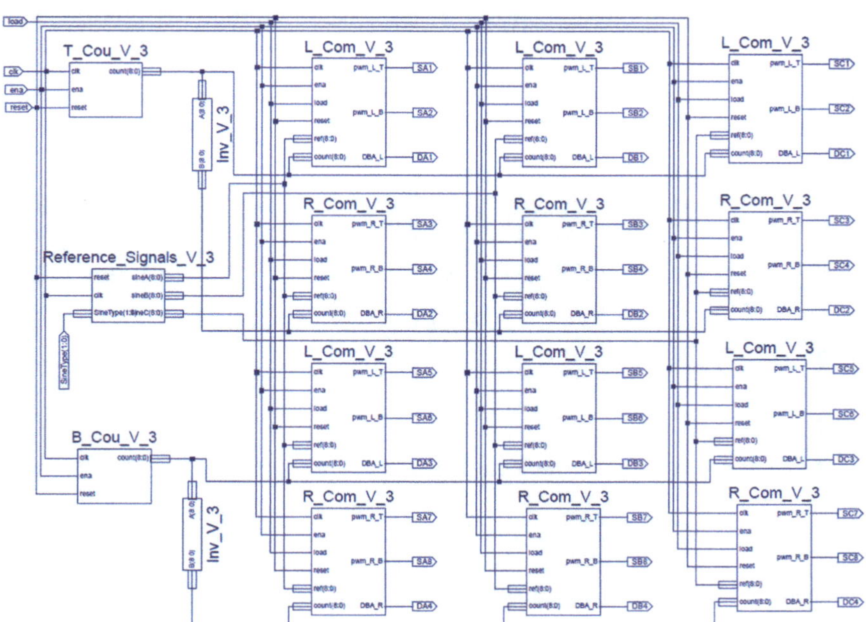

Fig. 5.33 Basic block diagram of FPGA-based 3-phase 5-level converter control circuit

5.5 Behavioral Simulation of Switching Controller

A test bench program file is created using VHDL code to verify the control circuit performance. The behavioral model of the complete switching controller is simulated by the ISim Simulator with the created test bench file. The gate pulse waveforms, deadtime wave forms, and other timing signals such as clock, load, enabling and reset are displayed on the waveform viewer. The gate pulses as well as deadtime waveforms are illustrated in Figs. 5.34, 5.35, and 5.36, each

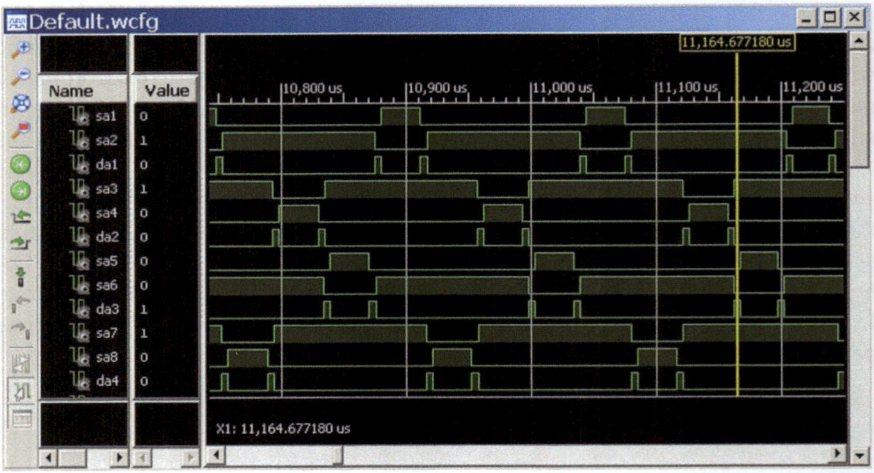

Fig. 5.34 Gate pulses and deadband signals for switches in phase A, where the narrow pulses represent the deadband (adjustable in software environment)

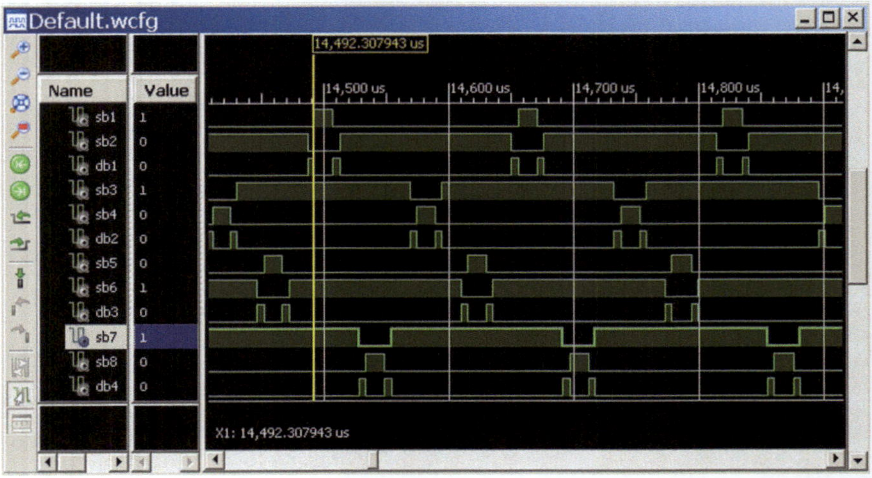

Fig. 5.35 Gate pulses and deadband signals for switches in phase B, where the narrow pulses represent the deadband (adjustable in software environment)

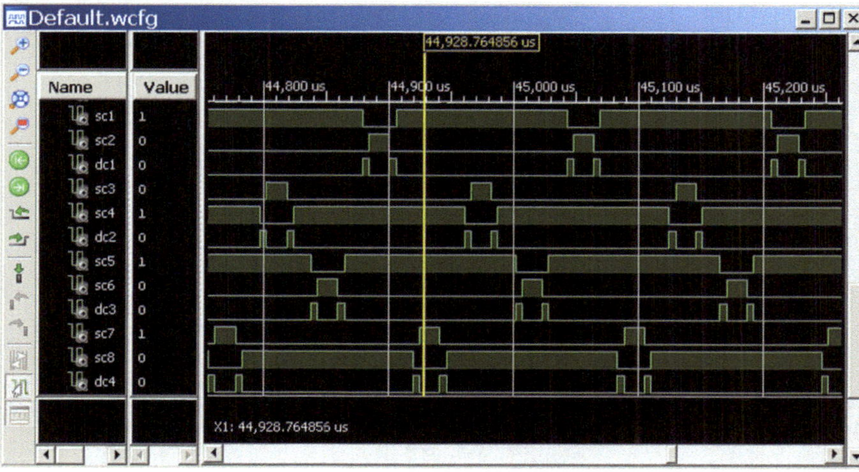

Fig. 5.36 Gate pulses and deadband signals for switches in phase C, where the narrow pulses represent the deadband (adjustable in software environment)

corresponding to one phase of the converter circuit. Through the waveform viewer, it is possible to analyze (by comparing with the MATLAB/Simulink results) the generated gate pulses. The deadtime is not considered in the MATLAB/Simulink system, but the pulse pattern verification can be done with this comparison.

5.6 Model Implementation

After analyzing the performance of the behavioral model of the complete switching controller with behavioral simulation, the model is ready for implementation. The schematic model is first synthesized using the Synthesized-XST tool. All input and output ports are assigned according to the available pin configuration of the Xilinx FPGA chip using the User Constraints-Floorplan Area/IO/Logic (PlanAhead) tool.

All output signals (24 gate signals and twelve deadband signals) are assigned to the first row connectors of the Hirose 100-pin FX2 connector on the Xilinx Spartan-3E development board used for testing. Most of all the connectors in the second row are connected to the ground. There are eight small LEDs (LD0–LD7) parallel connected with eight pins of the top row. Therefore, these signals should be assigned to the pins whose output requires monitoring. The board has four slide switches labeled as SW0–SW3. When a switch is in the UP or ON position, it connects the respective FPGA pin to 3.3 V (i.e., a logic high) and when in the DOWN or OFF position, it connects the FPGA pin to the ground (i.e., a logic low). The input control signals, such as the load, enabling, and reference type, are assigned to these switches. The board also has four momentary push button switches. Pressing a push button connects the associated FPGA pin to 3.3 V (i.e., a logic high). The I/O pin assignment pattern is summarized in Table 5.3.

Table 5.3 I/O pin configuration

Signal name	Direction	Pin number
SineType [0]	Input	L13
SineType [1]	Input	L14
Ena	Input	N17
Load	Input	H18
Clk	Input	C9
Reset	Input	K17
SA1	Output	C14
SA2	Output	B14
DA1	Output	A16
SA3	Output	E12
SA4	Output	E11
DA2	Output	B16
SA5	Output	A14
SA6	Output	B13
DA3	Output	E13
SA7	Output	F11
SA8	Output	C11
DA4	Output	C4
SB1	Output	A13
SB2	Output	F12
DB1	Output	B11
SB3	Output	D11
SB4	Output	E9
DB2	Output	A11
SB5	Output	E7
SB6	Output	B6
DB3	Output	A8
SB7	Output	F9
SB8	Output	E8
DB4	Output	G9
SC1	Output	A6
SC2	Output	C5
DC1	Output	C3
SC3	Output	F8
SC4	Output	C7
DC2	Output	B4
SC5	Output	D5
SC6	Output	A4
DC3	Output	D14
SC7	Output	D7
SC8	Output	F7
DC4	Output	D10

There are two ways to configure the FPGA pins in the Xilinx ISE Design Suite 13.2 software environment. The first method is to enter the pin number in the site field in the I/O Port Properties tab when the I/O signal is selected. The I/O Port Properties-based I/O pins assignment is illustrated in Fig. 5.37.

Dragging into the Package view is an alternative way for the I/O pins assignment. In this method, a signal needs to be selected, dragged into the Package

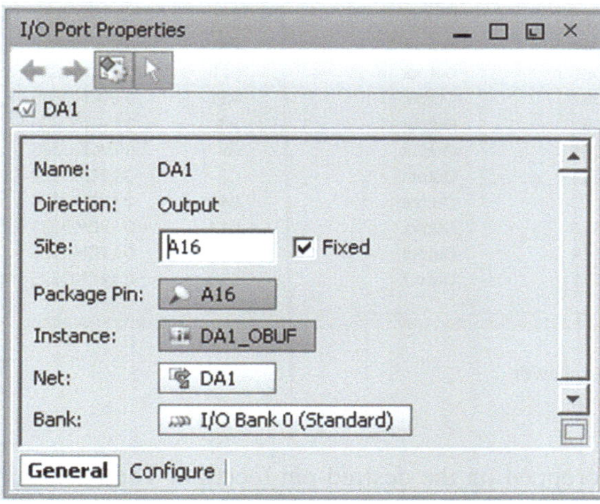

Fig. 5.37 Assigning I/O pins via I/O port properties

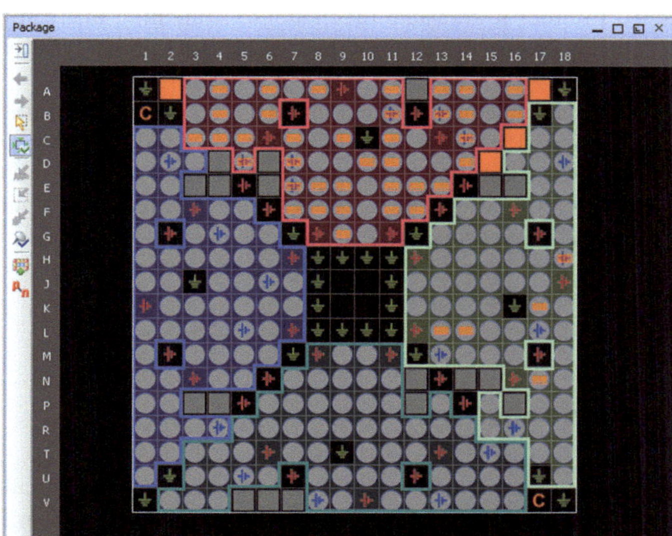

Fig. 5.38 Assigning I/O pins by dragging into package view

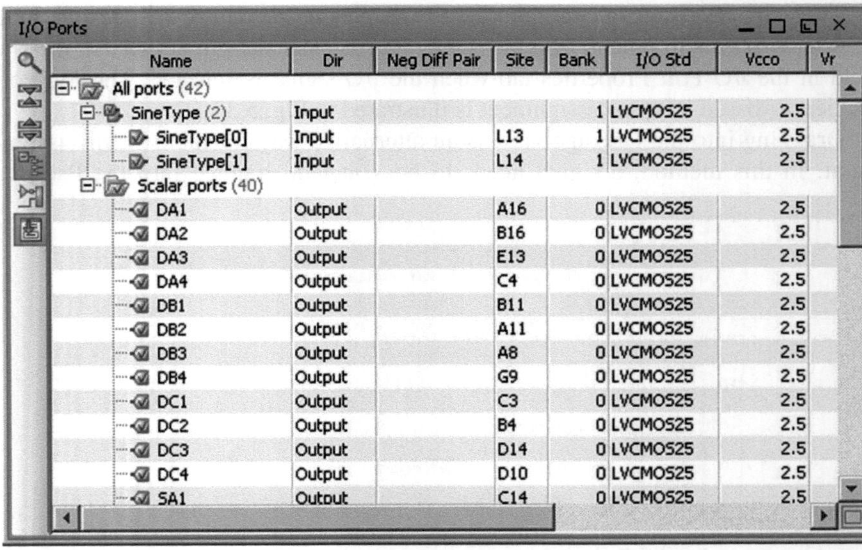

I/O Ports

Name	Dir	Neg Diff Pair	Site	Bank	I/O Std	Vcco	Vr
All ports (42)							
SineType (2)	Input			1	LVCMOS25	2.5	
SineType[0]	Input		L13	1	LVCMOS25	2.5	
SineType[1]	Input		L14	1	LVCMOS25	2.5	
Scalar ports (40)							
DA1	Output		A16	0	LVCMOS25	2.5	
DA2	Output		B16	0	LVCMOS25	2.5	
DA3	Output		E13	0	LVCMOS25	2.5	
DA4	Output		C4	0	LVCMOS25	2.5	
DB1	Output		B11	0	LVCMOS25	2.5	
DB2	Output		A11	0	LVCMOS25	2.5	
DB3	Output		A8	0	LVCMOS25	2.5	
DB4	Output		G9	0	LVCMOS25	2.5	
DC1	Output		C3	0	LVCMOS25	2.5	
DC2	Output		B4	0	LVCMOS25	2.5	
DC3	Output		D14	0	LVCMOS25	2.5	
DC4	Output		D10	0	LVCMOS25	2.5	
SA1	Output		C14	0	LVCMOS25	2.5	

Fig. 5.39 I/O ports viewer

view and then dropped on the desired pin location. The Package view is shown in Fig. 5.38. The I/O port viewer displays the ports with their configuration such as input/output and pin number as depicted in Fig. 5.39. Once the PlanAhead is closed by selecting the Exit tab, a ucf file is added to the project, which contains the constraints. In order to translate the model to something that can physically be mapped into the FPGA, i.e., xc3s500e-4fg320, the Implement Design tab is used.

Actually under the Implement Design tab, there are three main functions such as Translate, Map, and Place and Route. The Translate process merges all of the input netlists and design constraint information and creates a Xilinx native generic database (NGD) file. The Map process creates a native circuit description (NCD) file, which actually implements the logic functions in a device. The Place and Route process creates another NCD file to place and route the design by using the NCD file created during the Map process.

5.7 Design Verification

The design Summary viewer reports all the design issues such as overall summary (as shown in Fig. 5.40), timing constraints, pinout report (as shown in Fig. 5.41), errors and warnings and other information.

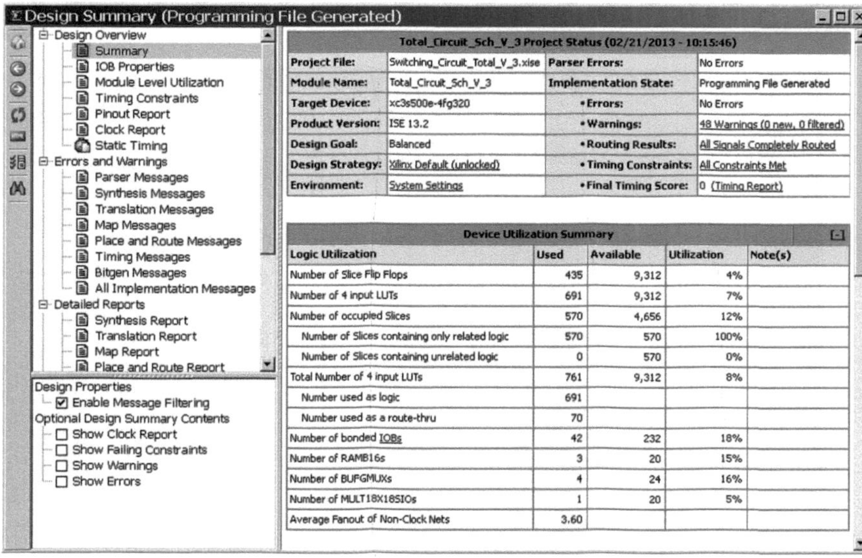

Fig. 5.40 Design summary/report viewer

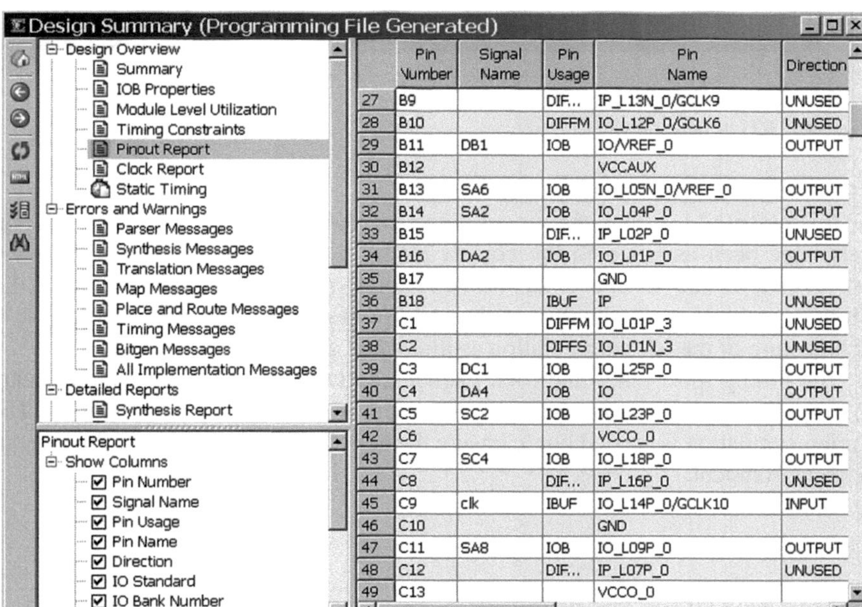

Fig. 5.41 Design summary/pinout report

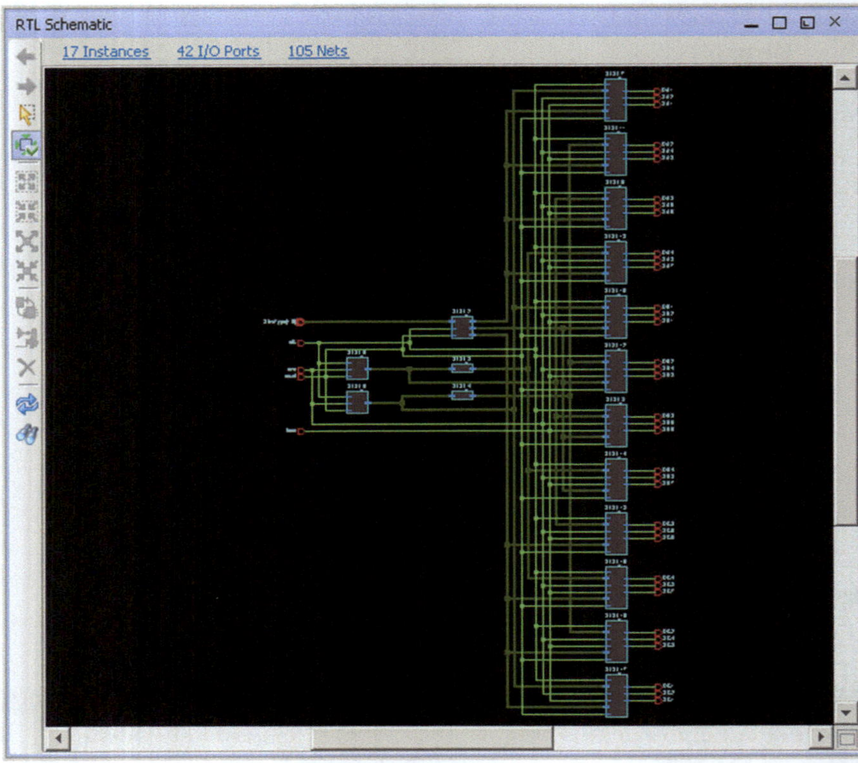

Fig. 5.42 RTL schematic viewer

It is essential to check the pinout report carefully (e.g., whether or not all signals have been assigned to the correct FPGA pins). The View RTL Schematic process views the schematic of the behavioral model as depicted in Fig. 5.42. Moreover, the View Technology Schematic process is used to view the physical architecture of the I/O configuration as illustrated in Fig. 5.43.

The timing (post-place and route) simulation uses the block and routing delay information to give a more accurate assessment of the behavior of the circuit. The timing simulation is carried out to ensure the actual device operation in the simulation environment.

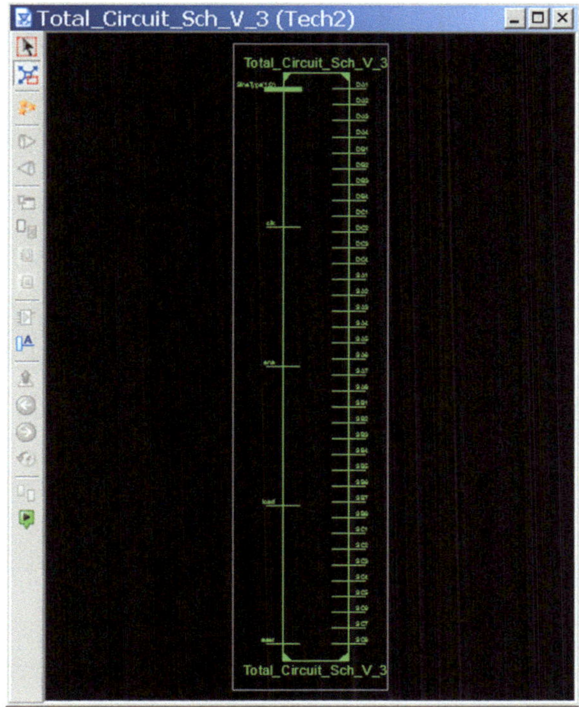

Fig. 5.43 View technology schematic

5.8 FPGA Programming

After analyzing the design, a configuration bitstream is created. The process tab Generate Programming File generates a bit file that can be used to program the FPGA. Before programming the FPGA, the configuration mode jumper settings must be set for JTAG programming. After connecting the FPGA board with the PC through a standard USB cable on the USB-based download/debug port of the test board, the power switch is turned on. Figure 5.44 shows a photograph of FPGA board with PC connection through USB cable. A green light-emitting diode (LED) lights up after a successful connection. The FPGA is programmed using the iMPACT tool under the Configure Target Device process. There is no need to include some configuration files for SPI ports or BPI PROM to this device. The configuration files assignment options for other ports of xcf04 s and xc2c64 should be bypassed. During this process, the device xc3s500e is selected to program the FPGA. The iMPACT process viewer is shown in Fig. 5.45. There are two indications of successful programming: a message Program Succeeded shows up in iMPACT process view and at the same time a yellow LED (underneath the J30 jumper) on the board lights up.

Fig. 5.44 Photograph of FPGA board with PC connection through USB cable

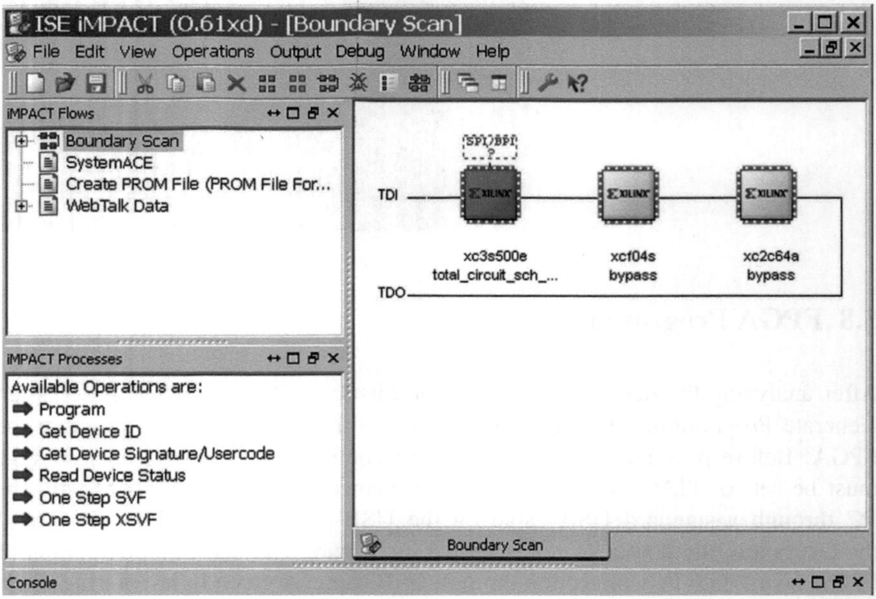

Fig. 5.45 iMPACT process view

5.9 Experimental Testing and Verifications

All FPGA I/Os that interface to the Hirose connector on the test board are in Bank 0 of the FPGA. The I/O Bank 0 supply is 3.3 V by default. Therefore, the PWM gate pulses from the FPGA are also 3.3 V, which is insufficient for the IGBT drivers used. Driver ICs TC4427A are used to increase the voltage level to the desired value

(TC4427A can support up to 18 V). Through Hirose 100-pin FX2 connector and ribbon cables, the generated 3.3-V gate pulses are applied to the interface circuit. The experimental setup of the switching controller for the 5-level 3-phase multilevel converter system is illustrated in Fig. 5.46. The gate pulses from the driver circuit can be used to switch the switching devices. The gate pulses for a 3-phase 5-level converter are illustrated in Figs. 5.47, 5.48, 5.49, 5.50, 5.51, and 5.52.

An Agilent Technologies DS06034A oscilloscope was used to observe the signal waveforms. This oscilloscope has four input channels, and the gate pulses for each H-bridge cell (i.e., four) are shown in a figure. The measured switching

Fig. 5.46 Photograph of the FPGA-based switching controller section

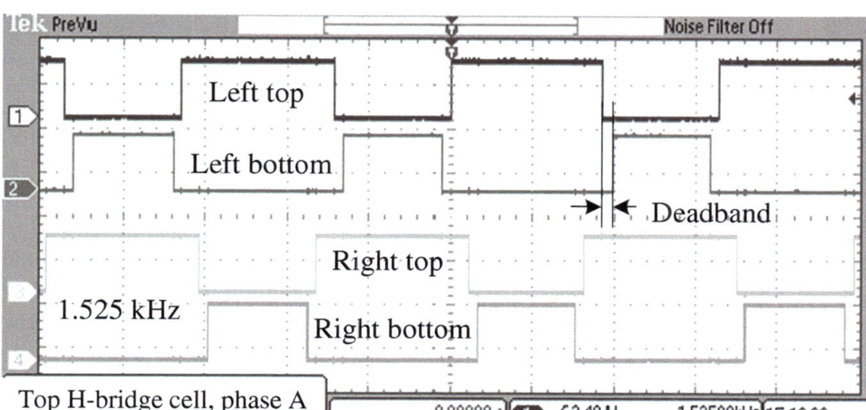

Fig. 5.47 Measured gate pulses for the *top* H-bridge cell in phase A: *1* left top, *2* left bottom, *3* right top, and *4* right bottom

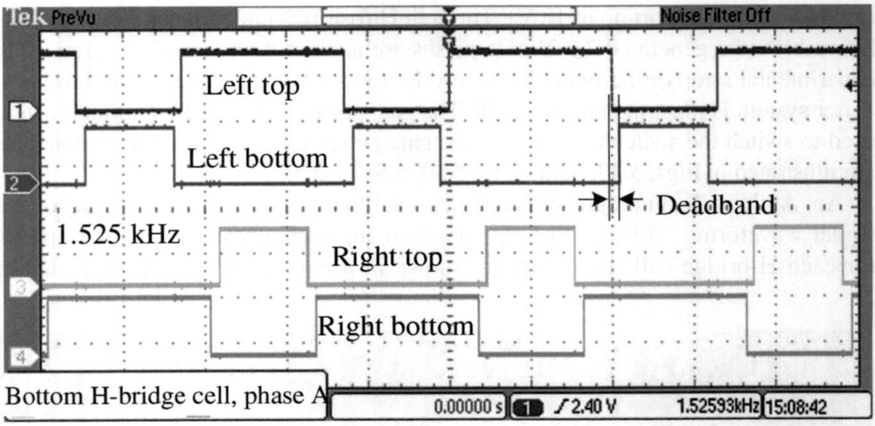

Fig. 5.48 Measured gate pulses for the *bottom* H-bridge cell in phase A: *1* left top, *2* left bottom, *3* right top, and *4* right bottom

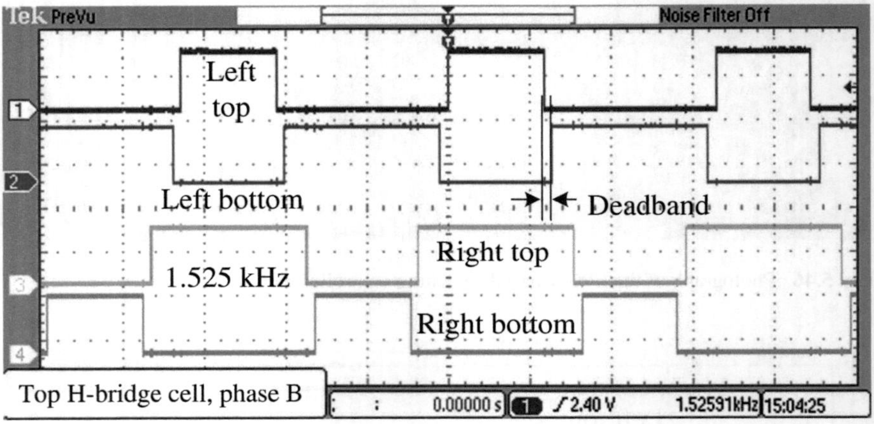

Fig. 5.49 Measured gate pulses for the *top* H-bridge cell in phase B: *1* left top, *2* left bottom, *3* right top, and *4* right bottom

frequency of the switching signals is 1.525 kHz, which precisely matches the theoretical result. Pulse patterns are also verified with theoretical results as well as MATLAB/Simulink and Xilinx ISE simulation results, and they are found highly consistent. Gate pulses are applied to the switching devices of the converter, and

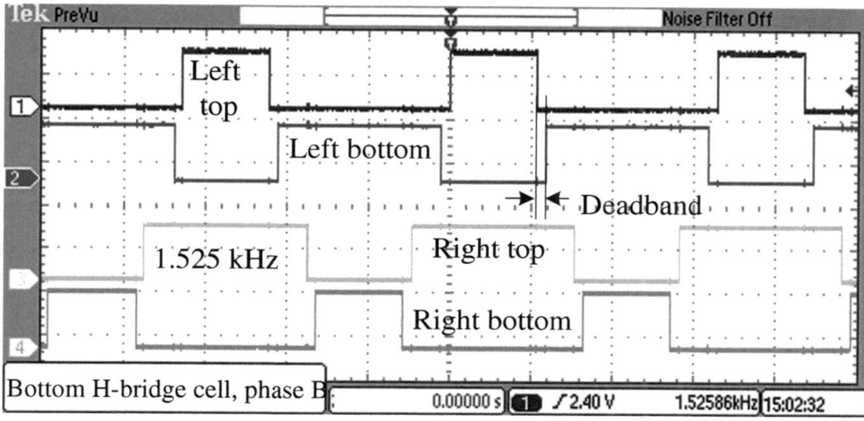

Fig. 5.50 Measured gate pulses for the bottom H-bridge cell in phase B: *1* left top, *2* left bottom, *3* right top, and *4* right bottom

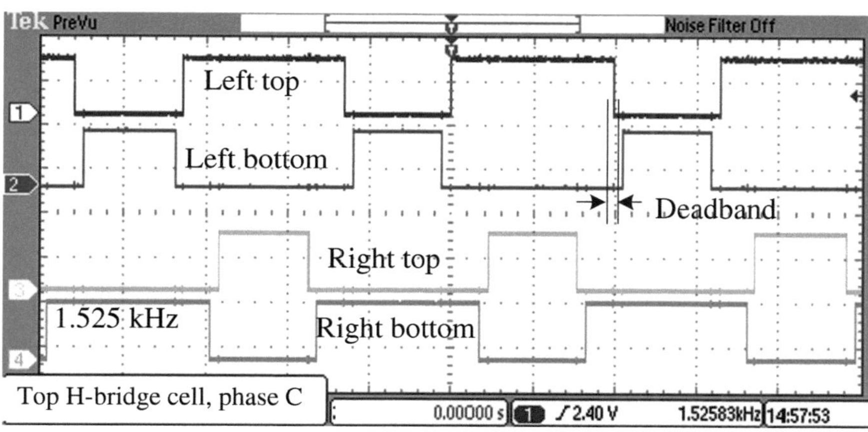

Fig. 5.51 Measured gate pulses for the *top* H-bridge cell in phase C: *1* left top, *2* left bottom, *3* right top, and *4* right bottom

the output phase voltage is measured using oscilloscope as shown in Fig. 5.53, and it is found satisfactory. The phase voltage waveforms were also found highly consistent with MATLAB simulation results. Figure 5.54 shows simulated phase voltage waveform of a 5-level converter.

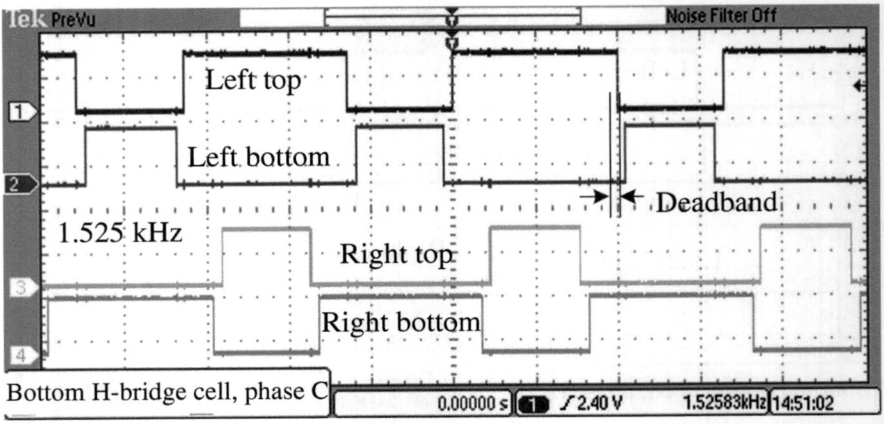

Fig. 5.52 Measured gate pulses for the *bottom* H-bridge cell in phase C: *1* left top, *2* left bottom, *3* right top, and *4* right bottom

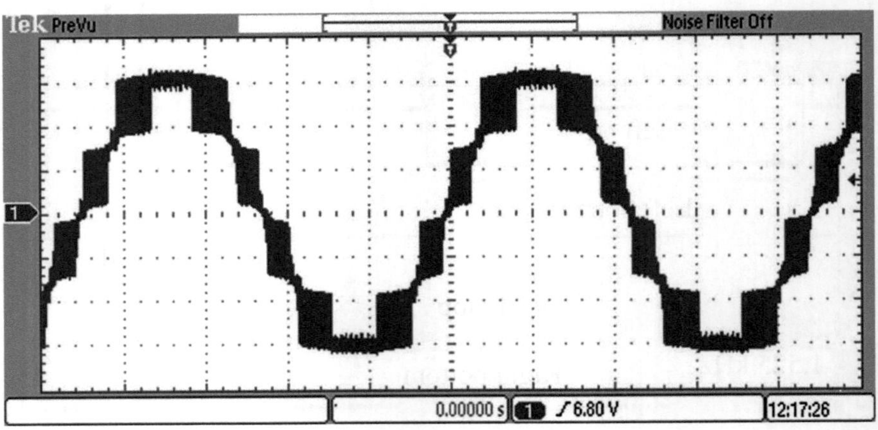

Fig. 5.53 Measured phase voltage of a prototype 5-level converter

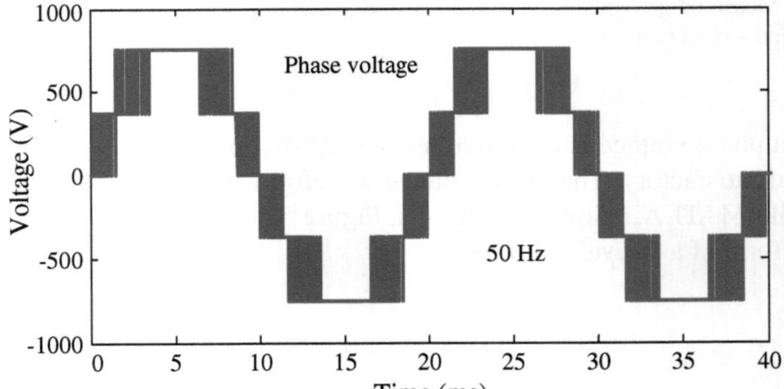

Fig. 5.54 Simulated phase voltage of a prototype 5-level converter

5.10 Summary

In order to develop switching circuits for multilevel converters, the FPGA may be the natural choice because of its high-speed operation and capacity to handle multiple PWM signals. An integrated fully digital switching controller has been developed for a 3-phase 5-level converter system using an FPGA development board, which has some user-friendly onboard switches and buttons for I/O. Through these switches and buttons, it is possible to control the operation of the switching controllers such as reset, enabling, load, type of reference signal, and many more. During the behavioral modeling of the switching controller, the MATLAB/Simulink model and its responses have been considered as the reference. Finally, the switching circuit has been implemented for experimental verification. The experimental results are found highly consistent with theoretical and MATLAB/Simulink results. This switching scheme and design technique can be used for any other converter with minor modification to the software environment.

References

1. Islam MR, Guo Y, Zhu J (2014) A high-frequency link multilevel cascaded medium-voltage converter for direct grid integration of renewable energy systems. IEEE Trans Power Electron 29(8):4167–4182
2. Texas Instruments, fixed-point digital signal processor, Application Note, SPRS051H (2004)
3. Xilinx, Spartan-6 family overview, product specification, DS160 (2011)
4. Ekas P (2007) FPGAs rapidly replacing high-performance DSP capability. DSP-FPGA.COM MAGAZINE. [Online] Available at: http://dsp-fpga.com/article-id/?2051. Accessed on 10 Feb 2013
5. Islam MR, Guo Y, Zhu J (2014) A review of offshore wind turbine nacelle: technical challenges, and research and developmental trends. Renew Sustain Energy Rev 33:161–176
6. Islam MR, Guo Y, Zhu J (2014) A multilevel medium-voltage inverter for step-up-transformer-less grid connection of photovoltaic power plants. IEEE J Photovoltaics 4(3):881–889
7. Islam MR, Guo Y, Zhu J (2012) A novel medium-voltage converter system for compact and light wind turbine generators. In Proceedings of Australasian universities power engineering conference (AUPEC2012), Bali, Indonesia, pp 1–6, 26–29 Sept 2012
8. Islam MR, Guo Y, Zhu J (2013) Multiple-input-multiple-output medium frequency-link based medium voltage converter for direct grid connection of photovoltaic arrays. In: Proceedings of International Conference on Electrical Machines and Systems (ICEMS2013), Busan, Korea, pp 202–207, 26–29 Oct 2013
9. ALTERA, All Development Kits. [Online] Available at: http://www.altera.com/products/devkits/kit-dev_platforms.jsp. Accessed on 13 July 2013
10. XILINX, Boards and Kits. [Online] Available at: http://www.xilinx.com/products/boards-and-kits/index.htm. Accessed on 01 Apr 2013
11. Rodriguez J, Lai JS, Peng FZ (2002) Multilevel inverters: a survey of topologies, controls and applications. IEEE Trans Ind Electron 49(4):724–738
12. Islam MR, Guo Y, Jafari M, Malekjamshidi Z, Zhu J (2014) A 43-level 33 kV 3-phase modular multilevel cascaded converter for direct grid integration of renewable generation systems. In: IEEE innovative smart grid technologies conference (ISGT) Asia, Kuala Lumpur, Malaysia, 20–23 May 2014

13. Islam MR, Guo Y, Zhu J (2014) Multilevel converters for step-up transformer-less direct grid integration of renewable generation units with medium voltage smart microgrids. Large scale renewable power generation: advances in technologies for generation, transmission and storage. Springer, Berlin, Heidelberg, pp 127–149
14. Suchenek M, Starecki T (2012) Programmable pulse generator based on programmable logic and direct digital synthesis. Rev Scientific Instruments 83(12):124704-1–124704-4
15. Li Z, Zhou W, Zhou H, Zhang X, Zhao J (2013) The optimization of super-high resolution frequency measurement techniques based on phase quantization regularities between any frequencies. Rev Scientific Instruments 84(2):025106-1–025106-5
16. Tierney J, Rader CM, Golg B (1971) A digital frequency synthesizer. IEEE Trans Audio Electroacoust AU-19:48–57
17. Islam MR, Guo Y, Zhu J (2011) H-bridge multilevel voltage source converter for direct grid connection of renewable energy systems. In: IEEE power and energy society innovative smart grid technology conference (ISGT 2011 Asia), Perth Australia, pp 1–7, 13–16 Nov 2011

Chapter 6
Experimental Validation of 1-kV Modular Multilevel Cascaded Converter with High-Frequency Magnetic Link

Abstract In this chapter, the high-frequency magnetic-link modular multilevel cascaded (MMC) converter is experimentally validated for grid integration of renewable power generation systems. The high-frequency magnetic link generates multiple isolated and balanced DC supplies for all the H-bridge inverter cells of the MMC converter. To verify the feasibility of the converter, a scaled down 1-kV laboratory prototype test platform with a 5-level MMC converter is developed in this chapter. The design and implementation of the prototyping, test platform, and experimental results are analyzed and discussed. The methods/techniques for component selection, converter fabrication, and experimentation are also applicable to other converter applications.

Keywords Modular multilevel cascaded (MMC) · Power converters · Prototype · Test platform · Component selection · Experimental validation

6.1 Introduction

Recently, the development of the medium-voltage converter, especially for off-shore and remote area renewable power generation applications, has attracted significant attention [1, 2]. In 2008, a multi-coil modular permanent magnet generator (isolated stator coils) was proposed to eliminate the step-up transformers of wind turbine generators [3]. This was validated by a 230-V test rig. This multiwinding generator generates multiple DC supplies for the modular multilevel cascaded (MMC) converter, which requires a special winding arrangement (still at a heavy weight and large volume due to the power frequency operation) and complicated control strategies. A multicoil generator-based wind turbine system is shown in Fig. 6.1. An improved control strategy was proposed and verified on almost the same generator converter system [4, 5]. In 2010, another approach to eliminate the step-up transformers of wind turbine generators was proposed [6]. A few 6-phase generators are placed in the wind turbine nacelle to generate multiple DC supplies

M.R. Islam et al., *Power Converters for Medium Voltage Networks*,
Green Energy and Technology, DOI 10.1007/978-3-662-44529-7_6

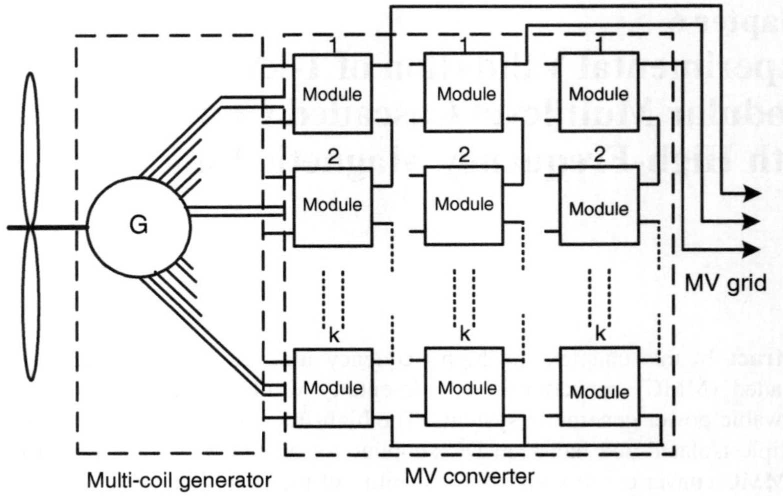

Fig. 6.1 Multi-coil generator-based wind turbine power generation systems [3, 7]

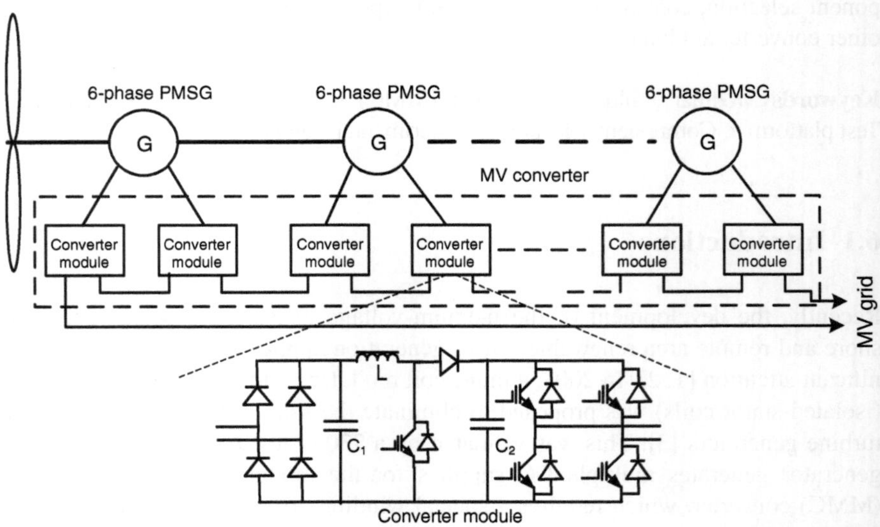

Fig. 6.2 Multiple generator-based wind turbine system: single-phase layout [6, 7]

for the MMC converter. All the generators are driven by the same wind turbine, and each stator winding generates an isolated source for all the H-bridge inverter cells of the MMC converter. The MMC converter generates medium-voltage AC output, which can be connected to the medium-voltage network directly. A multiple generator-based wind turbine system is shown in Fig. 6.2.

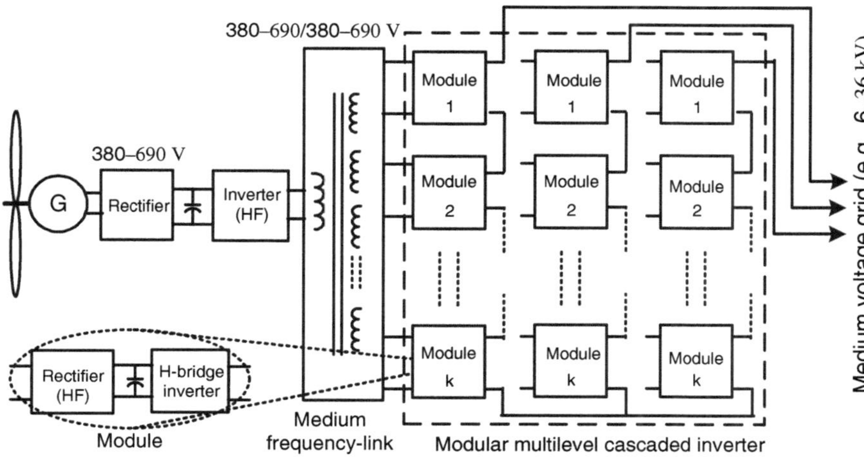

Fig. 6.3 Wind turbine generator system for direct medium-voltage grid integration [1]

However, these approaches require special modular generators or multiple traditional generators to generate isolated multiple DC supplies for all the H-bridge inverter cells of the MMC converter and introduce electrical isolation problems between the generator and grid [8]. On the other hand, compared with the conventional transformers operated at the power frequency, the high-frequency magnetic links have much smaller and lighter magnetic cores and windings, and thus much lower costs. They have, therefore, already been widely used in the low-voltage applications [9–11]. Operated at 1.2-kHz, the weight and size of a 3-MW, transformer can be less than 8 % of an equivalent 50-Hz unit [12]. As a result, the advanced magnetic material-based common high-frequency magnetic link may be the natural choice to generate multiple isolated and balanced DC supplies for all the H-bridge inverter cells of the MMC converter. To verify the feasibility of the new concept of generating balanced multiple sources from a single source with a high-frequency magnetic link, a comprehensive electromagnetic analysis was reported in [13, 14].

In this chapter, a high-frequency magnetic-link MMC medium-voltage converter is experimentally validated for step-up-transformer-less direct grid integration of renewable sources. The common magnetic link generates multiple isolated and balanced DC supplies for all the H-bridge inverter cells of the MMC converter from a single or multiple renewable sources. The grid electrical isolation and voltage imbalance problems are solved through the common high-frequency magnetic link. Figure 6.3 shows the basic block diagram of the proposed high-frequency magnetic-link MMC converter for wind turbine generator systems [1].

The available variable voltage as well as frequency AC power converts to DC power through a controller rectifier circuit, which also controls the rotor speed of the wind turbine generator. The DC power then converts to constant magnitude and frequency high-frequency AC power with a high-frequency H-bridge inverter. The single-input and multiple-output high-frequency magnetic links generate multiple isolated balanced sources for all the H-bridge inverter cells. Each module contains a dedicated rectifier circuit, which converts high-frequency AC power to DC power. The DC power works as the DC-link voltage of the respective H-bridge inverter cell. To ensure fixed grid voltage, a constant output voltage of the inverter is maintained. The inverter output is supplied to a primary winding of a multi-winding high-frequency magnetic link. Each secondary winding is connected with the H-bridge inverter through a bridge rectifier. The number of primary windings depends on the number of renewable sources, and the number of secondary windings depends on the number of levels of the MMC converter. The grid electrical isolation and voltage imbalance problems are solved through the high-frequency magnetic link. Figure 6.4 shows the detailed power converter circuit for wind turbine generators with a high-frequency magnetic-link MMC converter.

A few modules are cascaded on each phase, and each module consists of a bridge rectifier and an H-bridge inverter. The power density, switching loss, control complexity, and semiconductor cost are the significant factors, affecting the design of high-frequency magnetic-link-based DC/DC power conversion system. Different circuit topologies were proposed in recent years to improve the performance of the DC/DC power converters [15–17]. The dual active bridge (DAB) converter topology has received a lot of attention due to the lower switching losses [18–20]. In the DAB converter topology, a total of four additional switches are needed for each DC/DC conversion stage, which not only increases the system cost but also increases control complexity as opposed to topologies, which use a simple diode bridge rectifier at the output. The semiconductor cost and control complexity become critical for medium-voltage converter systems, where several parallel DC/DC converters are necessary. For example, a 19-level system requires 27 parallel DC/DC converters [7]. On the other hand, very short recovery time (25 ns) high-performance fast recovery power diodes are now available on the market [21]. Thus, the fast recovery diode rectifier-based DC/DC converter topology is considered for the proposed system. The proposed converter can be used with most of the commercial wind turbine generators: permanent magnet synchronous generator (PMSG), squirrel cage induction generator (SCIG), and wound rotor synchronous generator (WRSG). In WRSG-based systems, to achieve variable speed operation, the systems may use an extra excitation circuit, which feeds the excitation winding of WRSG. Therefore, the controlled rectifier can be replaced by an uncontrolled rectifier. The uncontrolled rectifier may also be used in the PMSG-based wind turbine generator system, where the converter may be equipped with a step-up chopper circuit. The step-up chopper adapts the rectifier voltage to the DC-link voltage of the inverter. Controlling the inductor current in the step-up chopper can control the generator torque and speed [22].

The high-frequency magnetic-link MMC medium-voltage converter-based grid integration system will have the following advantages: (i) no requirement for special or multiple generators for the wind turbine generator systems, (ii) a wide range of MPPT operation for PV systems, (iii) an inherent DC-link voltage

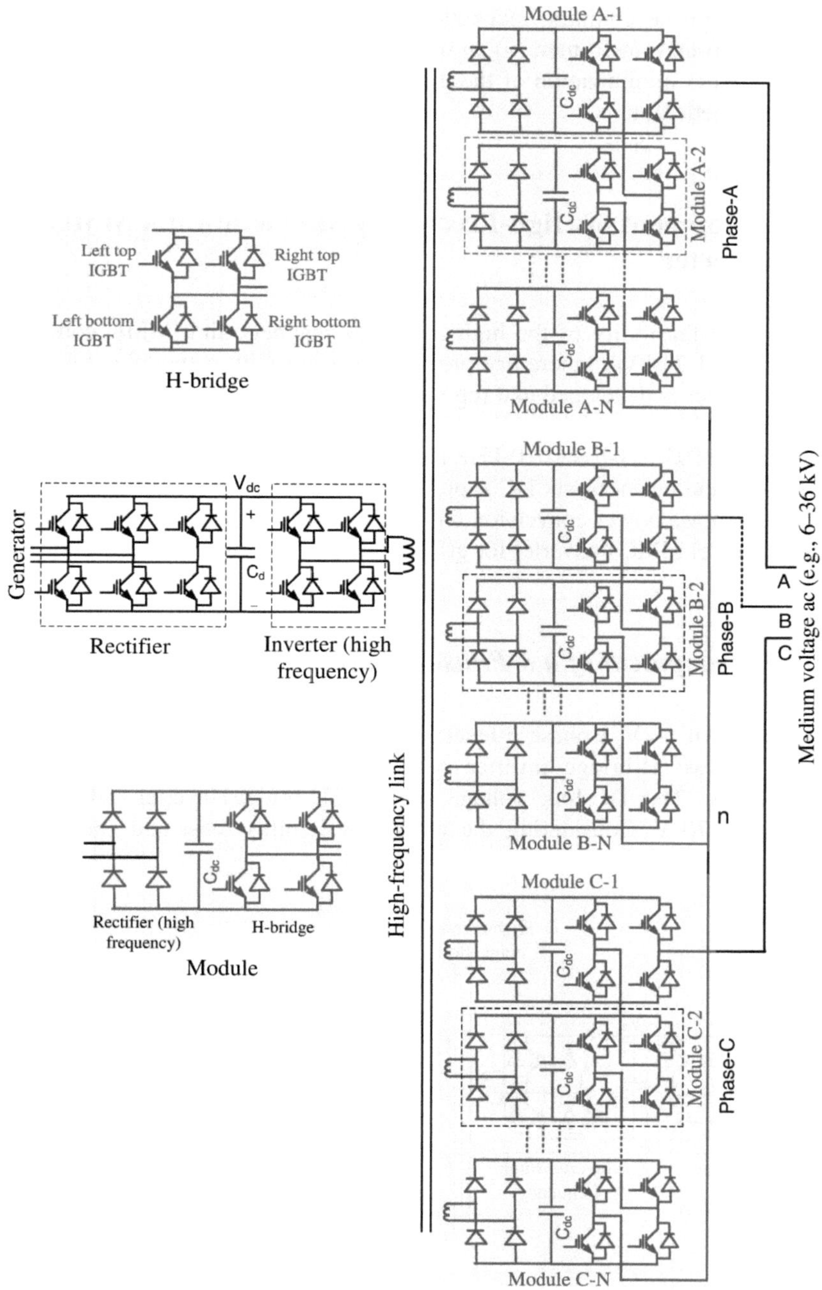

Fig. 6.4 Detailed power conversion circuit of high-frequency magnetic-link MMC converter for step-up-transformer-less direct grid connection of wind turbine generators

balance due to the common magnetic link, (iv) direct gird connection without using the step-up transformer, (v) an overall compact and lightweight system, and (vi) an inherent minimization of the grid isolation problem through the high-frequency magnetic link.

6.2 Development of High-Frequency Magnetic-Link MMC Converter

To verify the feasibility of the high-frequency magnetic-link MMC converter, a scaled down 1.73-kVA laboratory prototype system with a modular 5-level cascaded converter is developed and reported in this chapter, which converts a 210 V DC (rectified generator voltage) into 3-phase 1 kV rms 50 Hz AC. Metglas amorphous alloy 2605SA1-based 10-kHz magnetic link is also developed to generate the isolated and balanced six DC supplies for the 3-phase 5-level MMC converter. Figure 6.5 shows power conversion circuit of high-frequency magnetic-link-based 3-phase 5-level MMC converter for grid connection of wind turbine generators.

6.2.1 Development of a 3-Phase 5-Level MMC Converter

A scaled down 1-kV 3-phase 5-level MMC is developed, where each phase leg consists of two H-bridge inverter cells. Table 6.1 summarizes the converter specifications. The DC-link voltage of each H-bridge inverter cell is calculated at 367.70 V. Considering the market availability, cost and suitability for

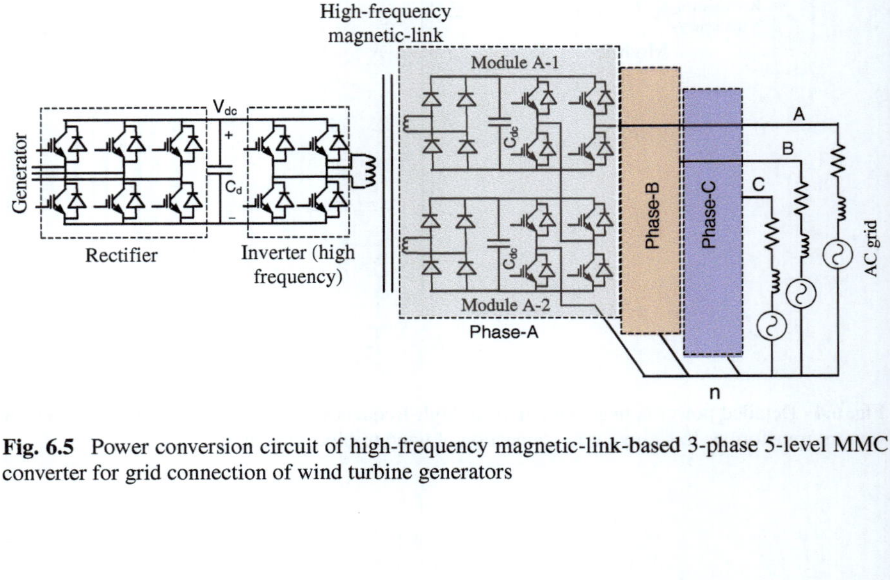

Fig. 6.5 Power conversion circuit of high-frequency magnetic-link-based 3-phase 5-level MMC converter for grid connection of wind turbine generators

Table 6.1 Converter specifications in the experiment

Technical data	Value
Line voltage (kV)	1
Line current (A)	1
Carrier frequency (kHz)	1.525
Output frequency (Hz)	50
Number of levels	5
DC-link voltage (V)	367.70
Number of phases	3
Carrier shifting technique	Phase shifted
Modulation scheme	SPWM and THPWM

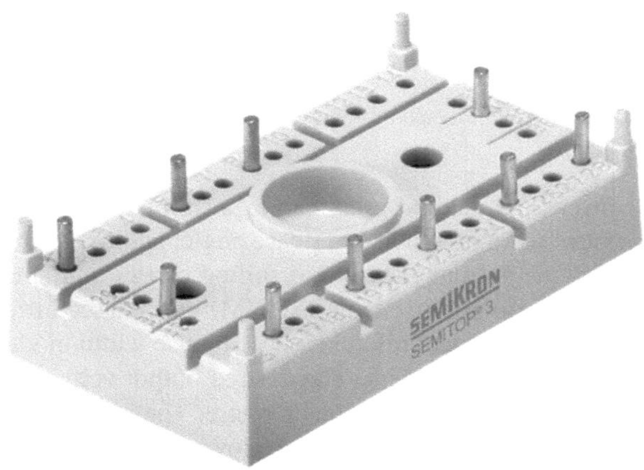

Fig. 6.6 A Photograph of Semikron IGBT modules SK30GH123

prototyping, the Semikron compact insulated gate bipolar transistor (IGBT) modules SK30GH123 have been collected for the development of the 3-phase 5-level MMC converter. The module consists of four N-channel 1.2-kV IGBTs in H-bridge connection. This module is specially designed for the power inverter, which recommends a switching speed of up to 20 kHz. Figure 6.6 shows a photograph of Semikron IGBT module SK30GH123. The Semikron driver SKHI 20opA is a dual driver for the half-bridge inverter cell, which may drive the IGBT with commutation voltage up to 1.2 kV. The driver has some special features. It offers short-circuit protection with soft turnoff, error memory, output signal with external or automatic reset, and it supports DC-bus voltage up to 800 V.

The driver circuits require 15-V DC supplies. The Semikron dual IGBT driver-isolated power supply SKHI PS1 has been considered. It has the capability to supply energy for up to seven IGBTs with additional transformers. The IGBT module and the driver and protection circuit (with power supply and filtering capacitors of the respective rectifier circuit) are mounted on the top of the heat sink. Figure 6.7 shows a photograph of the laboratory prototype 3-phase 5-level 1-kV MMC

Fig. 6.7 A photograph of 3-phase 5-level 1-kV MMC converter

converter. A field programmable gate array (FPGA)-based control circuit is used to control the magnitude and phase angle, and to ensure the power quality and stability of the system. The phase-shifted carrier and sinusoidal reference-based pulse width modulation scheme (SPWM) and phase-shifted carrier and third harmonic injected sinusoidal reference-based pulse width modulation scheme (THPWM) are used in the experiment. The desired output voltage level can be generated by cascading more modules on each phase. The high-level converter implies elimination of the step-up transformer and a lower total harmonic distortion (THD) with lower switching frequency, thereby eliminating the output filters and reducing the running cost. The high-level number attainability also allows a lower level DC-link voltage requirement for each H-bridge module that eliminates boosters.

6.2.2 Design of Driver Circuit

The Semikron driver SKHI 20op requires 15-V gate pulses. The generated gate pulses by Xilinx FPGA XC3S500E are 3.3 V, and these require stepping-up before feeding to the driver SKHI 20op. Figure 6.8 shows a photograph of Xilinx Spartan-3E starter kit. An extra driver circuit is required to interface the FPGA and driver SKHI 20op. The microchip dual high-speed power metal oxide field effect transistor (MOSFET) driver TC4427A is used to design the extra driver circuit. The IC has some special features: high output current (e.g., 1.5 A), generated gate pulse voltage of up to 18 V, short delay time (e.g., 40 ns), low output impedance (e.g., about 7 Ω), and low supply current (e.g., about 4 mA for logic 1 input and 400 μA for logic 0 input). Xilinx Spartan-3E board has 100-pin header

Fig. 6.8 A photograph of Spartan-3E starter kit

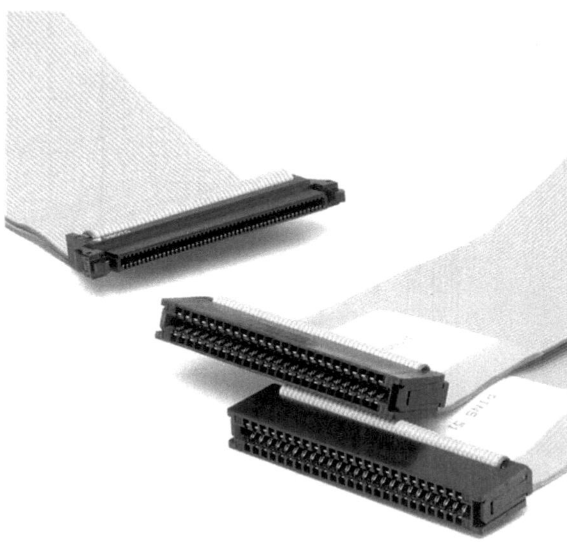

Fig. 6.9 A photograph of ribbon cable and header socket

to interface with peripherals. Therefore, 100-pin socket with proper ribbon cable is required for the connection of FPGA board with driver circuit. Figure 6.9 shows the ribbon cable connection with 100-pin socket. Figure 6.10 shows the detailed circuit diagram of the extra driver circuit. A photograph of the extra driver circuit is shown in Fig. 6.11. The 3-phase 5-level MMC converter consists of 24 IGBTs, and the switching controller evolves 24 gate signals. Each microchip dual high-speed power MOSFET driver TC4427A can drive two gate signals. Therefore, in total, 12 drivers are used. Each driver SKHI 20op handles two gate signals: one for the top IGBT of a half-bridge leg and the other for the bottom IGBT. The driver also requires DC supply with proper ground connection. A separate header

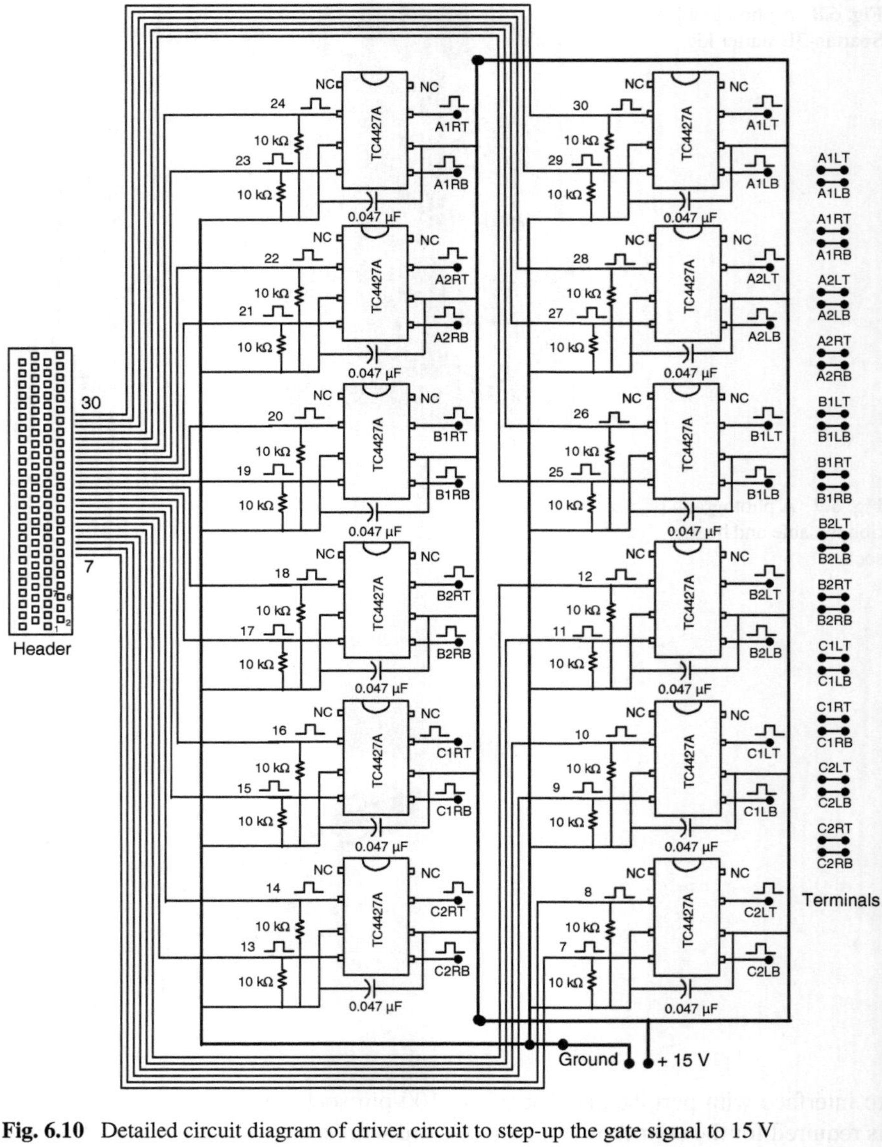

Fig. 6.10 Detailed circuit diagram of driver circuit to step-up the gate signal to 15 V

and socket set, each set for a half-bridge leg with ribbon cable are used to handle the gate signals and DC supplies of the driver circuits. A special switching control signal distribution board is designed for proper management. The gate signals, DC supply, and ground connection of the distribution board are depicted in Fig. 6.12. A photograph of the switching and control signal distribution board is shown in Fig. 6.13.

Fig. 6.11 A photograph of extra driver circuit to step-up the gate signal to 15 V

6.2.3 Development of High-Frequency Rectifiers with Fast Recovery Diodes

In order to minimize the switching loss, control complexity, and semiconductor cost, the diode bridge rectifier is considered to convert high-frequency alternating quantity to DC quantity with fast recovery diodes. Each module is associated with a separate rectifier. In total, six isolated bridge fast recovery rectifiers are required, as the system consists of six H-bridge inverter cells. Figure 6.14 shows the circuit diagram of the high-frequency rectifiers. The available average voltage is about 367 V. Table 6.2 summarizes the technical information of the module rectifier. A photograph of the fast recovery rectifier section is shown in Fig. 6.15.

The input-voltage waveform of this rectifier is similar to the alternating train of pulses. The rectified output waveform can be compared with the DC/DC chopper circuit output. The IXYS DSEE15-12CC superfast recovery (25 ns) dual diode module is considered for the development of the module rectifier of the medium-voltage converter. If we assume the capacitor discharge rate will remain constant at the DC level, the peak-to-peak ripple voltage, $V_{\mathrm{r(p\text{–}p)}}$ can be approximated by a waveform as shown in Fig. 6.16, which has a peak-to-peak value of ΔV_o and a time period of T_{r} and is centered around the DC level. If I_{dc} is the output rectifier average current, f_{s} is the source voltage frequency, V_{dc} is the output rectifier DC output voltage, D_{off} is the diode off time or capacitor discharge time factor, and γ_{r} is the ripple factor, then the peak-to-peak and rms output voltage can be deduced as

$$V_{\mathrm{r(p\text{–}p)}} = \frac{I_{\mathrm{dc}} D_{\mathrm{off}} T_{\mathrm{r}}}{C} = \frac{I_{\mathrm{dc}} D_{\mathrm{off}}}{C f_{\mathrm{r}}} = \frac{I_{\mathrm{dc}} D_{\mathrm{off}}}{2 C f_{\mathrm{s}}} \tag{6.1}$$

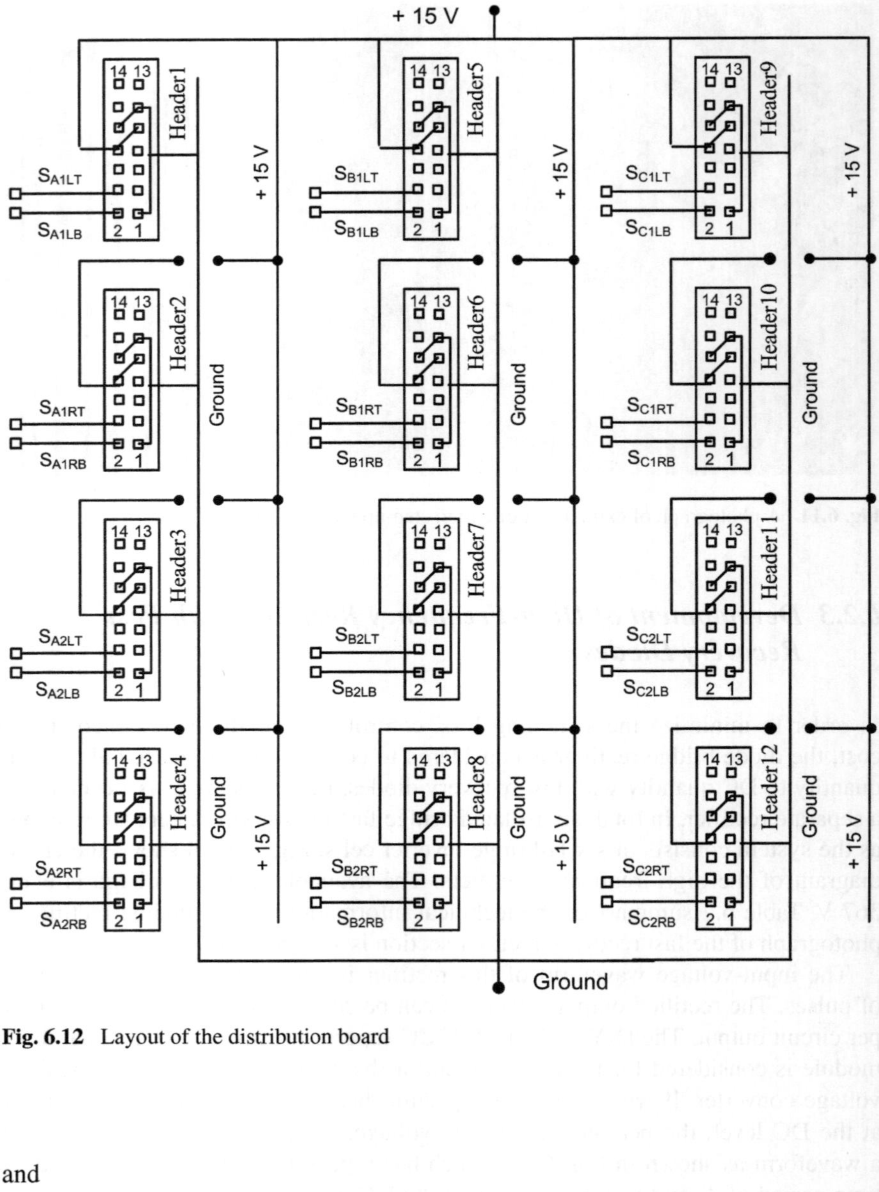

Fig. 6.12 Layout of the distribution board

and

$$V_{r(rms)} = \frac{V_{r(p-p)}}{2\sqrt{3}} = \frac{I_{dc}D_{off}}{4\sqrt{3}Cf_s}. \tag{6.2}$$

Hence, the ripple factor can be deduced as [1]

$$\gamma_r = \frac{V_{r(rms)}}{V_{dc}} = \frac{I_{dc}D_{off}}{4\sqrt{3}Cf_sV_{dc}} \tag{6.3}$$

Fig. 6.13 A photograph of the switching and control signal distribution board

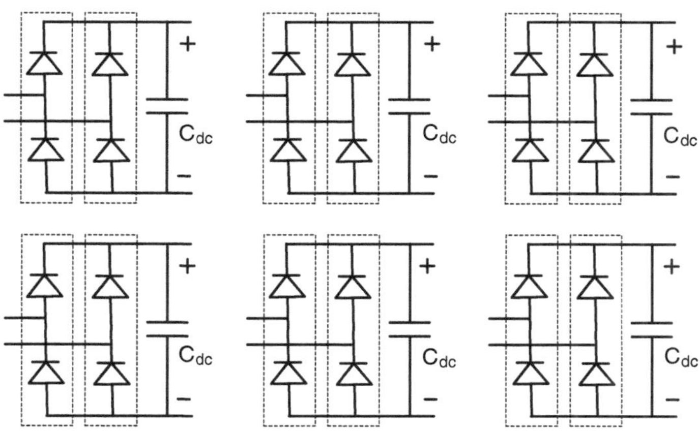

Fig. 6.14 High-frequency rectifier circuits with fast recovery diodes

Table 6.2 Specification of module rectifier

Technical data	Value (V)
Unit voltage (rms)	375
Average output voltage	367
PIV of the diodes	375

From (6.3), the minimum value of the capacitor can be deduced as

$$C = \frac{I_{dc} D_{off}}{4\sqrt{3}\gamma_r f_s V_{dc}} F. \tag{6.4}$$

Fig. 6.15 A photograph of the fast recovery rectifier section

Fig. 6.16 Capacitor voltage of the fast recovery rectifier [1]

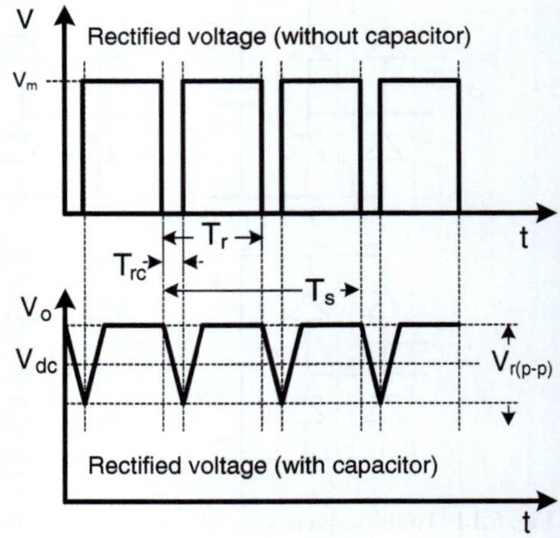

If $I_{dc} = 1$ A, $f_s = 10$ kHz, $V_{dc} = 367$ V, $D_{off} = 2$ %, and the ripple factor is 1 %, the minimum value of capacitor, C, for each module rectifier can be calculated as

$$C = \frac{1 \times 0.02}{4\sqrt{3} \times 0.01 \times 10 \times 10^3 \times 367} = 78.66\,\text{nF}.$$

The 3188EE152T400APA1 aluminum electrolytic capacitor is used for the module rectifier circuit.The maximum power handled by a power diode and the temperature of the diode junction are related since the power dissipated by the device causes an increase in temperature at the junction of the device. The junction temperature, T_J, case temperature, T_C, and ambient (air) temperature, T_A, are related by the device heat-handling capacity and can be presented in terms of thermal-electric analogy as

$$\theta_{JA} = (\theta_{JC} + \theta_{CS} + \theta_{SA}) \tag{6.5}$$

and

$$T_J = P_D\theta_{JA} + T_A, \tag{6.6}$$

where θ_{JA} is the total thermal resistance (junction to ambient), θ_{JC} is the transistor thermal resistance (junction to case), θ_{CS} is the insulator thermal resistance (case to heat sink), θ_{SA} is the heat-sink thermal resistance (heat sink to ambient), and P_D is the power dissipation. Using (6.5) and (6.6), the heat-sink thermal resistance can be deduced as [1]

$$\theta_{JC} + \theta_{CS} + \theta_{SA} = \frac{T_J - T_A}{P_D}$$

$$\theta_{SA} = \frac{T_J - T_A}{P_D} - (\theta_{JC} + \theta_{CS}). \tag{6.7}$$

From the data sheet of DSEE15-12CC, we obtain $\theta_{JC} = 1.60$ °C/W, $\theta_{CS} = 0.5$ °C/W, $T_J = 170$ °C, and P_D is about 2 W. Therefore, the heat-sink thermal resistance can be calculated as

$$\theta_{SA} = \frac{170 - 40}{2} - (1.6 + 0.5) = 62.9\,^{\circ}\mathrm{C/W}.$$

Fischer Elektronik FK 245 MI 247 O clip-on heat sinks without soldering lug are considered for this project. The heat sink is made of copper material with a thermal resistance of 20.2 °C/W. For proper contact with semiconductor devices, the Electrolube HTC10S non-silicone heat transfer compound is used on the contact surface area of the heat sinks.

6.2.4 Development of High-Frequency Inverter

A full-bridge high-frequency inverter is developed by using the Semikron SK30GH123 compact insulated gate bipolar transistor (IGBT) module to generate a high-frequency square wave primary excitation voltage of the high-frequency magnetic link. The Semikron SKHI 20opA with SKHI PS1 is used as an isolated driver for the SK30GH123 module. Figure 6.17 shows a photograph of the high-frequency inverter. The Texas Instruments high-speed PWM controller UC3825BN is used to generate the switching signals. Figure 6.18 shows a photograph of high-speed PWM controller UC3825BN. A simplified version of the switching signal generator circuit using high-speed PWM controller UC3825BN is shown in Fig. 6.19. The Texas Instruments high-speed PWM controller

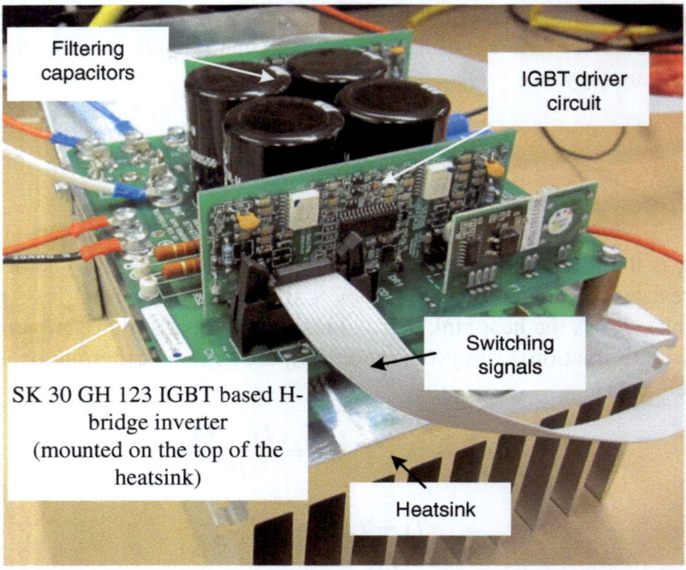

Fig. 6.17 A photograph of high-frequency inverter; generating high-frequency square wave primary excitation voltage of the high-frequency magnetic link [23]

Fig. 6.18 A photograph of Texas Instruments high-speed PWM controller UC3825BN

UC3825BN is a high-performance PWM controller with low start up current, accurate oscillator frequency, leading edge blanking, latched fault logic, full-cycle soft start, restart delay after fault, and many other features. The developed dedicated switching circuit is shown in Figs. 6.20 and 6.21.

The oscillator of the UC3825BN is a sawtooth that utilizes two pins; one for the timing resistor R_T and the other for the timing capacitor C_T. The resistor programs the charging current to the timing capacitor via an internal current mirror

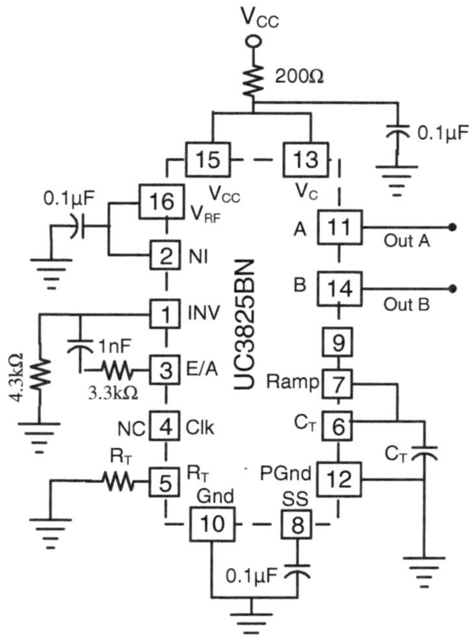

Fig. 6.19 Simplified circuit of the switching signal generator using Texas Instruments high-speed PWM controller UC3825BN [23]

Fig. 6.20 A photograph of control circuit using Texas Instruments high-speed PWM controller UC3825BN

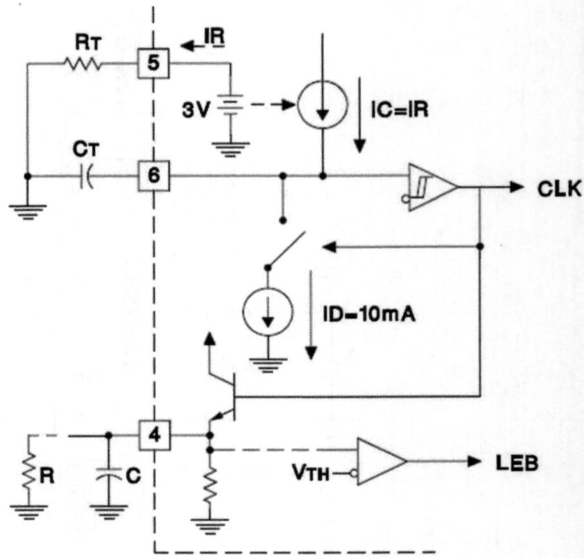

Fig. 6.21 Basic internal circuit of the Texas Instruments high-speed PWM controller UC3825BN (oscillator section only)

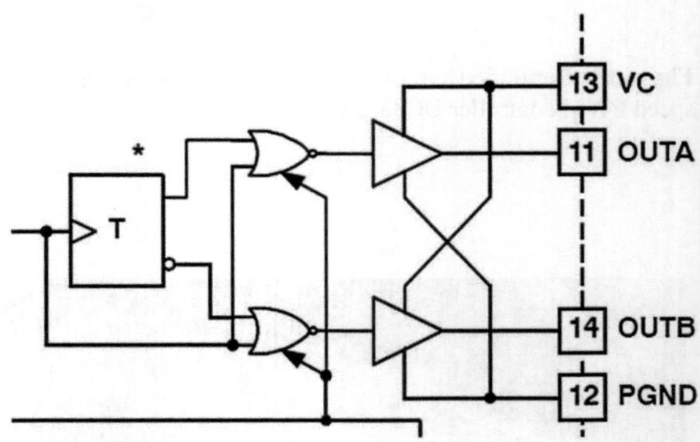

Fig. 6.22 Alternative two output topology

with high accuracy. The maximum signal high time is determined by the rising capacitor voltage, whereas dead time is determined by the timing capacitor discharge. Figure 6.21 shows the internal circuit diagram of the Texas Instruments high-speed PWM controller UC3825BN (oscillator section only).

Based on the desired maximum duty cycle, $D_{\max}$ the timing resistor can be calculated as

$$\begin{aligned} R_{\mathrm{T}} &= \frac{3V}{10\,\mathrm{mA} \times (1 - D_{\max})} \\ &= \frac{3V}{10\,\mathrm{mA} \times (1 - 0.90)} = 3 \times 10^{3}\Omega \text{ or } 3\,\mathrm{k}\Omega. \end{aligned} \tag{6.8}$$

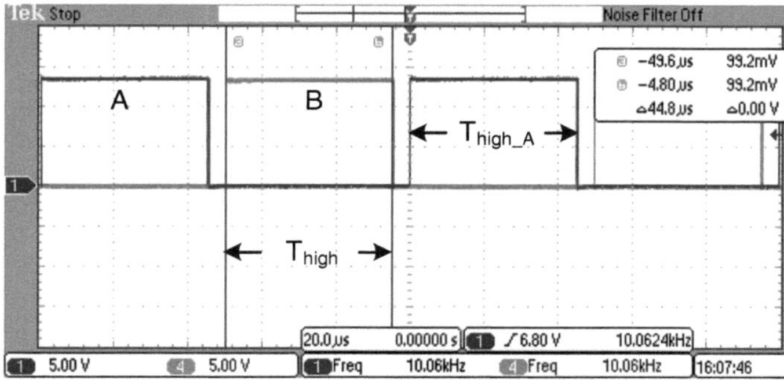

Fig. 6.23 Measured switching signals; on-time calculation

Two alternative signals are generated through a T flip-flop, and output signal frequency is half that of oscillator frequency, as shown in Fig. 6.22. Therefore, doubled oscillator frequency (compared to the output) is required for designing the timing capacitor.

Based on the calculated value of the timing resistor and desired maximum duty cycle, the timing capacitor can be calculated as

$$
\begin{aligned}
C_\mathrm{T} &= \frac{1.6 \times D_{\max}}{R_\mathrm{T} \times f} \\
&= \frac{1.6 \times 0.90}{3 \times 10^3 \times 20 \times 10^3} = 2.4 \times 10^{-8}(\mathrm{F})
\end{aligned}
\tag{6.9}
$$

The complete switching signal generator circuit with UC3825BN IC is fabricated and tested in the laboratory. The gate pulses were measured and compared with the simulation results. The measured pulses were found to be highly consistent with the simulation results. The signal high time (switch on) and dead time (switch off) are measured using cursors of the oscilloscope as shown in Figs. 6.23 and 6.24, and the duty cycle is also calculated. The percentage of duty cycle, $D_{\%}$, can be calculated as

$$
\begin{aligned}
D_{\%} &= \frac{T_\mathrm{high}}{T_\mathrm{high} + T_\mathrm{dead}} \times 100 \\
&= \frac{44.8}{44.8 + 4.8} \times 100 = 90.30\ \%
\end{aligned}
\tag{6.10}
$$

The expected time period, T, and frequency, f_H, of the medium frequency inverter can also be calculated as

$$
\begin{aligned}
T &= T_\mathrm{high\_A} + T_\mathrm{dead\_A} + T_\mathrm{high\_B} + T_\mathrm{dead\_B} \\
&= 44.8\,\mu\mathrm{s} + 4.8\,\mu\mathrm{s} + 44.8\,\mu\mathrm{s} + 4.8\,\mu\mathrm{s} = 99.20\,\mu\mathrm{s}
\end{aligned}
\tag{6.11}
$$

$$
\therefore f_\mathrm{H} = \frac{1}{T} = \frac{1}{99.2 \times 10^{-6}} = 10.08\ \mathrm{kHz}.
$$

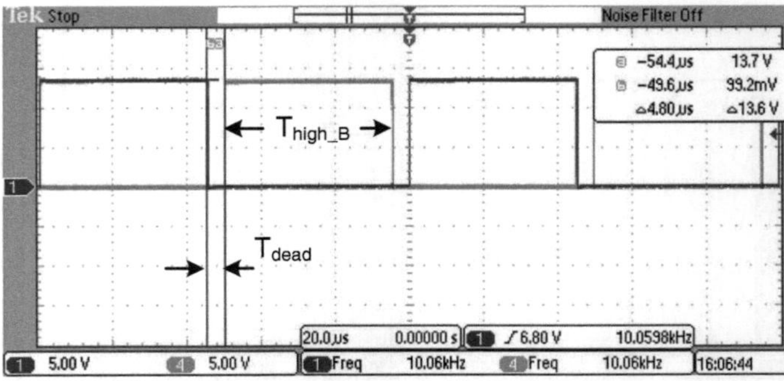

Fig. 6.24 Measured switching signals; dead-time calculation

6.2.5 Development of High-Frequency Magnetic Link

With the advent of new power semiconductor devices, different high magnetic saturation and low power loss soft magnetic materials are conceived to reduce the weight and volume of conventional power transformers. The grain-oriented silicon sheet steels, which are commonly used as the core material of power frequency transformers, are not suitable for high-frequency applications because of the heavy eddy current loss [24]. The soft ferrites have been widely used in medium- and high-frequency inductors and transformers due to the low price and general availability. Because of the low saturation flux density [24] (only 0.3–0.5 T), which would make the transformer bulky, they are not suitable for large power applications. On the other hand, the amorphous alloy and nanocrystalline materials have excellent magnetic characteristics for high-frequency applications, such as high permeability, high saturation flux density, and relatively low core losses. Two commercially available amorphous and nanocrystalline materials are Metglas and Finemet, respectively, both manufactured by Hitachi Metals, Japan. Although Finemet has lower specific core loss than Metglas, its saturation flux density (about 1 T) is much lower than that of Metglas, which is 1.56 T. Until now, many kinds of soft magnetic alloys with high magnetic flux density combined with low core loss have been developed [25, 26]. Taking into account the flux density, specific core loss, cost, and availability, we chose Metglas 2605SA1 stripe of 30 μm thickness and 25 mm width as the core material. The other parameters of 2605SA1 include mass density of 7.18 g/cm^3, saturation flux density of 1.56 T, and specific core loss of 180 W/kg, at 10 kHz sinusoidal excitation of 1 T. Figure 6.25 shows a photograph of the Metglas alloy 2605SA1 sheet-based multi-output high-frequency magnetic link. For small skin depth and proximity effect, the number of layers as well as the conductor diameter should be kept as small as possible. Moreover, the insulated strands should be twisted or braided together to equalize the flux linkages throughout the conductors. To achieve this so as to reduce the winding loss, a Litz wire with small number of layers should be always used in a high-frequency magnetic link. Figure 6.26 shows a photograph of Litz wire.

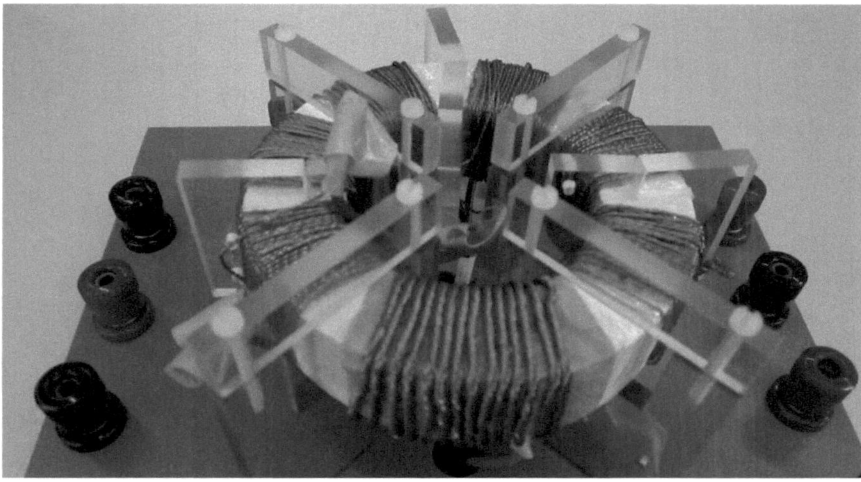

Fig. 6.25 Photograph of the Metglas amorphous alloy 2605SA1-based high-frequency magnetic link [2]

Fig. 6.26 A photograph of Litz wire

6.3 Experimental Testing and Performance Analysis

To verify the feasibility of the high-frequency magnetic-link MMC converter-based power conversion system, a scaled down laboratory prototype 1.73-kVA test platform is developed with a 5-level MMC converter, which converts 210 V DC (rectified generator voltage) into 3-phase 1 kV rms 50-Hz AC. The Semikron compact IGBT module SK30GH123 with Semikron-isolated driver SKHI 20opA is used to develop the MMC converter. The 10-kHz frequency magnetic link with single primary and six secondary windings is developed to generate the balanced isolated six DC supplies for a 3-phase 5-level MMC converter. Figure 6.27 shows a photograph

Fig. 6.27 A photograph of the experimental setup of high-frequency magnetic-link-based MMC converter (1 kV, 1.73 kVA)

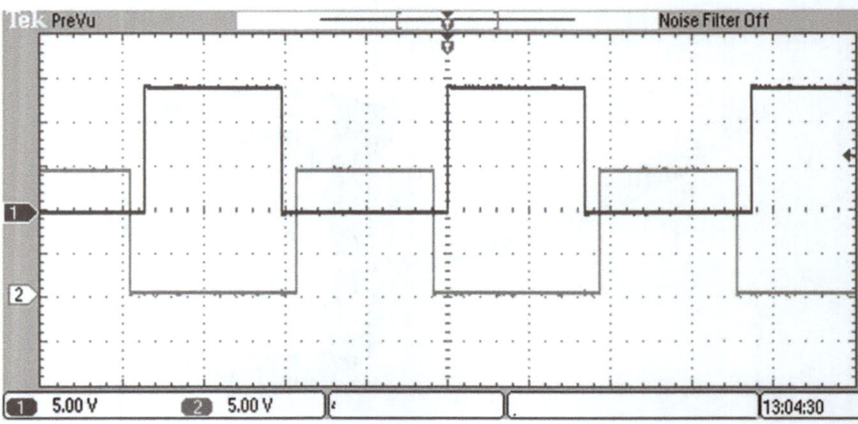

Fig. 6.28 Measured gate pulses of the high-frequency inverter

of the experimental test platform. The gate pulses for the high-frequency inverter were measured, as shown in Fig. 6.28, and compared with the simulation results.

The measured pulses were found to be highly consistent with the simulation results. The primary and secondary voltages of the high-frequency magnetic link were measured and the voltage transformation ratio was calculated, as listed in Table 6.3, and found to be highly consistent with the theoretical values. Figure 6.29 shows the voltage waveforms of the primary and secondary sides of the high-frequency magnetic link. Tektronix DPO 2024 digital phosphor oscilloscope with the P5200 high-voltage differential probe and Tektronix TCPA300 current probe were used to observe the voltage and current waveforms.

Table 6.3 Voltage transformation ratios against primary winding [27]

Windings	A	B	C	D	E	F
Ratios	1.781	1.780	1.779	1.780	1.780	1.781
Variation (%)	0.00	−0.05	−0.11	−0.05	−0.05	0.00

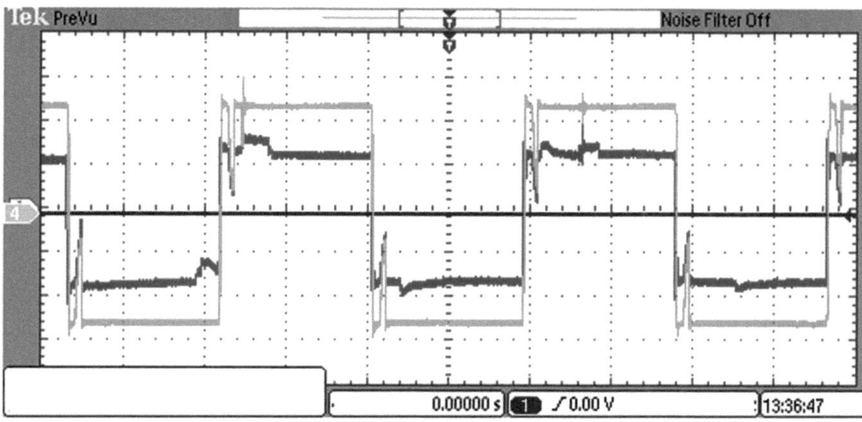

Fig. 6.29 Measured primary and secondary voltage waveforms of high-frequency magnetic link

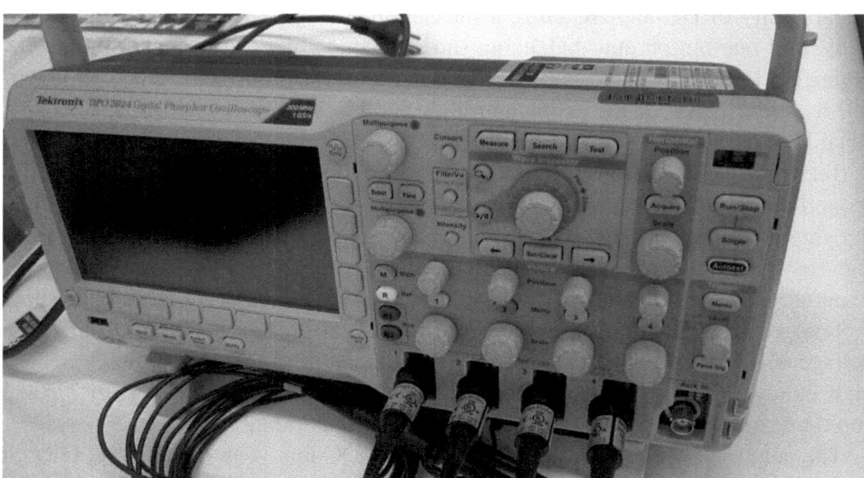

Fig. 6.30 A photograph of Tektronix digital phosphor oscilloscope DPO 2024

Figure 6.30 shows a photograph of Tektronix digital phosphor oscilloscope DPO 2024, which is a 200-MHz oscilloscope with four channels. The current probe TCPA300 can measure AC and DC currents. The probe converts the sensed

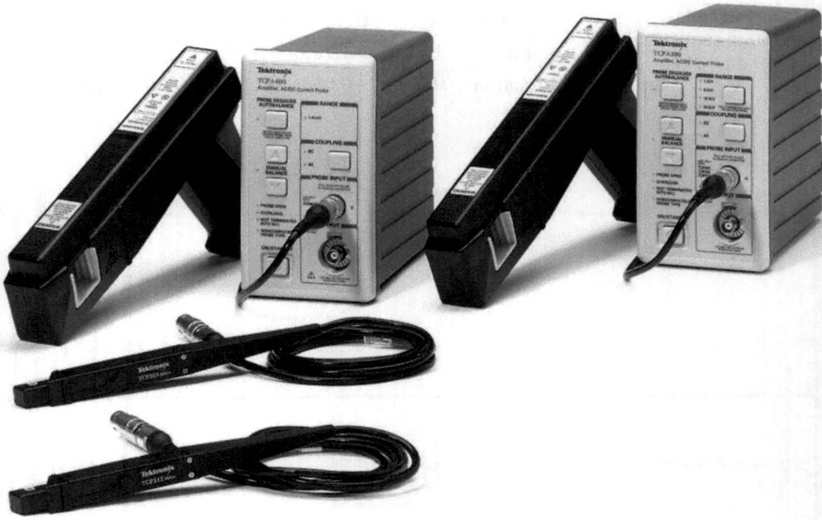

Fig. 6.31 A photograph of Tektronix current probes

AC or DC current into a proportional voltage signal that can be measured directly with an oscilloscope. The TCPA300 current probe has a few notable features, such as capability of DC and AC current measurements up to 750-A peak, high sensitivity, and one-button auto-balancing (no adjustment needed to match the probe). Figure 6.31 shows a photograph of Tektronix current probe TCPA300. To use the full dynamic range of the probe/amplifier combination, the oscilloscope must be capable of displaying a vertical scale factor of 1 mV/div to 1 V/div.

Tektronix high-voltage differential probe P5200 can be used with any oscilloscope to safely make measurements of floating circuits. The probe converts floating signals to low-voltage ground-referenced signals that can be displayed safely and easily on any ground-referenced oscilloscope. The P5200 high-voltage differential probe requires an oscilloscope with grounded inputs. Figure 6.32 shows a photograph of Tektronix high-voltage differential probe P5200. Tektronix high-voltage differential probes have a few remarkable features and benefits, such as high bandwidths up to 100 MHz, high-voltage-handling capacity up to 2,200 V common rms and up to 5,600 V differential, and over range indicator.

The outputs of each secondary voltage and the DC-link voltages (after the filter circuit) were found to be approximately equal, which can serve satisfactorily as the isolated and balanced DC sources for the MMC converter. The total losses (core loss plus copper loss) of all secondary windings of the Metglas amorphous alloy 2605SA1-based high-frequency magnetic link measured at different excitation frequencies ranging from 6 to 12 kHz are almost the same. Such a similarity of characteristics is obligatory to generate multiple balanced DC sources for all the H-bridge inverter cells of the MMC converter. A Voltech PW3000A universal power analyzer was used to measure the total

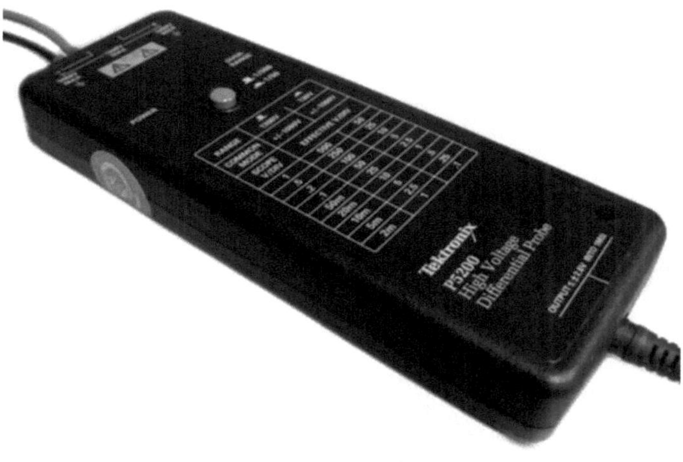

Fig. 6.32 A photograph of Tektronix high-voltage differential probe P5200

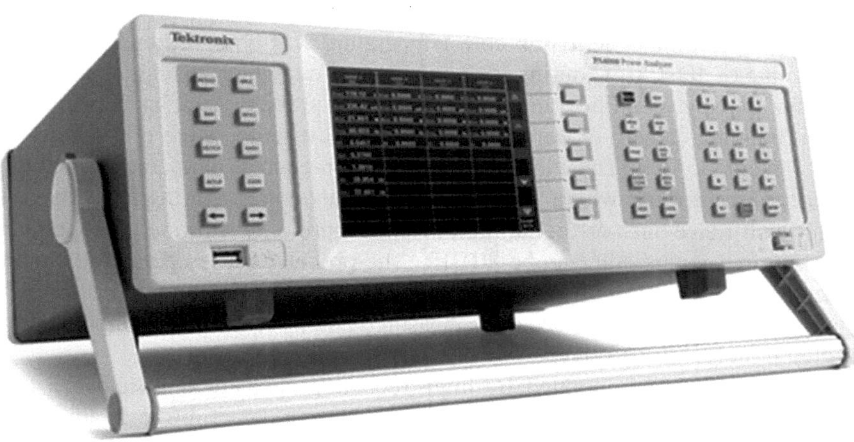

Fig. 6.33 A photograph of Tektronix power analyzer PA4000

loss. In March 2013, Voltech transferred all its power analyzer technology to Tektronix. Recently, Tektronix developed power analyzer PA4000, which delivers consistently accurate measurements, even with challenging power waveforms. Figure 6.33 shows a photograph of Tektronic power analyzer PA4000. The PA4000 supports input voltage and current up to 1,000 V rms and 30 A rms, respectively, with a measurement bandwidth up to 1 MHz. From the oscilloscope data, the copper loss of each winding at a different frequency ranging from 6 to 12 kHz was calculated by using the DC resistances (0.024 Ω for the primary and 0.16 Ω for the secondary), since the AC/DC resistance ratios, K_r, in this design is almost unity due to the use of Litz wires.

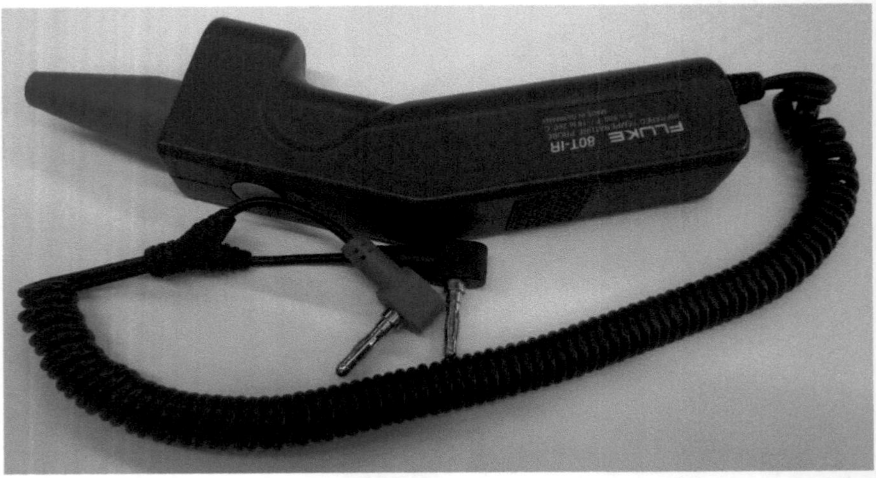

Fig. 6.34 A photograph of Fluke infrared temperature probe 80T-IR

High-frequency magnetic-link core temperature was observed and maintained in the range of 40–120 °C during the experimentation, which ensures stable maximum flux density. The Fluke infrared temperature probe 80T-IR and Flir infrared thermal imaging camera i7 were used to measure the temperature of the high-frequency magnetic link. The Fluke 80T-IR is a non-contact temperature measurement accessory for use with a test instrument capable of measuring DC voltage in the millivolt range such as digital multimeter. Figure 6.34 shows a photograph of Fluke infrared temperature probe 80T-IR. The probe has a temperature-handling capacity up to 260 °C, with a basic accuracy of 3 %. Temperature is measured by pointing the probe at the surface to be measured and by reading the temperature on the digital multimeter display.

Infrared radiation is emitted by high-frequency magnetic link above a temperature of −273 °C. Although the human eye cannot detect infrared radiation, a thermal imaging camera can, and takes picture of high-frequency magnetic link to show the amount of heat that is emitting. The Flir thermal imaging camera i7 produces non-contact temperature measurements as thermal images. The Flir i7 has a few remarkable features, such as high accuracy and sensitivity of 2 % and 0.1 °C, focus-free lens, different measurement modes (spot, area, and isotherm), and temperature range up to 250 °C. Figure 6.35 shows a photograph of Flir thermal imaging camera i7. With the reporting software included in the camera package, it is possible to create reports, analyze, and document the image easily.

The switching scheme is implemented with XC3S500E FPGA. The PWM gate pulses from FPGA are 3.3 V, which is insufficient for the IGBT drivers used. Driver IC TC4427A is used to increase the voltage level to the desired value (TC4427A can support up to 18 V). The gate pulses for an H-bridge module of a 5-level modular cascaded converter are illustrated in Fig. 6.36. The measured

Fig. 6.35 A photograph of Flir thermal imaging camera i7

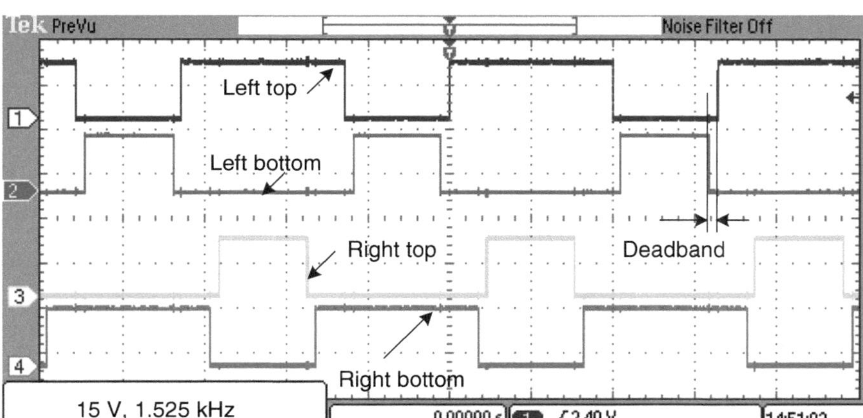

Fig. 6.36 Measured gate pulses of four IGBTs in the top module in phase A

switching frequency of the switching signals is 1.525 kHz, which precisely matches the theoretical result. Pulse patterns are also verified with theoretical results as well as MATLAB/Simulink and Xilinx ISE simulation results and they are found highly consistent. Gate pulses are applied to the switching devices of the

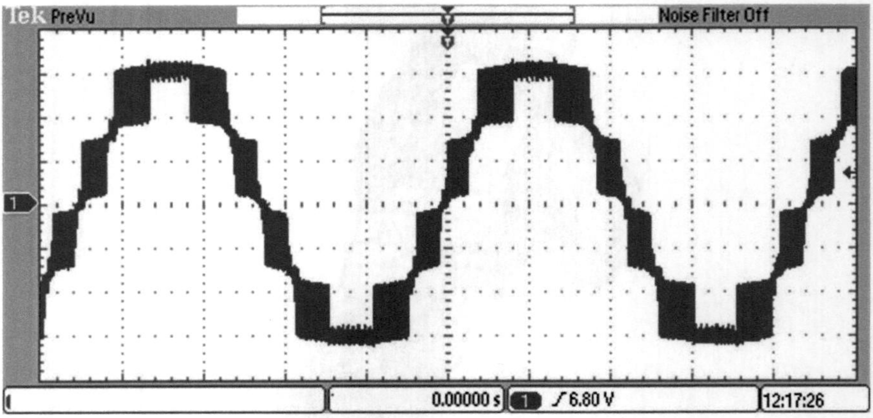

Fig. 6.37 Measured phase voltage of the prototype system (phase voltage before LC filter circuit)

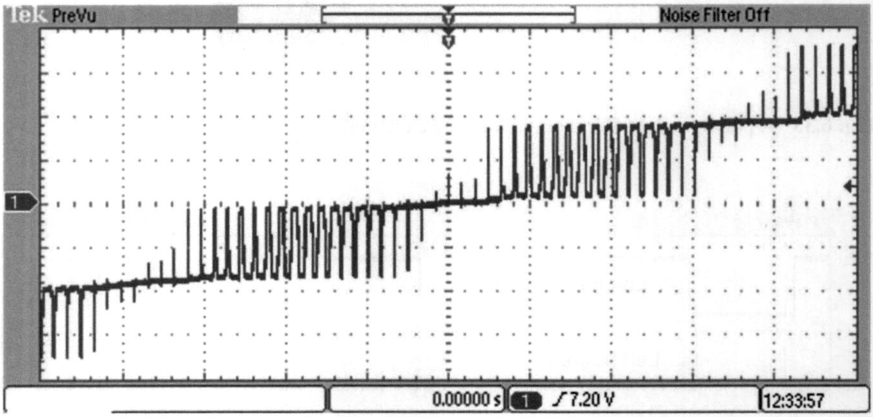

Fig. 6.38 Zoomed phase voltage of the prototype system (phase voltage before LC filter circuit; with 1.525-kHz PWM frequency

converter and the output phase voltage and line current are measured, and they are found to be satisfactory. Figure 6.37 depicts the output phase voltage of the prototype converter. Each level of the output voltage contains a number of PWM pulses. Figure 6.38 plots the zoomed (zoom factor = 8) output phase voltage of the prototype system. The line voltages of the prototype converter were measured; they were found to be highly consistent with the simulation results.

The SPWM and THPWM schemes were implemented practically and performances were verified by observing the line voltage waveforms, and they were found to be highly consistent with the theoretical and simulation results. The LC filter circuit is designed with the 3-mH MTE RL 00401 reactor and 6-μF RS

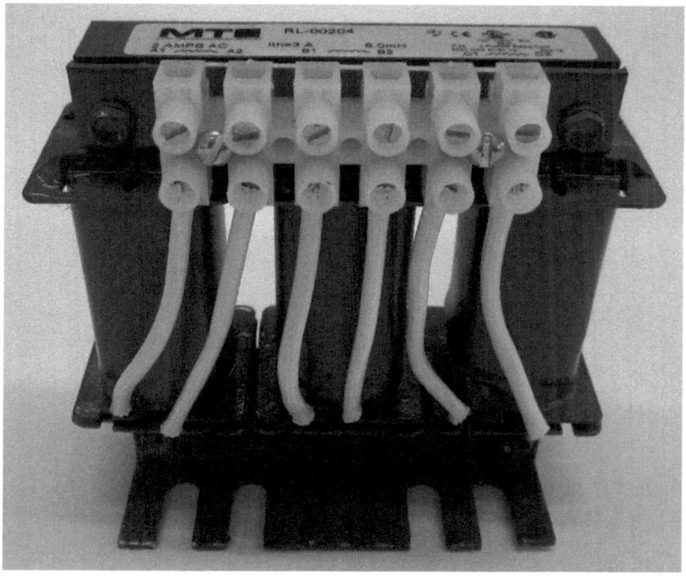

Fig. 6.39 A photograph of MTE reactor to design LC filter

Fig. 6.40 A photograph of RS AC capacitor

MR-P-MC-S-NF capacitors. Figure 6.39 shows a photograph of MTE reactor. The MTE reactors provide stable inductance, even at 150 % of their current rating; these reactors still have 100 % of their nominal inductance. The RS-metallized polypropylene motor run AC capacitors can be used to design LC line filter circuit. The RS MR-P-MC-S-NF is a 440-V AC screw mount 35-mm-diameter polypropylene film capacitor of 6 μF with a tolerance of 10 %. A few capacitors can be connected in series to obtain the required voltage rating. Figure 6.40 shows a

Fig. 6.41 A photograph of LC line filter

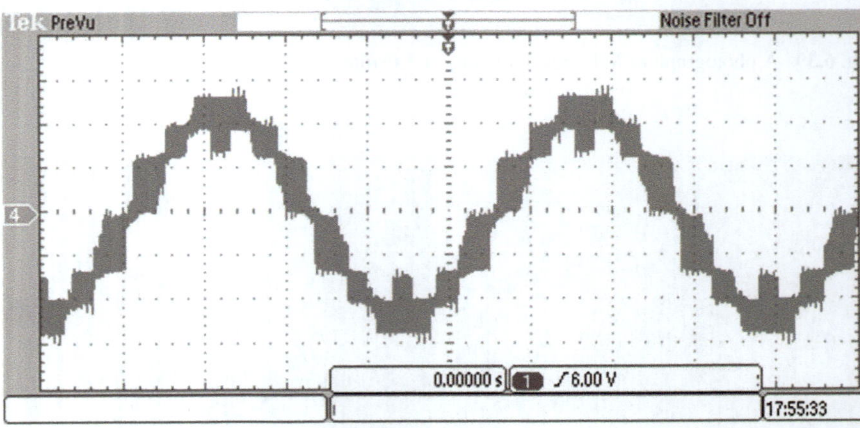

Fig. 6.42 Measured line voltage with SPWM of the prototype system before filter

photograph of RS capacitor MR-P-MC-S-NF. Figure 6.41 shows a photograph of LC line filter circuit of high-frequency magnetic-link MMC converter-based power conversion system.

In each phase, three capacitors are connected in series, which provides equivalent 2-μF capacitance with a rated voltage at 1.32 kV. The line voltage before the filter circuit with SPWM scheme is illustrated in Fig. 6.42, and this shows that the SPWM scheme gives a higher THD, which is about 27 %. The performances are also observed with the filter capacitors connected in Y and Δ. The line voltages after filter are shown in Fig. 6.43, when the filter capacitors were connected in Y. Significant improvement was observed when filter capacitors were connected in

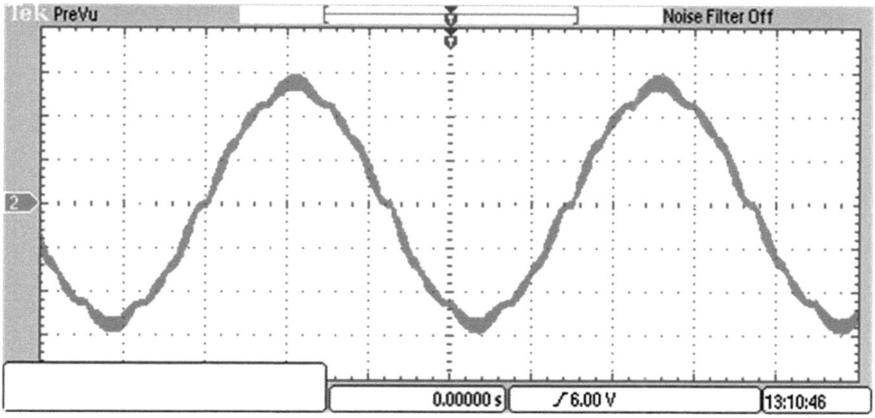

Fig. 6.43 Measured line voltage with SPWM of the prototype system after the filter when the filter capacitors are connected in Y

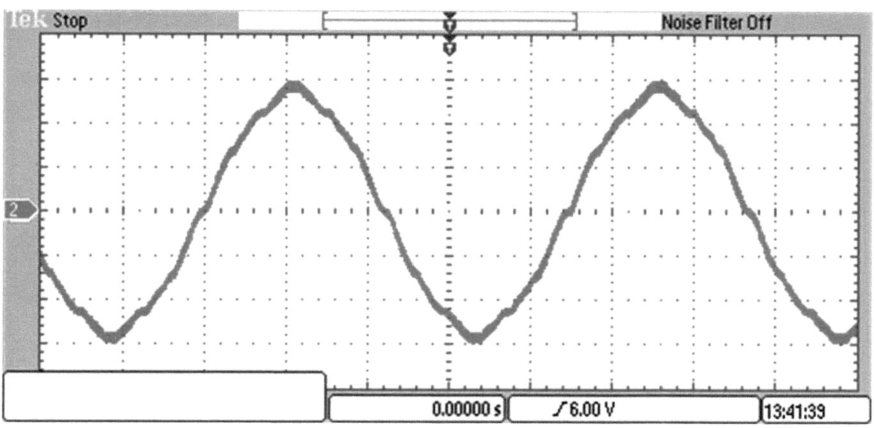

Fig. 6.44 Measured line voltage with SPWM of the prototype system after the filter when the filter capacitors are connected in Δ

Δ as the THD was reduced from 4.5 % to about 3.2 %. After filter circuit, the line voltages are depicted in Fig. 6.44, when the filter capacitors are connected in Δ.

The THPWM scheme provides much better results than the SPWM, which was found to be highly consistent with the simulation results. The line voltage waveform before the filter circuit with THPWM is illustrated in Fig. 6.45. Figure 6.46 plots the zoomed (zoom factor = 6) output line voltage with THPWM. The frequency spectrum of the measured line voltage (before the filter circuit) is shown in Fig. 6.47, which contains about 19 % THD. The line voltages were also measured after the filter circuit, and they were found to be consistent with the simulation

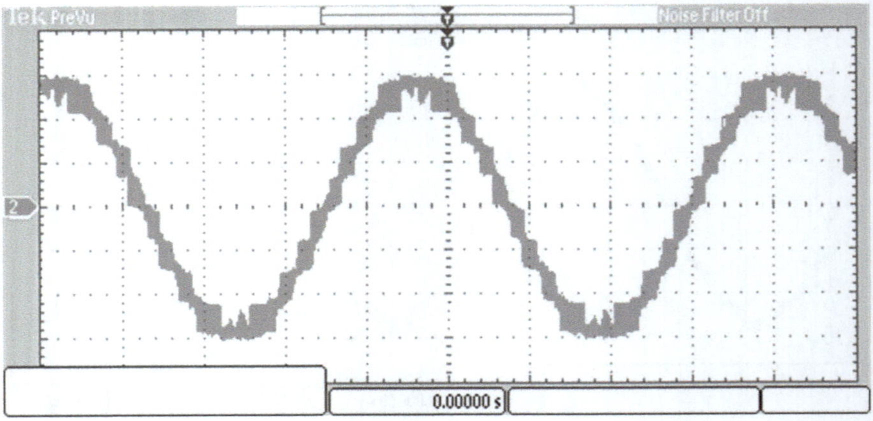

Fig. 6.45 Measured line voltage (before filter circuit) of the prototype 1-kV modular multilevel cascaded converter with THPWM

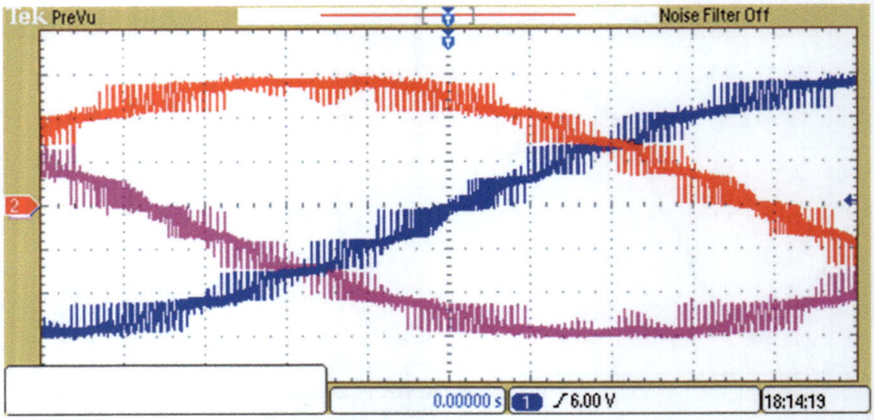

Fig. 6.46 Zoomed measured line voltage (before filter circuit) of the prototype 1-kV modular multilevel cascaded converter with THPWM

results. The measured line voltages after the filter circuit are shown in Fig. 6.48. Figure 6.49 plots zoomed 3-phase line voltage after the filter circuit with THPWM. As measured, after the filter circuit, the output voltage waveform contains about 2.75 % THD. The frequency spectrum of the line voltage after the filter circuit is shown in Fig. 6.50. The line currents were measured and compared with theoretical as well as simulation results, and they were found to be almost the same.

If P_{c_inv} is the conduction loss and P_{sw_inv} is the switching loss, the losses in the inverter section of the proposed converter can be described as

$$P_{loss\_inv} = P_{c\_inv} + P_{sw\_inv} \tag{6.12}$$

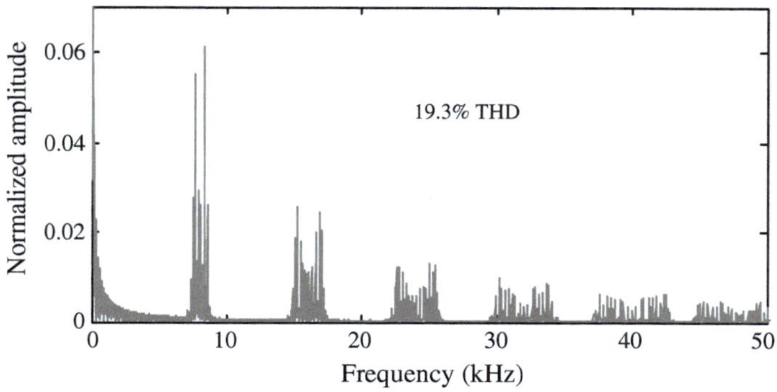

Fig. 6.47 Frequency spectrums of line voltages (before the filter circuit)

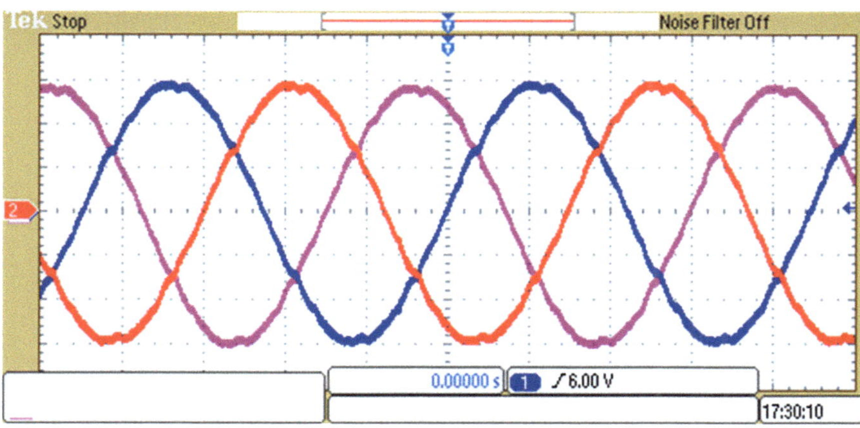

Fig. 6.48 Measured line voltage (after filter circuit) of the prototype 1-kV modular multilevel cascaded converter with THPWM

The switching losses in a two-level and multilevel inverters were approximated by [29]

$$P_{\mathrm{sw\_inv\_2}} = (\mathrm{AI_r} + \mathrm{BI_r^2})f_{\mathrm{c\_2}} \tag{6.13}$$

and

$$P_{\mathrm{sw\_inv\_}M} = (\mathrm{AI_r} + \mathrm{BI_r^2})f_{\mathrm{c\_M}}. \tag{6.14}$$

where $f_{\mathrm{c\_M}}$ and $f_{\mathrm{c\_2}}$ are the carrier frequencies of multilevel and 2-level inverters, respectively. If A and B are assumed to be the same for both inverter systems, the switching losses in a 2-level inverter would be $f_{\mathrm{c\_2}}/f_{\mathrm{c\_M}}$ times higher than that of a multilevel inverter.

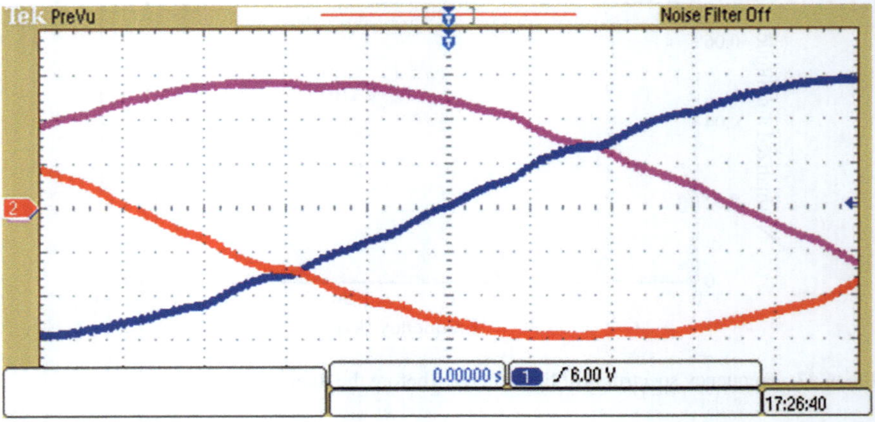

Fig. 6.49 Zoomed measured line voltage (after filter circuit) of the prototype 1-kV modular multilevel cascaded converter with THPWM

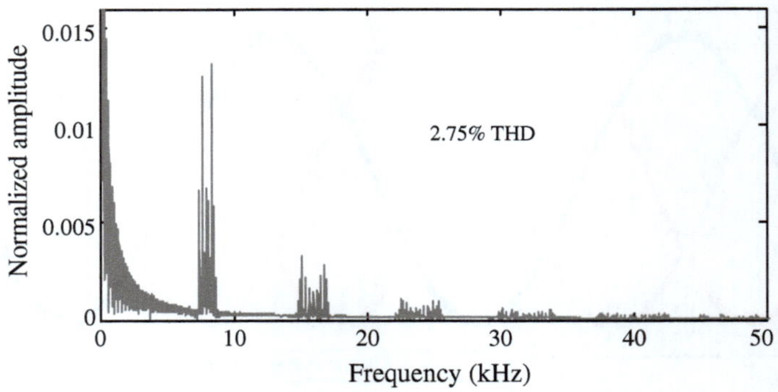

Fig. 6.50 Frequency spectrum of line voltages (after the filter circuit)

The conduction losses in an IGBT and in an antiparallel diode were represented by [28]

$$P_{\mathrm{c\_sw}} = \frac{1}{2} I_{\mathrm{r}} V_{\mathrm{t}} \left(\frac{1}{\pi} + \frac{m_a}{4} p_{\mathrm{f}} \right) + I_{\mathrm{r}}^2 R_{\mathrm{CE}} \left(\frac{\sqrt{3}}{8\sqrt{\pi}} + \frac{m_a}{3\pi} p_{\mathrm{f}} \right). \tag{6.15}$$

and

$$P_{\mathrm{c\_D}} = \frac{1}{2} I_{\mathrm{r}} V_{\mathrm{f}} \left(\frac{1}{\pi} - \frac{m_a}{4} p_{\mathrm{f}} \right) + I_{\mathrm{r}}^2 R_{\mathrm{AK}} \left(\frac{\sqrt{3}}{8\sqrt{\pi}} - \frac{m_a}{3\pi} p_{\mathrm{f}} \right). \tag{6.16}$$

where m_a is the amplitude modulation index, p_{f} the power factor of the current, I_{r} the device current, and V_{t} and V_{f} are the voltage drops at zero current condition, and R_{CE} and R_{AK} the forward resistances of IGBT and diode, respectively, which can be collected from the manufacturer's data sheets. The total conduction loss in a 3-phase 2-level inverter can be approximated as

$$P_{c\_inv\_2} = 6(P_{c\_sw} + P_{c\_D}). \tag{6.17}$$

The inverter section of the medium/high-frequency-link MMC converter consists of a series of H-bridge inverter cells in a cascaded connection. Therefore, the total conduction losses of an m-level inverter can be approximated as

$$P_{c\_inv\_M} = 6(m-1)(P_{c\_sw} + P_{c\_D}). \tag{6.18}$$

The device commutation voltage of m-level inverter is $(m-1)$ times lower than that of a device in the 2-level inverter. The on-state saturation voltage of an IGBT and forward voltage of a diode are highly dependent on device voltage ratings. For these reasons, although the multilevel inverter uses a large number of devices, the total conduction loss is similar to that of a 2-level inverter with the same power conditions. Due to lower switching losses, the total loss of a multilevel inverter is much lower than that of a 2-level inverter with the same power conditions.

The efficiency of each section and for the whole system of the high-frequency magnetic-link MMC converter can be calculated from

$$\eta = \frac{P_{out}}{P_{in}}. \tag{6.19}$$

and percentage of efficiency can be calculated from

$$\eta\% = \frac{P_{out}}{P_{in}} \times 100. \tag{6.20}$$

As calculated, the multilevel inverter section of the high-frequency magnetic-link MMC converter gives about 80 % efficiency at 20 % rated load. The efficiency reaches 90 % at about 50 % rated load. Almost 95 % efficiency was measured with the fully rated power condition. The overall efficiency of the whole system was also calculated. The overall efficiency of the whole system was much lower than that of the multilevel inverter section due to the significant power losses in the 10-kHz inverter, high-frequency magnetic link, and fast recovery rectifiers. About 58 % efficiency was evaluated with the 20 % rated power. The overall efficiency increases to 70 % at about 50 % rated power. The full load overall efficiency of the proposed system was measured at 76 %. Compared with the traditional 2-level converter (inverter section only), about 15 % lower efficiency was evaluated. Although the high-frequency magnetic-link MMC system gives 15 % lower efficiency than that of a 2-level converter, it is still similar to a 2-level converter-based traditional system, because the traditional system uses two main additional components, i.e., line filter, and the power frequency step-up transformer. The step-up transformers and harmonic neutralizing filters produce about 50 % of the total losses and occupy up to 40 % of the system volume [29]. With the high-frequency magnetic-link MMC converter, the elimination of the heavy and large step-up transformer and line filter will enable large cost savings in terms of the installation, running, and maintenance of renewable power generation systems. No matter what type of power transmission system, the medium-voltage converter is equally applicable in the high-voltage AC (HVAC) or high-voltage DC (HVDC) system.

6.4 Summary

The multiple secondary windings high-frequency magnetic link can be a good solution to provide multiple isolated and balanced DC supplies for the modular multilevel cascaded converter and to ensure electrical isolation between grid and renewable generation systems. The modern FPGA-based switching controller may solve the control complexity of MMC converters. In this chapter, the high-frequency magnetic-link MMC converter-based medium-voltage system has been validated by a scaled down 3-phase 1-kV system with a 5-level MMC converter topology. The same concept can be used to model the 11-kV system with 19-level MMC converter topology. The switching scheme and design technique can be used for any other converter with minor modifications in the software environment. Although the high-frequency magnetic-link MMC converter-based system does not improve the efficiency significantly, the elimination of the heavy and large-size step-up transformer and line filter will enable large cost savings in the installation, running, and maintenance of renewable generation systems. These design and implementation techniques would have great potential for the development of new medium-voltage converters for renewable generation systems and smart micro-grid applications.

References

1. Islam MR, Guo YG, Zhu JG (2014) A high-frequency link multilevel cascaded medium-voltage converter for direct grid integration of renewable energy systems. IEEE Trans Power Electron 29(8):4167–4182
2. Islam MR, Guo YG, Zhu JG (2014) A multilevel medium-voltage inverter for step-up-transformer-less grid connection of photovoltaic power plants. IEEE J Photovoltaics 4(3):881–889
3. Ng CH, Parker MA, Ran L, Tavner PJ, Bumby JR, Spooner E (2008) A multilevel modular converter for a large, light weight wind turbine generator. IEEE Trans Power Electron 23(3):1062–1074
4. Yuan X, Li Y, Chai J, Ma M (2009) A modular direct-drive permanent magnet wind generator system eliminating the grid-side transformer. In: Proceedings of the 13th European conference on power electronics and applications, Barcelona, Spain, 8–10 Sept 2009, pp 1–7
5. Yuan X, Chai J, Li Y (2012) A transformer-less high-power converter for large permanent magnet wind generator systems. IEEE Trans Sustain Energy 3(3):318–329
6. Deng F, Chen Z (2010) A new structure based on cascaded multilevel converter for variable speed wind turbine In: Proceedings of the 36th annual conference on IEEE industrial electronics society, Glendale, AZ, USA, 7–10 Nov 2010, pp 3167–3172
7. Islam MR, Guo YG, Zhu JG (2014) A review of offshore wind turbine nacelle: technical challenges, and research and developmental trends. Renew Sustain Energy Rev 33:161–176

8. Islam MR, Guo YG, Zhu JG (2014) A novel medium-voltage converter system for compact and light wind turbine generators In: Proceedings of Australasian universities power engineering conference (AUPEC2012), Bali, Indonesia, 26–29 Sept 2012, pp 1–6
9. Jia L, Mazumder SK (2012) A loss-mitigating scheme for DC/pulsating-DC converter of a high-frequency-link system. IEEE Trans Ind Electron 59(12):4537–4544
10. Huang R, Mazumder SK (2009) A soft-switching scheme for an isolated DC/DC converter with pulsating DC output for a three-phase high-frequency-link PWM converter. IEEE Trans Power Electron 24(10):2276–2288
11. Mazumder SK, Rathore AK (2011) Primary-side-converter-assisted soft-switching scheme for an AC/AC converter in a cycloconverter-type high-frequency-link inverter. IEEE Trans Ind Electron 58(9):4161–4166
12. Prasai A, Yim JS, Divan D, Bendre A, Sul SK (2008) A new architecture for offshore wind farms. IEEE Trans Power Electron 23(3):1198–1204
13. Islam MR, Guo YG, Zhu JG (2013) A medium frequency transformer with multiple secondary windings for medium voltage converter based wind turbine power generating systems. J Appl Phys 113(17):17A324–17A324-3
14. Islam MR, Guo YG, Lin ZW, Zhu JG (2014) An amorphous alloy core medium frequency magnetic-link for medium voltage photovoltaic inverters. J Appl Phys 115(17):17E710-1–17E710-3
15. Xu D, Chen M, Luo M (2004) Transformer secondary leakage inductance based ZVS dual bridge DC/DC converter. IEEE Trans Power Electron 19(6):1408–1416
16. Mezaroba M, Martins DC, Barbi I (2007) A ZVS PWM half-bridge voltage source inverter with active clamping. IEEE Trans Ind Electron 54(5):2665–2672
17. Chen G, Lee YS, Hui SYR, Xu D, Wang Y (2000) Actively clamped bidirectional flyback converter. IEEE Trans Ind Electron 47(4):770–779
18. Kheraluwala MN, Gascoigne RW, Divan DM, Baumann ED (1992) Performance characterization of a high-power dual active bridge. IEEE Trans Ind Appl 28(6):1294–1301
19. Krismer F, Kolar JW (2012) Closed form solution for minimum conduction loss modulation of DAB converters. IEEE Trans Power Electron 27(1):174–188
20. Krismer F, Kolar JW (2012) Efficiency-optimized high-current dual active bridge converter for automotive applications. IEEE Tran Ind Electron 59(7):2745–2760
21. Islam MR, Guo YG, Zhu JG (2011) H-bridge multilevel voltage source converter for direct grid connection of renewable energy systems. In Proceedings of IEEE power & energy society innovative smart grid technology conference, Perth, Australia, 13–16 Nov 2011, pp 1–7
22. Islam MR, Guo YG, Zhu JG (2013) Power converters for wind turbines: current and future development. In: Mendez-Vilas A (ed) Materials and processes for energy: communicating current research and technological developments, Energy book series-2013 edition. Spain, pp 559–571
23. Islam MR, Guo YG, Zhu JG, Lu H, Jin JX (2013) Medium frequency-link power conversion for high power density renewable energy systems. In: Proceedings of IEEE 2013 international conference on applied superconductivity and electromagnetic devices (ASEMD2013), 25–27 Oct 2013, Beijing, China, pp 102–106
24. Agheb E, Bahmani MA, Hoidalen HK, Thiringer H (2012) Core loss behavior in high frequency high power transformers-II: arbitrary excitation. J Renew Sustain Energy 4(3):033113–033113-11
25. Zhang Z, Sharma P, Makino A (2012) Role of Si in high Bs and low core-loss $Fe_{85.2}B_{10-x}P_4Cu_{0.8}Si_x$ nano-crystalline alloy. J Appl Phys 112(10):103902–103902-8
26. Makino A, Kubota T, Yubuta K, Inoue A, Urata A, Matsumoto H, Yoshida S (2011) Low core losses and magnetic properties of Fe85-86Si1-2B8P4Cu1 nanocrystalline alloy with high B for power applications. J Appl Phys 109(7):07A302–07A302-5
27. Islam MR, Guo YG, Zhu JG (2012) A medium-frequency transformer with multiple secondary windings for grid connection through H-bridge voltage source converters. In: Proceedings of international conference electrical machines and systems, Sapporo, Japan, 21–24 Oct 2012, pp 1–6
28. Tolbert LM, Peng FZ, Habetler TG (1999) Multilevel converters for large electric drives. IEEE Trans Ind Appl 35(1):36–44
29. Peng FZ, Lai JS, McKeever JW, Coevering JV (1996) A multilevel voltage—source inverter with separated DC sources for static var generation. IEEE Trans Ind Appl 32(5):1130–1138

Chapter 7
Design and Analysis of 11- and 33-kV Modular Multilevel Cascaded Converters

Abstract The high number of levels of modular multilevel cascaded (MMC) converters enables the direct connection of the renewable generation units to the medium-voltage grid and improvement of the output power quality. The component number and control complexity increase linearly with the increase in the number of levels. On the other hand, the distortion in generated output voltage and semiconductor cost of the converter decrease dramatically with the increase in the converter number of levels. As the number of levels increase, it is possible to use lower switching frequencies, even the fundamental switching frequency, which can significantly reduce the switching losses. Therefore, the optimal selection of the number of converter levels is important for the best performance/cost ratio of the medium-voltage converter systems and this is the central content of this chapter. In this chapter, an 11-kV system and a 33-kV system are designed and analyzed taking into account the specified system performance, control complexity, and cost and market availability of the power semiconductors. It is found that the 19-level and 43-level converters are the optimal choice for the 11- and 33-kV systems, respectively. Besides the design and analysis of medium-voltage converters, the traditional low-voltage converters with power frequency step-up transformers are also discussed.

Keywords Medium voltage · Modular multilevel cascaded · Number of levels · Design and optimization · Conventional converters/inverters · Renewable generations · Step-up transformers

7.1 Introduction

Different power electronic converters have been developed using the conventional topologies to fulfill the requirements of renewable generation systems. However, it is hard to connect the traditional converters to the grids directly, as the distortion in generated output voltages is high and a single switch cannot stand the grid voltage level [1, 2]. Therefore, the traditional low-voltage converters/inverters have typically used step-up transformer to step up the voltage to the grid voltage level.

M.R. Islam et al., *Power Converters for Medium Voltage Networks*,
Green Energy and Technology, DOI 10.1007/978-3-662-44529-7_7

7.1.1 *Traditional Converters for Wind Power Generation Systems*

Many power semiconductor vendors such as Semikron, ASEA brown boveri (ABB), IXYS, and Mitsubishi Electric produce devices specially designed for the diode rectifier-based converter and back-to-back converter (modular form) for wind turbine generator systems. All of the devices are in a single pack, which reduces the cost and complexity of the power conditioning system. Semikron developed modules IGDD6-4-426-D3816-E1F12-BL-FA and SKS 660F B6U+E1C+B6CI 250 V06 for the diode rectifier-based power conditioning systems [3, 4]. This type of power converter is normally used in a wound rotor synchronous generator (WRSG) or a permanent magnet synchronous generator (PMSG)-based wind power generation system instead of an induction generator. According to the internal circuit configuration, module IGDD6-4-426-D3816-E1F12-BL-FA is suitable for WRSG-based wind turbine generator systems and module SKS 660F B6U+E1C+B6CI 250 V06 is suitable for PMSG-based wind turbine generator systems. In a WRSG-based system, to achieve variable-speed operation, the system uses an extra excitation circuit, which feeds the excitation winding of WRSG. Figure 7.1 shows the Semikron's SEMIKUBE converter module IGDD6-4-426-D3816-E1F12-BL-FA. The internal electrical circuit of the Semikron's SEMIKUBE converter module IGDD6-4-426-D3816-E1F12-BL-FA is shown in Fig. 7.2. The PMSG-based wind turbine generator systems are equipped with a step-up chopper circuit. The step-up chopper adapts the rectifier voltage to the DC-link voltage of the inverter. Controlling the inductor current in the step-up chopper can control the generator torque and speed. The diode rectifier with step-up chopper-based power conditioning system is illustrated in Fig. 7.3. In this converter system, the grid-side inverter controls the active and reactive power delivered to the grid. Mitsubishi Electric developed the IGBT module CM100MXA-24S with this converter topology which can be used for wind turbine generator systems [5]. The module is recommended for the collector current of 100 A and the collector–emitter

Fig. 7.1 A photograph of Semikron's SEMIKUBE converter module IGDD6-4-426-D3816-E1F12-BL-FA

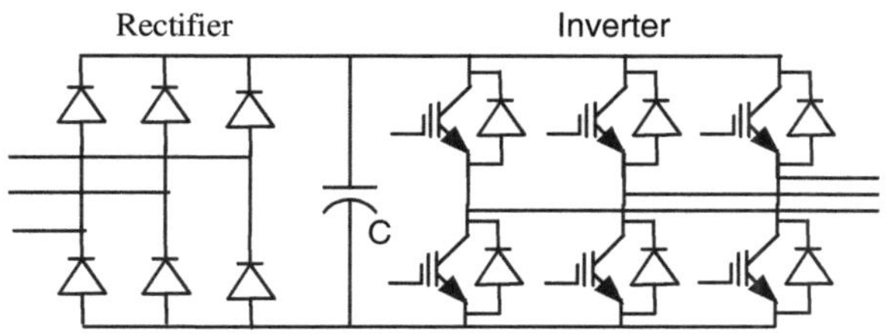

Fig. 7.2 Electrical circuit of Semikron's SEMIKUBE converter module IGDD6-4-426-D3816-E1F12-BL-FA

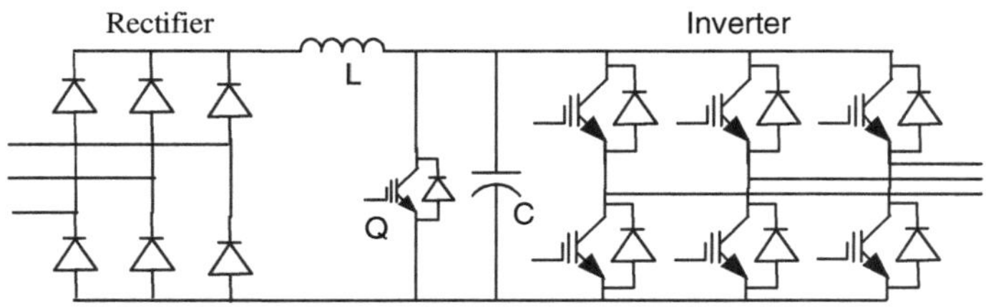

Fig. 7.3 Power conditioner circuit for PMSG-based wind turbine generator systems

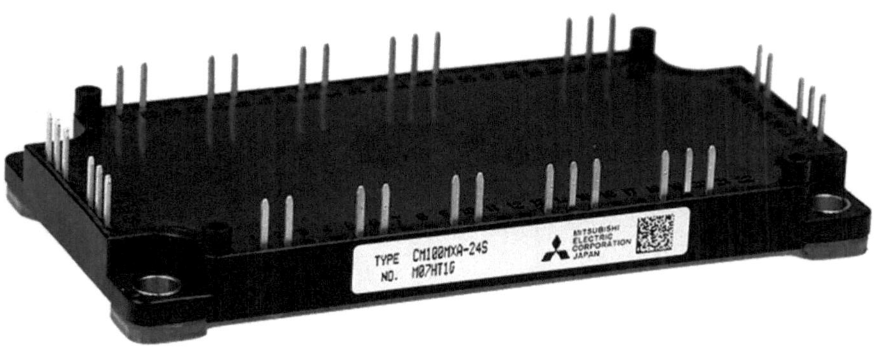

Fig. 7.4 A photograph of MITSUBISHI ELECTRIC IGBT module CM100MXA-24S

voltage of 1,200 V. Figure 7.4 shows a photograph of Mitsubishi Electric IGBT module CM100MXA-24S. By adding the DC-link capacitor and chopper inductor, the module can be used for both the WRSG- and PMSG-based wind turbine generator systems. Although it is a cost-effective choice, the power quality and unidirectional power flow capability of this converter system are critical to the selection.

When the diode rectifier and chopper circuit is replaced by controlled rectifier, the controlled rectifier may have the bidirectional power flow capability, which is not possible in the diode rectifier-based power conditioning system. Moreover, the controlled rectifier strongly reduces the input current harmonics and harmonic power losses. The grid-side converter enables the control of the active and reactive power flows to the grid and keeps the DC-link voltage constant, improving the output power quality by reducing the total harmonic distortion (THD). The generator-side converter works as a driver, controlling the magnetization demand and the desired rotor speed of the generator. The decoupling capacitor between grid-side converter and generator-side converter provides independent control capability of the two converters. Due to some special features, this converter topology has received great attention recently. Many power semiconductor manufacturers, such as Semikron, ABB, Hitachi, Siemens, IXYS, and Mitsubishi Electric, produce components in module forms, suitable for this converter, which makes the converter compact and lightweight. The back-to-back converter can be used for PMSG and squirrel cage induction generator (SCIG)-based wind power generation systems. Siemens employs back-to-back converter for power conditioning of SCIG-based wind turbine generator systems. Semikron's SEMISTACK RE module SKS B2 140 GDD 69/12 U-A11 MA PB is a back-to-back converter typically used for power conditioning of synchronous and doubly fed generator-based wind power systems, as well as in central solar PV inverters [6]. Figure 7.5 shows a photograph of SEMISTACK RE. The module's maximum ratings are the input/

Fig. 7.5 A photograph of Semikron SEMISTAK RE converter module SKS B2 140 GDD 69/12 U-A11 MA PB

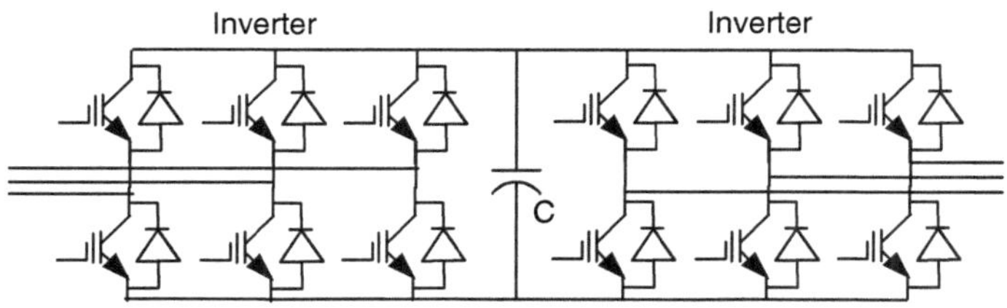

Fig. 7.6 Electrical power circuit of Semikron SEMISTAK RE converter module SKS B2 140 GDD 69/12 U-A11 MA PB

output current of 1,400 A, input/output voltage of 760 V, and switching frequency of 5 kHz, respectively. Up to four SEMITACK RE converters can be connected in parallel to support applications of capacity up to 6 MVA. SEMISTACK RE uses the latest generation of Semikron's SKiiP 4 intelligent power module family which integrates the power components, driver, and heat sink in a single case. Figure 7.6 shows the internal electrical circuit of SEMISTACK RE. In addition to bidirectional power flow capability, back-to-back converters have some other remarkable features. For example, the DC-link voltage can be boosted to a level higher than the amplitude of the grid line-to-line voltage in order to achieve full control of the grid current and the capacitor between the inverter and rectifier makes it possible to decouple the control of the two inverters, allowing the compensation of asymmetry on both the generator side and the grid side. On the other hand, the back-to-back converter has some critical disadvantages. The heavy and large DC-link capacitor increases the costs and reduces the overall lifetime of the system. Since the back-to-back converter consists of two inverters, the switching losses might be even more pronounced, and the combined control of the controlled rectifier and inverter is quite complicated.

7.1.2 Traditional Converters for Solar PV Power Plants

Semikron's SEMIKUBE modules IGD-8-326-E1F12-BH-FA and IGD-8-426-E1F12-BH-FA are three-phase inverters typically used for solar PV applications. Figure 7.7 shows a photograph of the Semikron's SEMIKUBE solution family, covering maximum power of 110, 220, 400, and 900 kW (from the left), for output voltages up to maximum 460 V. The SEMIKUBE is UL508c recognized and fulfills the EN 50178 standard. The IGBTs of SEMIKUBE may allow a maximum switching frequency of 15 kHz. Figure 7.8 shows the power circuit of SEMIKUBE.

The ABB central inverters are specially designed for the medium-scale PV power plants. The PVS800 version is a 3-phase inverter with a power capacity in the range of 100–1,000 kW [7]. Figure 7.9 shows a photograph of ABB central PV inverter PVS800. The transformer steps up the inverter output voltage from

Fig. 7.7 A photograph of Semikron's SEMIKUBE converter modules designed for 110-, 220-, 400-, and 900-kW (*from the left*) solar PV power converters

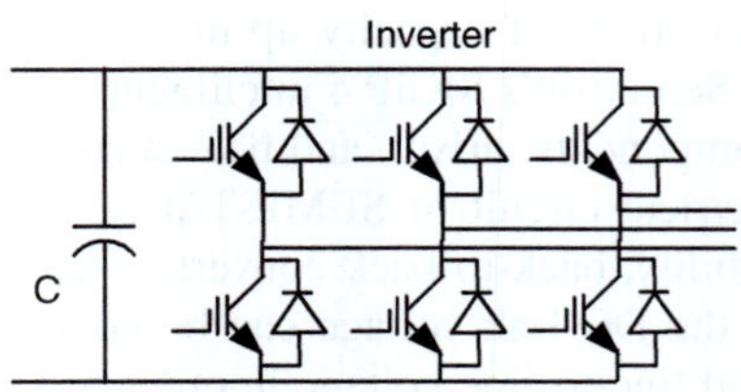

Fig. 7.8 Electrical equivalent circuit of Semikron's SEMIKUBE converter modules

Fig. 7.9 ABB central solar PV inverter PVS800

Fig. 7.10 Housing of ABB central solar PV inverter PVS800 with step-up transformer

300–400 V AC to the distribution grid voltage level (e.g., 6–36 kV). Figure 7.10 shows the assembly of PVS800 inverter with step-up transformer. Siemens also developed the SINVERT PVS inverter for the medium-scale PV power plants. The AC output voltage and power capacity of the PVS version inverters are in the range of 288–370 V and 500–630 kW, respectively [8]. Figure 7.11 shows a photograph

Fig. 7.11 Siemens PV inverter SINVERT PVS; the left two cabinets are the DC cabinets and the other two the AC cabinets

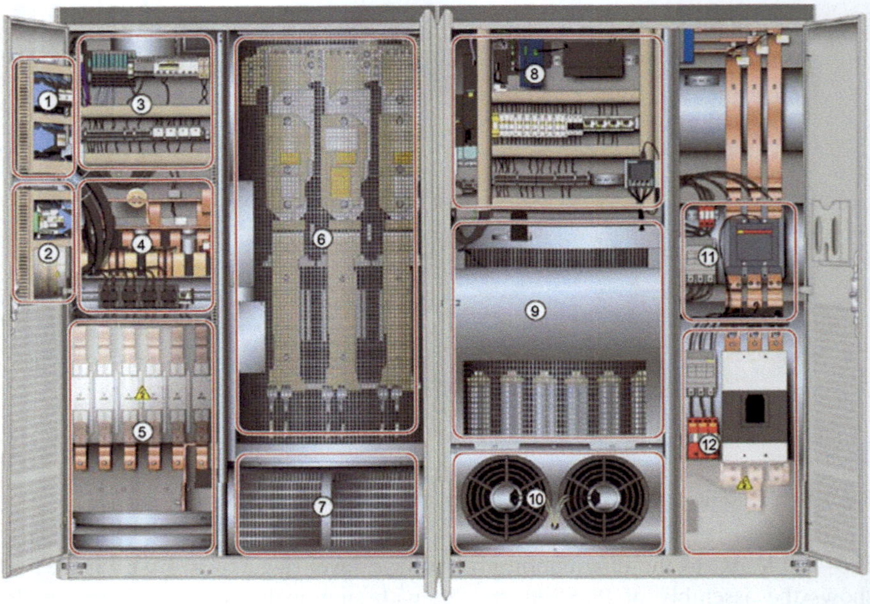

Fig. 7.12 Function units of Siemens PV inverter SINVERT PVS

of Siemens central PV inverter SINVERT PVS. The cabinet holds touch panel which consists of indicators and key-operated switch. Figure 7.12 shows the function units of Siemens PV inverter SINVERT PVS, where the markings of 1–12 indicates the modules for 1,000 V option, modules for PV array grounding option, modules for options, DC connectors, DC terminal compartment of the PV array and LV HRC fuses, inverter module (power unit), connection to AC cabinet, communication area, AC filter, cooling ventilators, AC contactor, and circuit breaker for isolating the AC system and overvoltage protection, respectively.

7.1.3 Power Transformers in Traditional Converter-Based Renewable Generation Systems

Installation of megawatt-scale renewable power generation unit, such as wind farms and solar PV power plants, is becoming common in recent years. The voltage levels of the wind generator and PV inverter are usually in the range of 380–690 V and 280–370 V, respectively [9–11]. To integrate the scattered wind turbine generators and solar PV arrays into a medium-voltage grid before the voltage step-up for long-distance transmission, a step-up transformer is typically used to step up the voltage. In a wind turbine generation system, this transformer is usually installed inside the nacelle together with other equipment, such as the generator and power converter at a height of about 80 meters, whereas in PV power plants, these transformers are typically installed together with low-voltage PV inverter.

Several vendors have been developing special transformers for renewable power generation systems aiming to reduce the size and weight with the consideration of environmental concern. ABB has been delivering worldwide vacuum cast coil transformers and liquid-filled transformers for wind farm and solar PV power plant applications, and has already installed more than 5,000 units for wind farm applications. The cast coil dry-type transformers are nonflammable and moisture proof, featuring a discharge-free solid isolation system. The volume and weight of a 2.5-MVA vacuum cast coil transformer are about 8.5 m^3 and 6,200 kg, and the no-load and full-load losses are 5.8 and 25 kW, respectively [1, 12]. Figure 7.13 shows a photograph of ABB cast coil dry-type three-phase transformer. However, the dry-type transformers can be sensitive to water, microcracks, temperature variations, and pollution which can block cooling ducts. ABB cast coil dry-type transformers are typically installed in different locations of wind turbines, such as in nacelle and tower. Figure 7.14 shows the possible locations to install the cast coil dry-type transformers.

In addition to the ABB's existing high-efficiency transformers, ABB has also introduced new liquid-filled amorphous metal (AM) transformers which can offer even more energy savings and are suitable for renewable power generation systems. Since

Fig. 7.13 ABB cast coil dry-type transformer

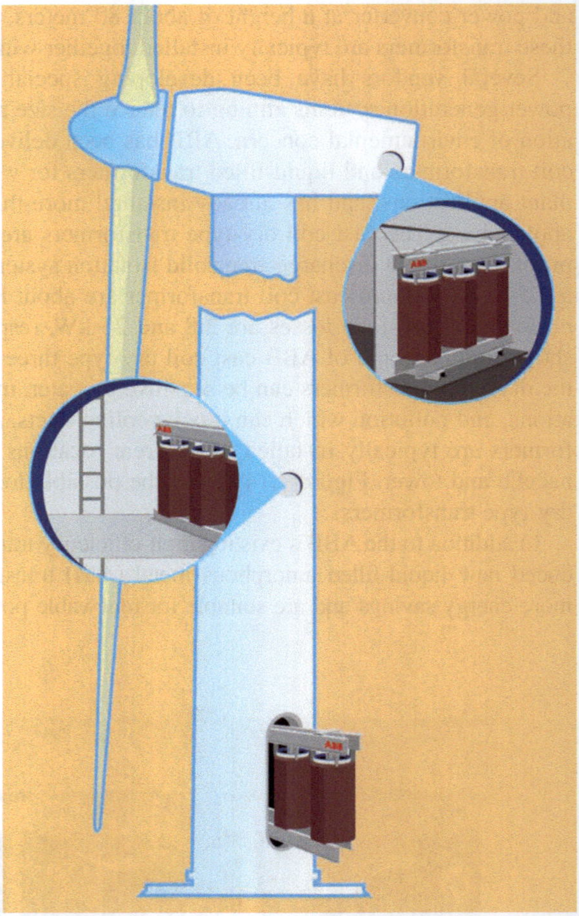

Fig. 7.14 Housing of ABB cast coil dry-type transformer

wind farms and solar PV power plants are intermittent power sources, it is of great importance to minimize transformer no-load losses. A liquid-filled 2-MVA transformer is about 5.7 m^3 in volume and 4,530 kg in weight with 870 kg of liquid as the coolant and insulator [1, 13]. The no-load and full-load losses are approximately 3.2 and 21 kW, respectively. Figure 7.15 shows a photograph of an ABB liquid-filled transformer. In offshore wind farms, this transformer is usually installed inside the nacelle of the wind turbine. Ground-mounted liquid-filled transformers are typically used for solar PV plants and onshore wind farm applications. Figure 7.16 shows a photograph of a ground-mounted liquid-filled transformer for onshore wind farm.

The SLIM transformer developed by Pauwels is compact in size and has low no-load losses, typically half of those of the dry-type transformers. For example, a 33/0.69-kV, 2.6-MVA SLIM transformer has a no-load loss of 2.6 kW and a full-load loss of 22.5 kW [1, 14, 15]. Figure 7.17 shows a photograph of a Pauwels SLIM transformer. This transformer is typically lifted through the tower door.

Fig. 7.15 ABB liquid-filled transformer

Figure 7.18 shows the transformer lifting technique. The conventional liquid-immersed transformer uses cellulose and mineral oil, whereas SLIM uses a high-temperature aramid insulation material called NOMEX and a silicone liquid. The use of about 900 kg or more silicone fluid as the coolant and insulator in these transformers ensures a high degree of fire safety (the flash points of mineral oil dielectric and silicone fluid are about 150 and 360 °C, respectively). These insulation materials are usually chosen on the basis of reliability, performance, safety, and environment concerns. Especially, the liquid degradation has received a lot of attention. For example, according to European requirements, more than 65 % of the liquids must be degraded in 28 days. This concern can be overcome by using fully biodegradable liquids. M&I Materials Ltd. developed the synthetic ester MIDEL 7131, which is the preferred dielectric fluid for transformers in environmentally sensitive locations and already recognized as "readily biodegradable." Pauwels developed a new transformer by optimally using the synthetic ester MIDEL 7131 in combination with the high-temperature NOMEX insulation system to obtain the compactness and reliability of the SLIM transformer. This resulted biodegradable liquid-based SLIM transformer is called Bio-SLIM. The volume and weight of a 20/0.69-kV, 2.3-MVA Bio-SLIM transformer are about

Fig. 7.16 Housing of liquid-filled transformer in onshore wind farm

Fig. 7.17 Pauwels SLIM transformer

Fig. 7.18 A Pauwels transformer being lifted through the door of a wind turbine tower during installation

Fig. 7.19 Pauwels Bio-SLIM transformer for offshore applications

4 m^3 and 5,040 kg, respectively [1, 14, 15]. The Bio-SLIM transformer is physically similar to SLIM, but painted blue to distinguish it from SLIM and to link to marine and water applications. Figure 7.19 shows a photograph of the Bio-SLIM transformer. Table 7.1 summarizes the data of various transformers.

Table 7.1 Transformer data

Transformer type	Rating (MVA)	Size (m^3)	Weight (kg)	No-load loss (kW)	Full-load loss (kW)	Liquid (kg)
Cast coil	2.50	8.50	6,200	5.80	25.00	–
Liquid filled	2.00	5.70	4,530	3.20	21.00	870
SLIM	2.30	4.00	5,040	2.60	22.50	900

Although industries have been trying to develop environmentally friendly compact and lightweight transformers for renewable generation systems, still the transformers are heavy and large. This heavy and large-size step-up transformer significantly increases the weight and volume of the nacelle as well as the mechanical stress of the tower. These penalties are critical in offshore applications, where the cost of installation and regular maintenance is extremely high. For example, an offshore wind farm requires about 20 % higher operating and maintenance cost compared with an onshore farm. Moreover, environmental concerns are still unsolved and can never be compromized.

7.1.4 Medium-Voltage Converters for Direct Grid Integration

With the traditional low-voltage converters, the conventional renewable generation systems possess the power frequency (i.e., 50 or 60 Hz) step-up transformers, filters, and boosters, resulting in not only large size, weight, and loss but also high cost and complexity of the system, both capital and operational [16]. Today, the industrial trend is to move away from these heavy and large-size passive components to power electronic systems that use more and more semiconductor elements controlled by a digital circuit. In such a way, smart operation can be ensured. In comparison with the conventional two-level converters, multilevel converters present lower switching losses, lower voltage stress on switching devices, and better harmonic performance. These remarkable features enable the connection of renewable energy systems directly to the grid without using large, heavy, and costly power transformers and also minimizing the input and output filter requirements. Because of some special features (e.g., the number of components scales linearly with the number of levels, and individual modules are identical and completely modular in constriction, hence enabling high-level attainability), the MMC converter topology can be considered as the best possible candidate for medium-voltage applications [17, 18]. The high number of levels means that medium-voltage attainability is possible to connect the renewable generation units to the medium-voltage grid directly and it is also possible to improve the output power quality. The component number and control complexity increase linearly with the increase in number of levels. Figure 7.20 plots the relationship between component number and complexity of different multilevel converters.

On the other hand, the distortion in generated output voltage and semiconductor cost of the converter decrease dramatically with the increase in the converter number of levels [9, 10, 19]. Figure 7.21 plots the THD (%) and semiconductor cost of different number of MMC converter levels for an 11-kV system.

Due to the unavailability of IGBT ratings, the 13- and 17-level converters used the IGBTs that are used in the 11- and 15-level converters, respectively. Hence, the semiconductor cost curve is up and down in nature. Moreover, lower switching frequency, even fundamental switching frequency, can be used with the high-level number converters, which can significantly reduce the switching losses of the converter [9]. Therefore, the optimal selection in the number of converter levels is important for the best performance/cost ratio of the medium-voltage converter systems and this is the central content of this chapter. In this chapter, two systems of 11 and 33 kV, respectively, are designed and analyzed taking into account the specified system performance, control complexity, and cost and market availability of the power semiconductors. It is found that the 19-level and 43-level converters are the optimal choice for the 11- and 33-kV systems, respectively. The design and analysis of the 11-kV MMC converter and 33-kV MMC converter systems are presented in detail in the following sections.

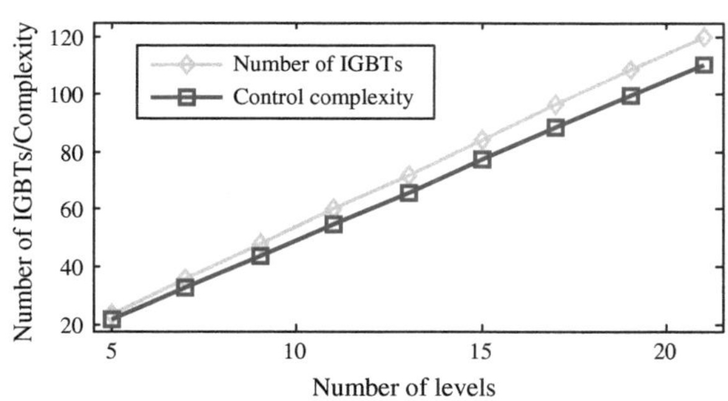

Fig. 7.20 Number of IGBTs/control complexity versus converter level numbers

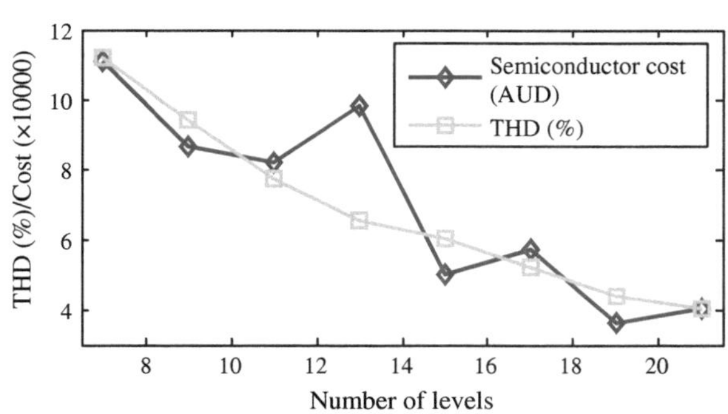

Fig. 7.21 THD (%) and semiconductor cost versus converter level numbers

7.2 Design and Analysis of 11-kV Converter Systems

The total nominal DC-link voltage for an 11-kV 2-level converter is 16,176 V. Each H-bridge inverter cell commutation voltage of a 5-level converter is about 4,044 V. The highest voltage rating of a commercially available IGBT is 6.5 kV, which is recommended for a maximum voltage of 3,600 V. Therefore, the 5-level or lower-level converter cannot be used to design the 11-kV converter. Each H-bridge inverter cell commutation voltage of a 7-level topology-based 11-kV converter is 2,696 V which may be supported by the 6.5-kV IGBT. Accordingly, at least 7-level topology is required to design an 11-kV converter. The output power quality of a 21-level converter is good enough to feed into the 11-kV AC grid directly. The cheap 1.7-kV IGBT can be used to design the 21-level converter. Therefore, 7-level to 21-level MMC topologies are considered for an 11-kV converter system. If V_{com} is the commutation voltage of respective commutation cells, the device voltage utilization factor (DVUF) can be calculated from

$$\mathrm{DVUF} = \frac{V_{com}}{V_{com@100FIT}}. \tag{7.1}$$

Table 7.2 summarizes the commutation voltage and DVUF of different level MMC converters. A higher DVUF is essential for cost-effective design, since the semiconductor cost is a significant figure in medium-voltage converter applications. From Table 7.2, it can be seen that only a few converters have high DVUFs. Considering the availability of the power semiconductor devices, one can find only a few converter topologies of 9, 11, 15, 19, and 21 levels may give good DVUFs. Although the 7-level circuit topology can be used to develop an 11-kV converter, the available commercial IGBTs are not well suited. Therefore, devices of higher ratings are required, which would yield a very poor DVUF of about 75 %. A low DVUF means the use of unnecessarily high-cost semiconductors. In general, to use the active switching devices cost-effectively, a converter must have a DVUF of 90 % or above. As listed in the table, only two converter systems give more than 90 % DVUF. In order to ensure a cost-effective design, the converters of 9, 11, 15, 19, and 21 levels for 11-kV systems were considered for further analysis. The lowest number of levels of an MMC converter is

Table 7.2 DVUF with different level number; 11-kV system

No. of levels	V_{com} (V)	Rated device voltage (kV)	$V_{com@100FIT}$ (V)	DVUF (%)
7	2,696	6.50	3,600	75
9	2,022	4.50	2,250	90
11	1,618	3.30	1,800	90
13	1,348	3.30	1,800	75
15	1,156	2.50	1,200	96
17	1,011	2.50	1,200	84
19	898	1.70	900	99
21	809	1.70	900	90

3, and such a converter consists of only one H-bridge inverter cell on each phase leg. Each additional cascaded H-bridge inverter cell contributes two voltage levels to the output voltage waveform. Therefore, the numbers of levels are only of odd values.

7.2.1 Design and Analysis of 9-Level 11-kV Converter

The circuit diagram of a 9-level MMC converter is shown in Fig. 7.22. There are 48 active switching devices in the three-phase circuit, and each phase contains 16 devices. Each phase leg consists of 4 H-bridge inverter cells in cascaded connection. The line voltage of a 9-level MMC converter is illustrated in Fig. 7.23. The line voltage THDs of an 11-kV 9-level converter are calculated and found as

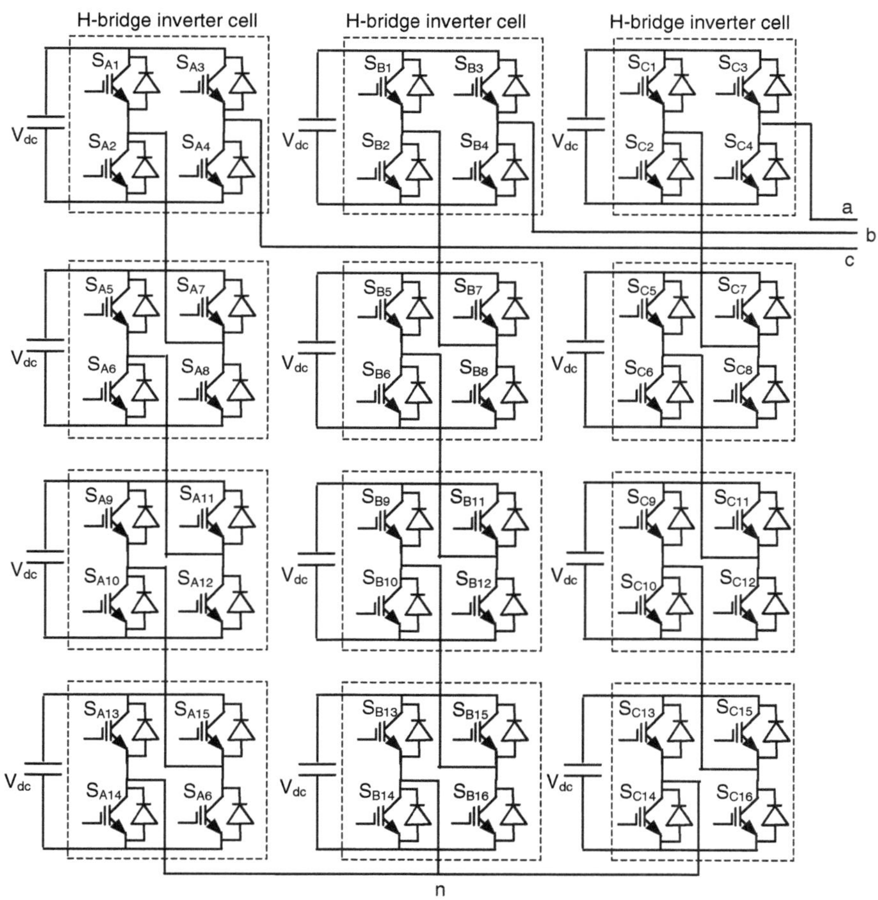

Fig. 7.22 Three-phase 9-level MMC converter

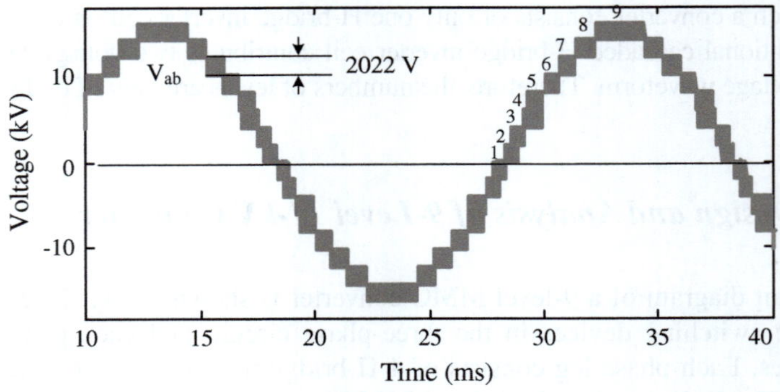

Fig. 7.23 Line voltage of 9-level MMC converter

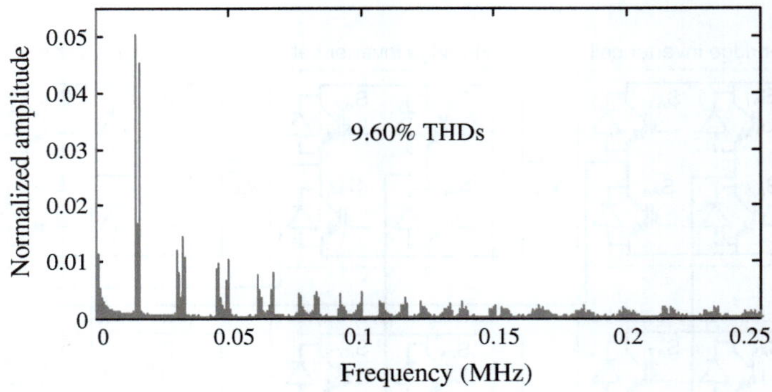

Fig. 7.24 Line voltage frequency spectrum of 9-level MMC converter

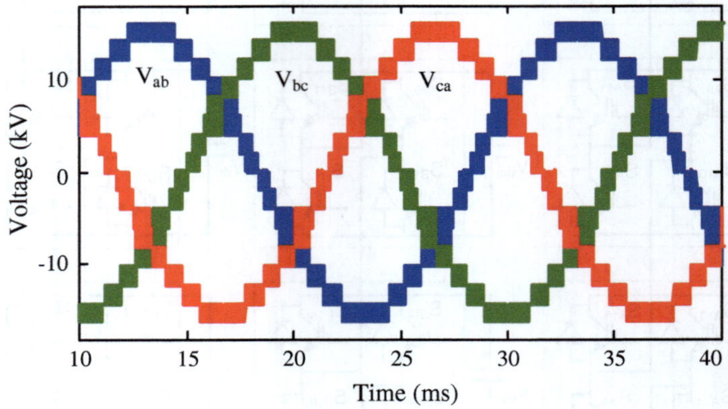

Fig. 7.25 Three line voltages of 9-level MMC converter

9.60 %. The frequency spectrum of line voltage is depicted in Figs. 7.24 and 7.25 shows the three line voltages of three-phase converter system.

Table 7.3 Commercially available 4.5-kV-rated IGBTs

IGBT modules	Rated current (A)	Brand
CM400HB-90H	400	Mitsubishi Electric
CM600HB-90H	600	Mitsubishi Electric
5SNA 0650J450300	650	ABB
5SNA 1200G450350	1,200	ABB
FZ800R45KE3_B5	800	Infineon

A total of 44 arithmetic and logic operations (ALOs) are required in the MATLAB/Simulink environment to control the 3-phase 9-level MMC converter with the phase-shifted carrier-based modulation scheme. The switching controller should have at least 22 dedicated PWM channels. Different vendors, such as ABB, Infineon, and Mitsubishi Electric, commercially developed 4.5-kV-rated IGBTs. Table 7.3 tabulates the possible 4.5-kV IGBT modules for design of the 9-level 11-kV converter. Figure 7.26 shows a photograph of Mitsubishi Electric 4.5-kV IGBT module CM400HB-90H, whose current rating is 400 A. A 4.5-kV-rated 48 IGBT costs about AUD 86,400.00.

The conduction losses of a 9-level MMC converter can be approximated by

$$P_{c\_9} = 8 \times 6(P_{c\_sw} + P_{c\_D}). \tag{7.2}$$

Compared with a 2-level converter, the loss from (7.2) appears to be 8 times higher, but the total conduction loss of a 9-level converter is similar to that of the 2-level converter because the device commutation voltage is 8 times lower than that of the switching devices in the 2-level converter. The conduction loss highly depends on the collector–emitter saturation voltage, $V_{CE(sat)}$, and the diode

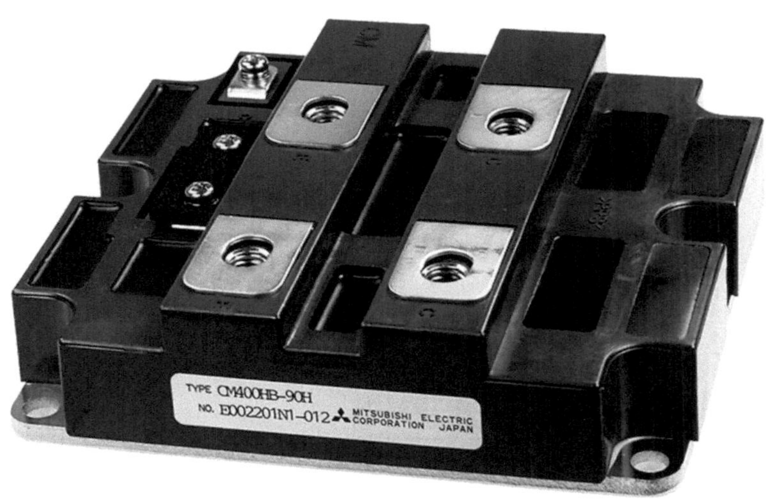

Fig. 7.26 Photograph of Mitsubishi Electric 4.5-kV 400-A IGBT CM400HB-90H

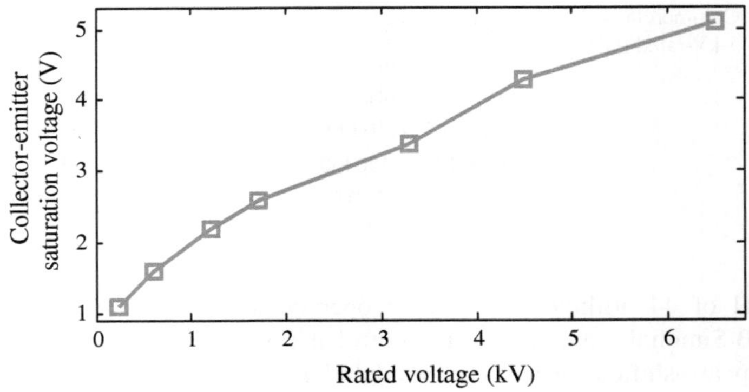

Fig. 7.27 Collector–emitter saturation voltages of 600-A-rated IGBTs [10]

forward voltage, which in turn depend on the voltage rating of switching devices. Figure 7.27 plots the collector–emitter saturation voltages of different voltage rated Mitsubishi Electric IGBTs [20].

7.2.2 *Design and Analysis of 11-Level 11-kV Converter*

The DC supply voltage rating of an 11-kV 11-level MMC converter is 1,618 V. The available 3.3-kV IGBT may be used to design the 11-kV 11-level converter, because this IGBT is recommended for 1,800 V maximum applications. With these IGBTs, 90 % DVUF can be ensured, which is important to conform to the high performance/cost of the power converter. Figure 7.28 plots the line voltage waveform of an 11-level 11-kV MMC converter. There are 20 steps in the peak-to-peak line voltage waveform, and each step contributes 1,618 V to the line voltage of 32,360 V peak to peak. The frequency spectrum of the line voltage of an

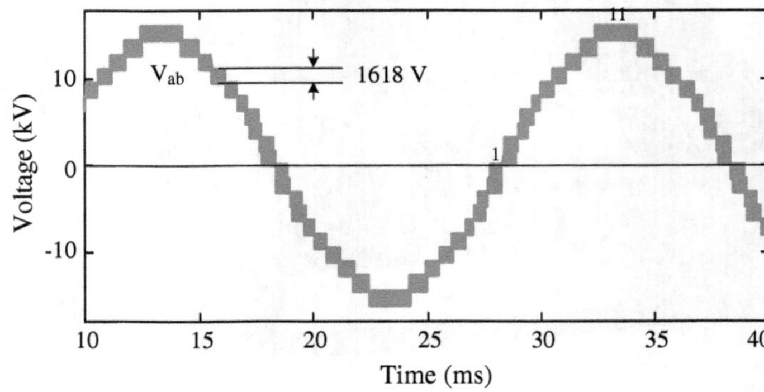

Fig. 7.28 Line voltages of 11-level MMC converter

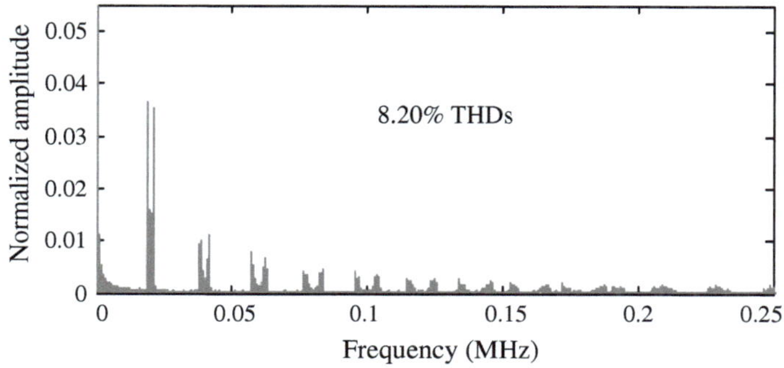

Fig. 7.29 Line voltage frequency spectrum of 11-level MMC converter

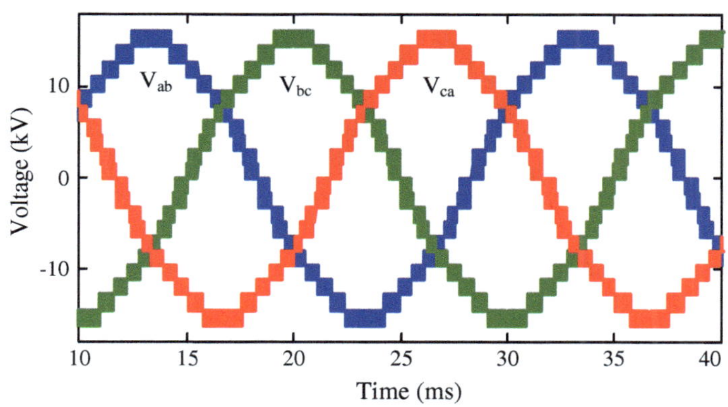

Fig. 7.30 Line voltages of 11-level MMC converter

11-level MMC converter is depicted in Fig. 7.29. Figure 7.30 shows the three line voltages of three-phase converter system.

In total, 60 IGBTs are in the 3-phase 11-level MMC converter circuit. Each IGBT price is about AUD 1,369.31, such that the cost of 60 IGBTs is about AUD 82,159.00. A few vendors, such as Mitsubishi Electric, ABB, and Infineon, commercially developed 3.3-kV IGBT modules. Table 7.4 summarizes the possible 3.3-kV IGBT modules to design the 11-level 11-kV converter. Figure 7.31 shows the photograph of Infineon 3.3-kV IGBT module FZ800R33KL2C. Since the switching control circuits consist of 55 ALOs, at least 30 PWM channels are required to control the 3-phase 11-level converter. The single DSP cannot handle the PWM pulses as required by the 3-phase 11-level converter. The parallel operation of several DSPs makes the control circuit hard to implement. The modern FPGA may be an appropriate option to design the control circuit for multilevel converters.

Table 7.4 Commercially available 3.3-kV-rated IGBTs

IGBT modules	Rated current (A)	Brand
CM400HG-90H	400	Mitsubishi Electric
5SNA 1000N330300	1,000	ABB
FZ400R33KL2C_B5	400	Infineon
FZ800R33KL2C	800	Infineon
FZ800R33KF2C	800	Infineon

Fig. 7.31 Infineon 3.3-kV IGBT module FZ800R33KL2C

7.2.3 Design and Analysis of 15-Level 11-kV Converter

The 15-level converter topology is also well suited with the available commercial semiconductors. Each phase leg consists of seven H-bridge inverter cells, 28 IGBTs in a phase, and 84 in total in the 3-phase circuit. The DC supply voltage rating of an 11-kV 15-level converter is 1,155 V. The commercially available 2.5-kV IGBT may be used to design the 11-kV 15-level converter, because this IGBT is recommended for 1,200 V maximum applications. Figure 7.32 plots

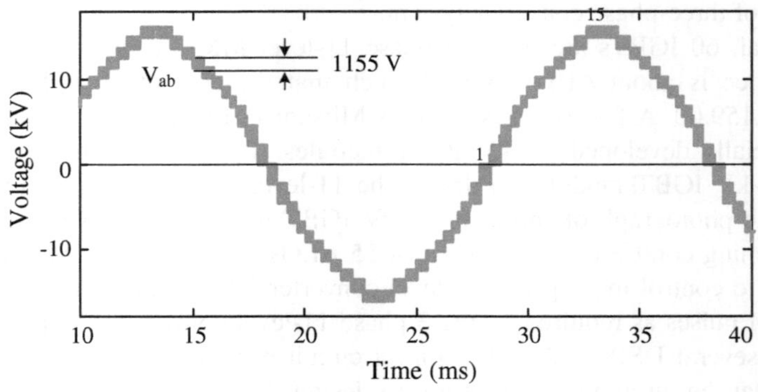

Fig. 7.32 Line voltage of 15-level MMC converter

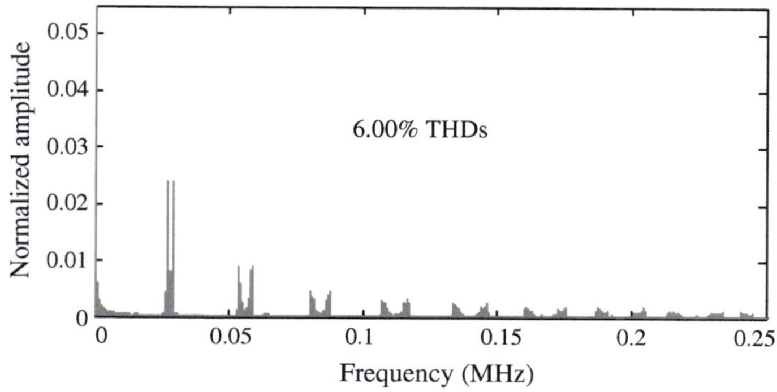

Fig. 7.33 Line voltage frequency spectrum of 15-level MMC converter

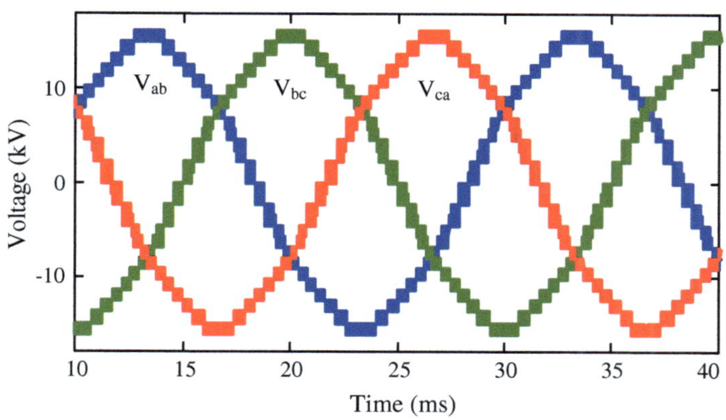

Fig. 7.34 Line voltages of 15-level MMC converter

the line voltage waveform of a 15-level MMC converter. There are 28 steps in the peak-to-peak line voltage waveform, and each step contributes 1,155 V to the line voltage. The frequency spectrum of the line voltage of a 15-level MMC converter is depicted in Fig. 7.33. Figure 7.34 shows the three line voltages of three-phase 15-level 11-kV converter. Quite a few vendors, such as ABB and Mitsubishi Electric, commercially developed 2.5-kV IGBT modules. Table 7.5

Table 7.5 Commercially available 2.5-kV-rated IGBTs

IGBT modules	Rated current (A)	Brand
5SNA 1500E250300	1,500	ABB
CM400DY-50H	400	Mitsubishi Electric
5SNA 1000E250100	1,200	ABB
CM800HA-50H	800	Mitsubishi Electric
CM1200HA-50H	1,200	Mitsubishi Electric

Fig. 7.35 ABB 2.5-kV IGBT module 5SNA 1200E250100

summarizes possible 2.5-kV IGBT modules for designing the 15-level 11-kV converter. Figure 7.35 shows the photograph of ABB 2.5-kV IGBT module 5SNA 1200E250100, whose rated current is 1,200 A.

7.2.4 Design and Analysis of 19-Level 11-kV Converter

Figure 7.36 shows the circuit diagram of a 19-level converter. Each phase leg of a 19-level converter consists of 9 H-bridge inverter cells, i.e., 36 IGBTs in a phase and 108 in total in the 3-phase circuit. The DC-link voltage rating of an 11-kV 19-level converter is about 899 V. Figure 7.37 shows voltage waveform across an IGBT in 19-level 11-kV converter.

The available 1.7-kV IGBT may be used to design the 11-kV 19-level converter. This IGBT is cheap and matured in terms of technology and is recommended for 900 V maximum applications. With this converter topology, the highest achievable DVUF is almost 100 %. Although a large number of IGBTs are required for a 19-level converter, the total semiconductor cost is about AUD 36,670.00, which is only about 42 % of a 9-level converter with the same power rating. Each H-bridge inverter cell in cascaded multilevel converter generates output voltage as 2-level converter. Figure 7.38 shows the voltage waveform of an H-bridge inverter cell in a 19-level 11-kV converter. There are 18 and 36 steps in the peak-to-peak phase and line voltage waveforms, and each step contributes 899 V to the phase or line voltage. Figure 7.39 shows the phase voltage waveform of a 19-level 11-kV converter. Figure 7.40 plots the line voltage waveform of a 19-level MMC converter. Due to the small step size, the output power quality is suitable for filter-less grid connection. The frequency spectrum of the line voltage of a 19-level MMC converter is depicted in Fig. 7.41. Figure 7.42 shows the three line voltages of the three-phase 19-level 11-kV

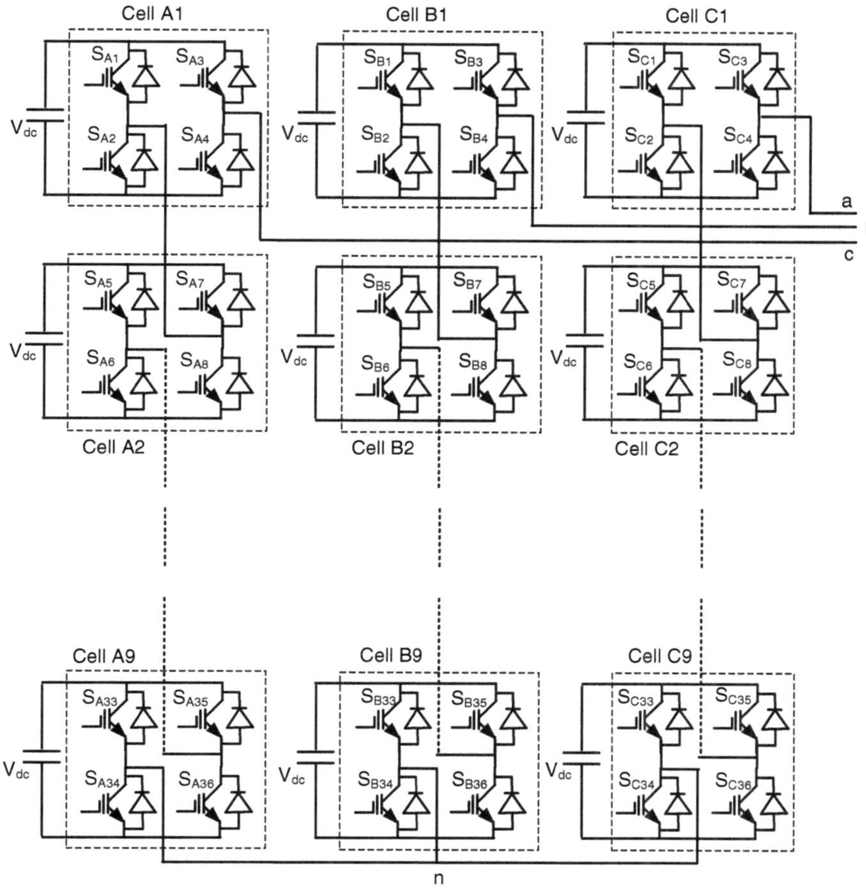

Fig. 7.36 Three-phase 19-level MMC converter [1]

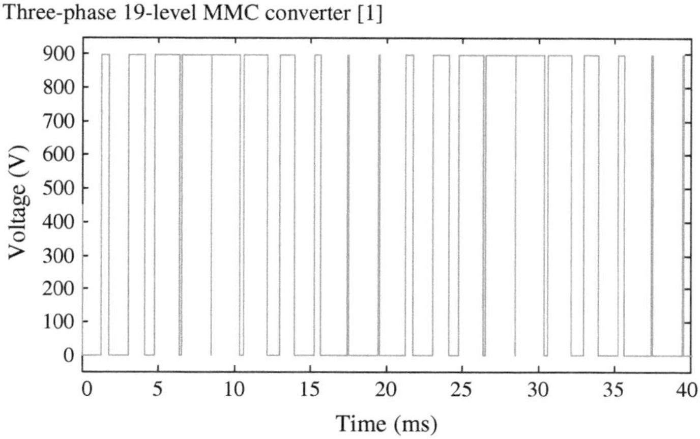

Fig. 7.37 Voltage waveform across an IGBT in 19-level MMC converter

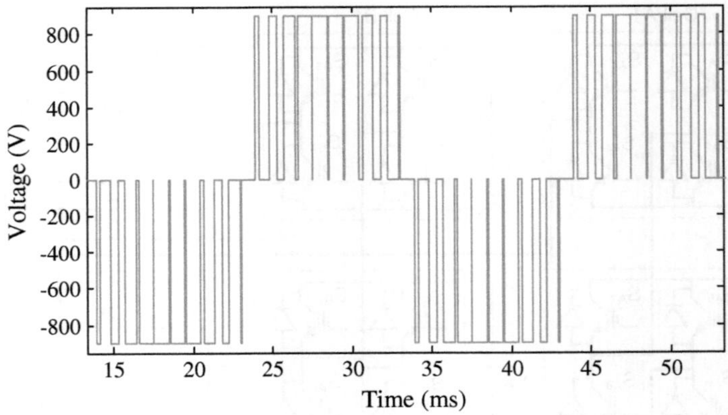

Fig. 7.38 Voltage waveform of an H-bridge inverter cell in 19-level MMC converter

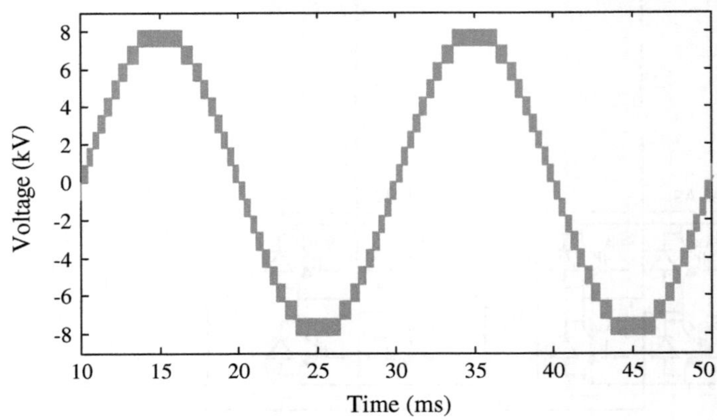

Fig. 7.39 Phase voltage of 19-level MMC converter

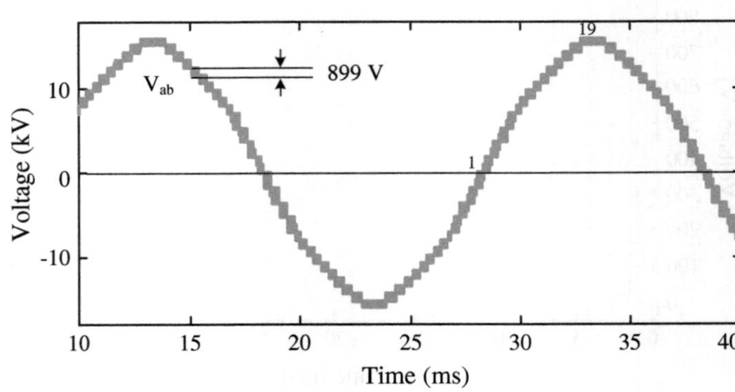

Fig. 7.40 Line voltage of 19-level MMC converter

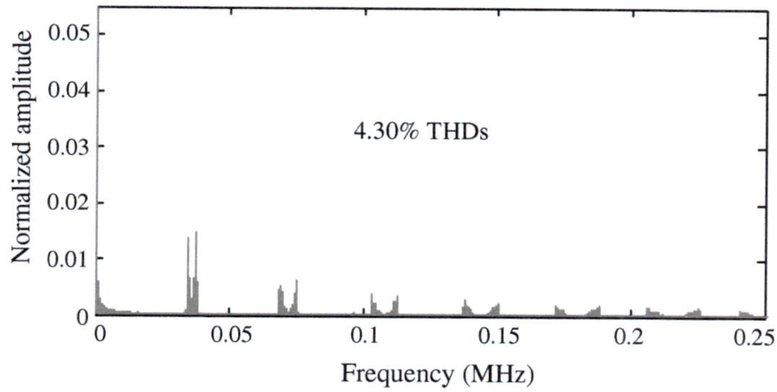

Fig. 7.41 Line voltage frequency spectrum of 19-level MMC converter

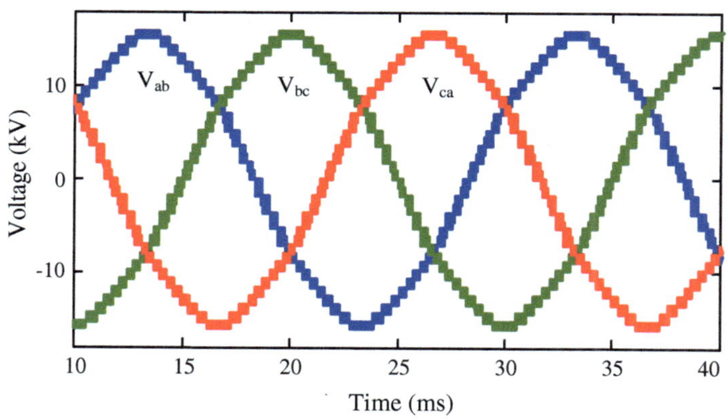

Fig. 7.42 Three line voltages of three-phase 19-level MMC converter

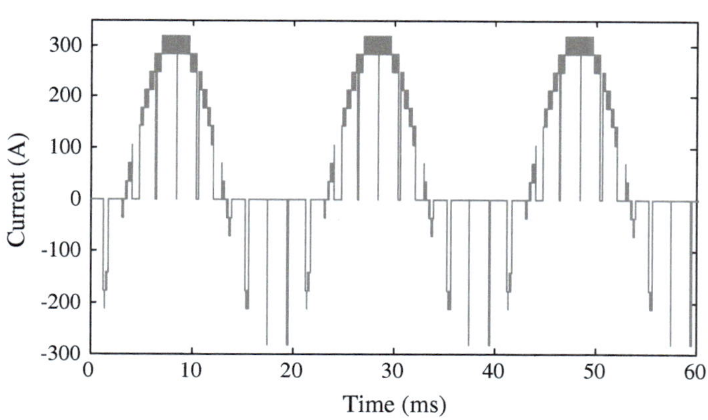

Fig. 7.43 Current waveform in an IGBT of 19-level MMC converter

Table 7.6 Commercially available 1.7-kV-rated IGBT modules

IGBT modules	Rated current (A)	Brand
CM500HA-34A	500	Mitsubishi Electric
SKM 600GA176D	400	Semikron
SKM 800GA176D	600	Semikron
5SNE 0800M170100	800	ABB
FZ400R17KE4	400	Infineon
FZ600R17KE4	600	Infineon
FZ1200R17HE4	1,200	Infineon
CM600DY-34H	600	Mitsubishi Electric

Fig. 7.44 Semikron 1.7 IGBT module SKM 600GA176D

converter. The current waveform of an IGBT in the 19-level 11-kV converter is shown in Fig. 7.43. There are a few vendors, such as Semikron, ABB, Mitsubishi Electric, and Infineon, who commercially developed 1.7-kV-rated IGBT modules. Table 7.6 summarizes the possible 1.7-kV IGBT modules for designing the 19-level 11-kV converter. Figure 7.44 shows the photograph of Semikron 1.7 IGBT module SKM 600GA176D, whose current rating is 400 A.

7.2.5 *Design and Analysis of 21-Level 11-kV Converter*

The DC-link voltage rating of an 11-kV 21-level converter is about 809 V. The 1.2-kV IGBT or lower-rated devices cannot be used and at least 1.7-kV IGBT is required to design the 11-kV 21-level converter. With this converter topology,

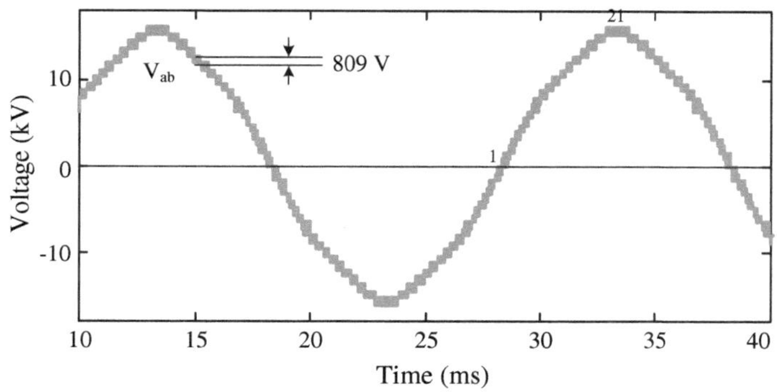

Fig. 7.45 Line voltage of 21-level MMC converter

the highest achievable DVUF is 90 %, which is much lower than that of the 19-level converter. In total, 120 IGBTs are used in a 21-level converter, where 10 H-bridge inverter cells are cascaded on each phase leg. The total semiconductor cost is about AUD 40,744.00, which requires about 11 % more cost than that of the 19-level converter. Figure 7.45 shows the line voltage waveform of a 21-level converter, and its frequency spectrum is illustrated in Fig. 7.46. The THDs are calculated and found to be 4.25 %, which means only 1.1 % improvement compared with the 19-level converter. The control scheme consists of 110 ALOs to drive 120 active switching devices. The control complexity of an 11-kV 21-level converter is about 11 % more than that of a 19-level converter. Figure 7.47 shows the three line voltages of three-phase 21-level 11-kV converter.

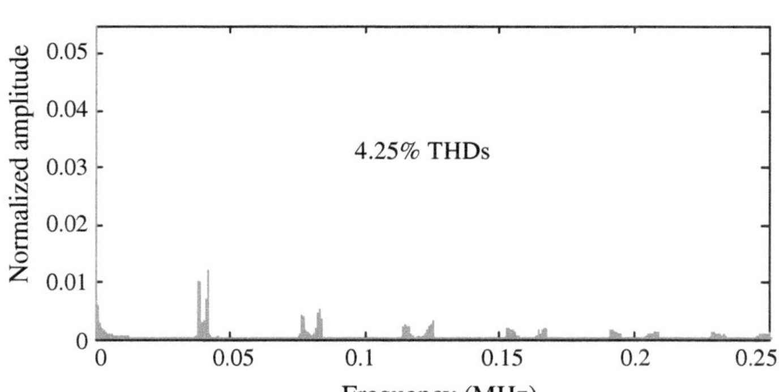

Fig. 7.46 Line voltage frequency spectrum of 21-level MMC converter

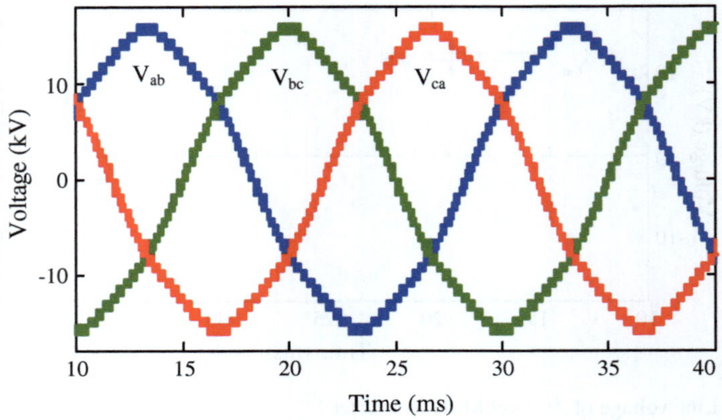

Fig. 7.47 Three line voltages of 21-level MMC converter

7.2.6 Selection of Number of Levels for 11-kV Converter Systems

The number of ALOs for the switching section and cost of semiconductors are calculated as tabulated in Table 7.7. The number of ALOs is used to compare the complexity of the converters. The THDs are calculated through the MATLAB/Simulink environment. The price data quoted for the semiconductor devices were collected from the Galco Industrial Electronics and Farnell catalogs, and quotations from different vendors where devices were chosen from the same family so that it was possible to meet the requirements.

Using (3-29), the normalized index values are calculated to obtain the overall performance/cost of the converters. Table 7.8 tabulates the normalized index values of Table 7.7. Based on Table 7.8, the performance indicators are plotted as shown in Fig. 7.48. For the 11-kV converter, the total index value is the lowest

Table 7.7 Converter comparison for an 11-kV system [9, 10]

No. of levels	9	11	15	19	21
IGBTs	48	60	84	108	120
THDs (%)	9.60	8.20	6.00	4.30	4.25
Cost (AUD)	86,400	82,159	47,066	36,670	40,744
ALOs	44	55	77	99	110

Table 7.8 Performance indexes of 11-kV converters [9, 10]

No. of levels	9	11	15	19	21
THD	1.00	0.74	0.33	0.01	0.00
IGBTs cost	1.00	0.92	0.21	0.00	0.09
Complexity	0.00	0.17	0.50	0.83	1.00
Total index	2.00	1.83	1.04	0.84	1.09

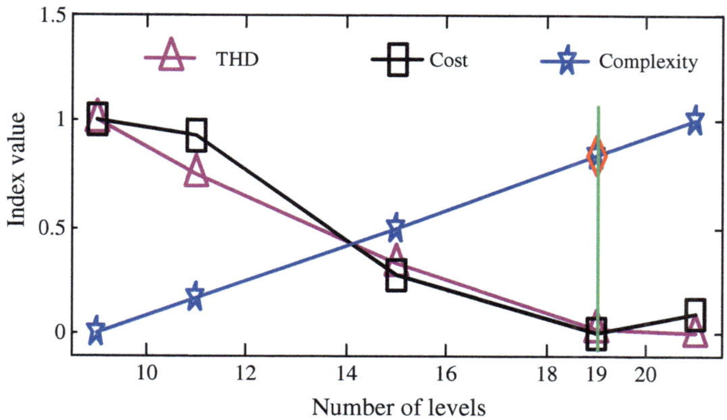

Fig. 7.48 Graphical representation converter performance parameters [9, 10]

at level 19, because there is no significant output power quality improvement and semiconductor cost reduction for inverters with more than 19 levels. Moreover, the component number and control complexity increase linearly with the increase in the number of levels. Therefore, the 19-level topology is the optimum for the 11-kV converter systems. The output power quality of a 19-level converter is good enough to feed the converter output directly into the medium-voltage grid directly (i.e., without using step-up transformer or line filter circuit). Figure 7.49 shows the total indexes of different 11-kV converters. Lower total index value means lower converter cost and complexity and better quality of output power.

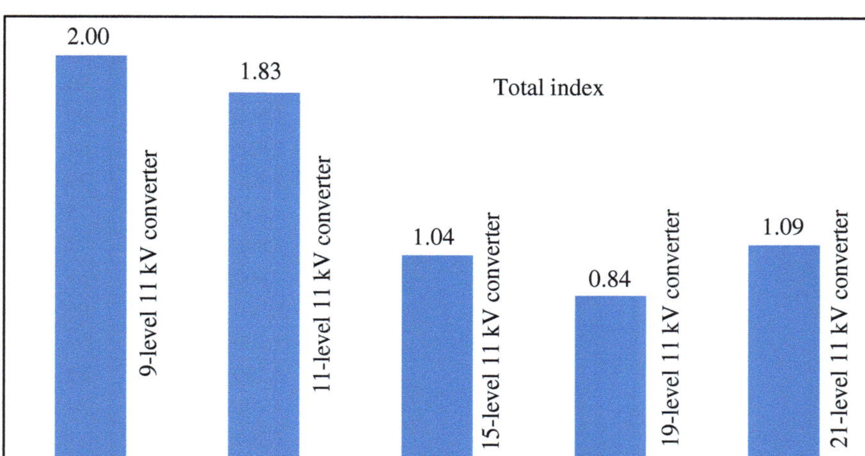

Fig. 7.49 Total performance indexes of different 11-kV MMC converters

7.3 Design and Analysis of 33-kV Converter Systems

The total nominal DC-link voltage for a 33-kV 2-level converter is 48,538 V. Each H-bridge inverter cell communication voltage of a 13-level converter is about 4,044 V. The highest voltage rating of a commercially available IGBT is 6.5 kV, which is recommended for a maximum voltage of 3,600 V. Therefore, the converter should have 13 or more levels for direct connection to a 33-kV grid. Each H-bridge inverter cell communication voltage of a 15-level topology-based 33-kV converter is 3,467 V which may be supported by the 6.5-kV IGBT. Accordingly, at least 15-level topology is required to design a 33-kV converter. While the output power quality of 55-level converter is good enough to feed into the 33-kV AC grid directly, the cheap 1.7-kV IGBT can be used to construct the 55-level converter. Therefore, MMC topologies 15 level to 55 level are considered for a 33-kV converter system. The DVUFs are tabulated for converters of different level numbers in Table 7.9.

A high DVUF is essential for a cost-effective design, since the semiconductor cost is the significant figure in MV inverter applications. From Table 7.9, it can

Table 7.9 DVUFs of 33-kV MMC converters of different levels

No. of levels	V_{com} (V)	Rated device voltage (V)	$V_{com@100FIT}$ (V)	DVUF (%)
15	3,467	6,500	3,600	96
17	3,034	6,500	3,600	85
19	2,697	6,500	3,600	75
21	2,427	6,500	3,600	68
23	2,206	4,500	2,250	98
25	2,022	4,500	2,250	90
27	1,867	4,500	2,250	83
29	1,734	3,300	1,800	96
31	1,618	3,300	1,800	90
33	1,517	3,300	1,800	84
35	1,428	3,300	1,800	79
37	1,348	3,300	1,800	75
39	1,277	3,300	1,800	71
41	1,214	3,300	1,800	65
43	1,156	2,500	1,200	96
45	1,103	2,500	1,200	92
47	1,055	2,500	1,200	88
49	1,011	2,500	1,200	84
51	971	2,500	1,200	81
53	934	2,500	1,200	78
55	899	1,700	900	100

be seen that only some converters have high DVUFs. In order to ensure a cost-effective design, the converters with level numbers of 15, 23, 29, 43, and 55 for a 33-kV system were considered for further analysis in the following sections. The converters with DVUFs of more than 95 % are assumed to satisfactorily utilize the active switching devices.

7.3.1 Design and Analysis of 15-Level 33-kV Converter

The 15-level converter topology may be considered for designing a 33-kV converter with the available commercial semiconductors. Each phase leg consists of seven H-bridge inverter cells with 28 IGBTs in a phase and 84 in total in the 3-phase circuit. The DC supply voltage for a 33-kV 15-level converter is 3,467 V. The available 6.5-kV IGBT may be used for constructing the 33-kV 15-level converter, because this IGBT is recommended for application of 3,600 V in maximum. Figure 7.50 plots the phase voltage waveform of a 33-kV 15-level MMC converter. There are 28 steps in the peak-to-peak line voltage waveform, and each step contributes 3,467 V to the line voltage. Figure 7.51 shows the line voltage waveforms of the 15-level 33-kV converter. The frequency spectrum of the line voltage of a 15-level MMC converter is depicted in Fig. 7.52.

ABB, Infineon, and Mitsubishi Electric commercially developed 6.5-kV IGBT modules with various power ratings. Table 7.10 summarizes some possible 6.5-kV IGBT modules for constructing medium-voltage converters. ABB and Mitsubishi Electric are the market leaders to supply 6.5-kV IGBT module, who developed 6.5-kV IGBT modules with a current rating of a few hundred amperes. Figure 7.53 shows the ABB 6.5-kV IGBT module 5SNA 0400J650100.

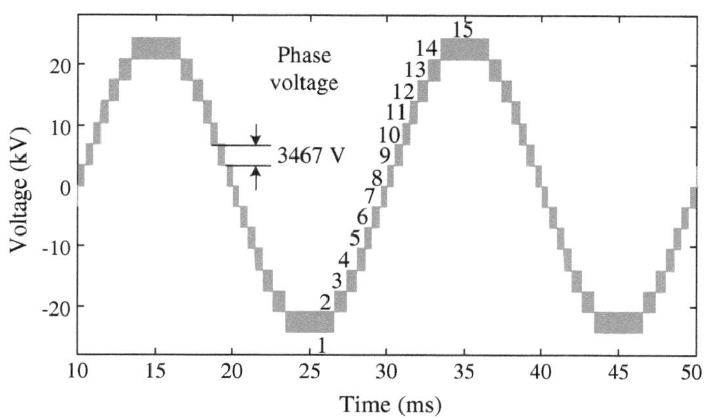

Fig. 7.50 Phase voltage of 33-kV 15-level MMC converter

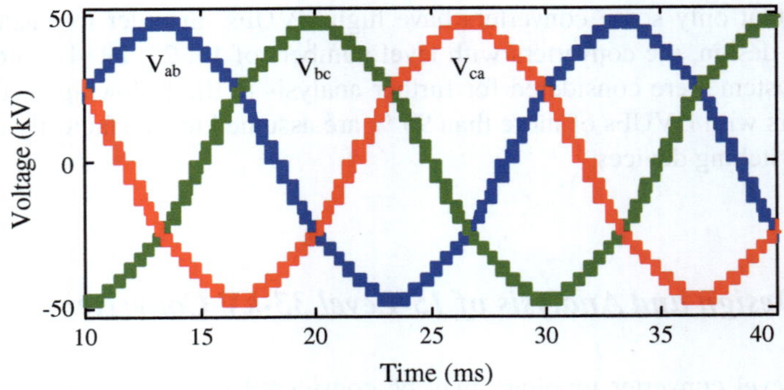

Fig. 7.51 Three line voltages of three-phase 33-kV 15-level MMC converter

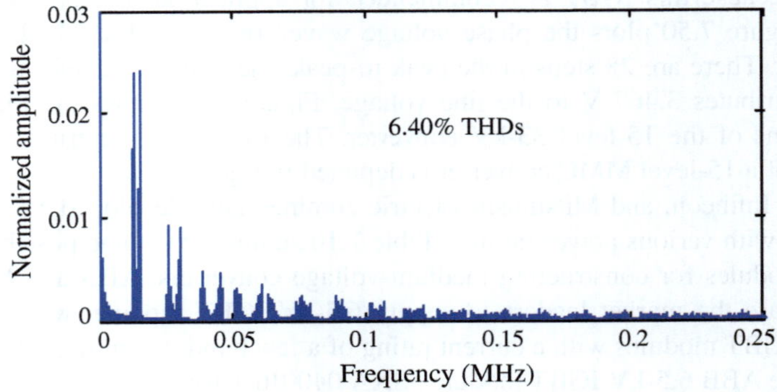

Fig. 7.52 Line voltage frequency spectrum of 33-kV 15-level MMC converter

Table 7.10 Commercially available 6.5-kV-rated IGBTs

IGBT modules	Rated current (A)	Brand
CM400HG-130H	400	Mitsubishi Electric
CM600HG-130H	600	Mitsubishi Electric
5SNA 0400J650100	400	ABB
5SNA 0750G650300	750	ABB
FZ400R65KE3	400	Infineon
FZ600R65KE3	600	Infineon

Fig. 7.53 Photograph of ABB 6.5-kV IGBT module 5SNA 0400J650100

7.3.2 Design and Analysis of 23-Level 33-kV Converter

The power circuit of a 23-level 3-phase MMC converter is shown in Fig. 7.54.

The H-bridge inverter DC-link voltage of a 33-kV 23-level converter is 2,206 V. The available 4.5-kV IGBT may be used to construct the 33-kV 23-level converter because this IGBT is recommended for application of 2,250 V in maximum. An about 98 % DVUF can be obtained with the 4.5-kV IGBTs. In total, 11 H-bridge inverter cells are in each phase leg. In order to drive such a converter, 11 phase-shifted carriers are required and adjacent carriers are shifted by (180/11) degrees. Figure 7.55 shows the phase-shifted carrier-based switching control scheme for the 23-level MMC converter. The scheme involves about 121 ALOs. The phase voltage waveform of the 23-level 33-kV converter is shown in Fig. 7.56. The line voltage waveforms of the 23-level 33-kV converter are depicted in Fig. 7.57. The line peak-to-peak voltage consists of 44 voltage levels, and each level contributes 2,206 V to the peak-to-peak line voltage. The line voltage contains 4.54 % THDs, which moderately satisfy the grid requirements. The frequency spectrum of the 23-level 33-kV converter is illustrated in Fig. 7.58.

The cost of 132 IGBTs is about AUD 237,600.00, which is about 8.8 % lower than that of a similar 15-level converter. Compared with a 15-level 33-kV converter, about 29 % improvement of THDs is obtained.

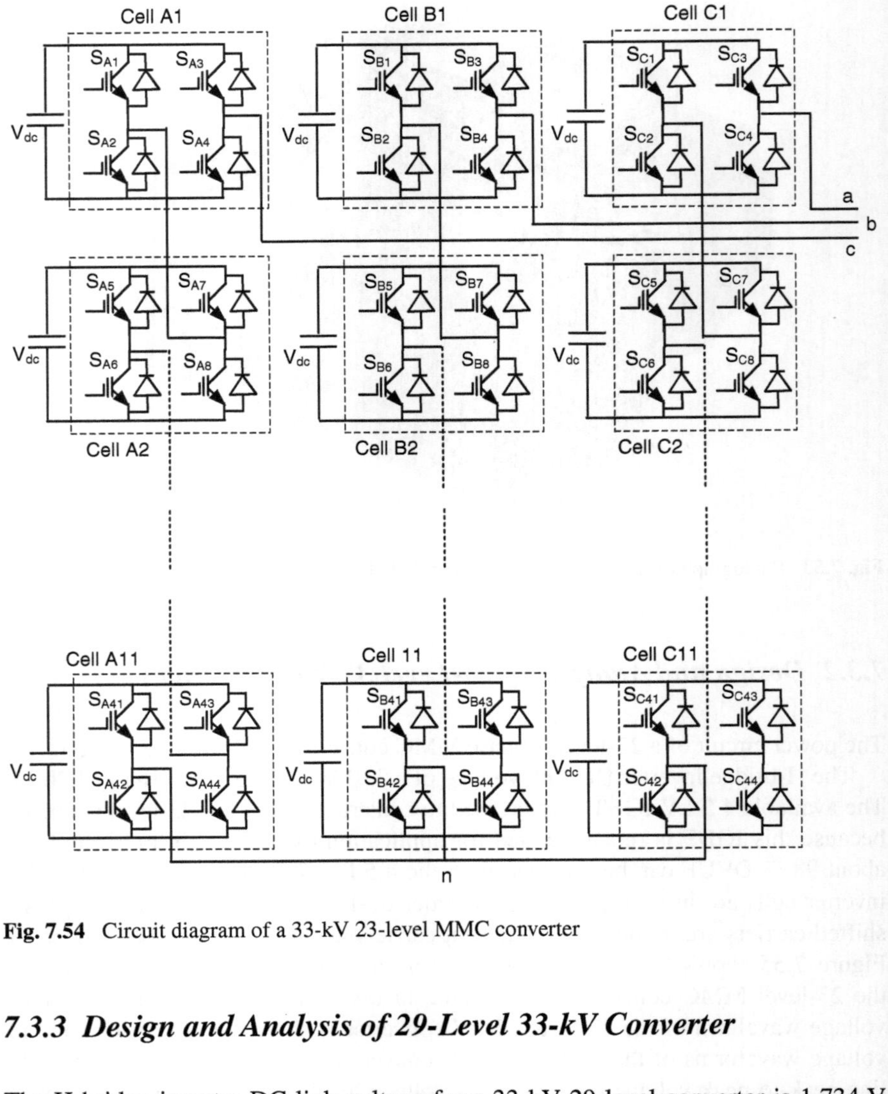

Fig. 7.54 Circuit diagram of a 33-kV 23-level MMC converter

7.3.3 Design and Analysis of 29-Level 33-kV Converter

The H-bridge inverter DC-link voltage for a 33-kV 29-level converter is 1,734 V. The available 3.3-kV IGBT may be used to construct the 33-kV 29-level converter, because this IGBT is recommended for 1,800 V maximum applications. About 96 % DVUF can be obtained with the 3.3-kV IGBTs. The device's dv/dt is about 50 % lower than that of the devices in the 15-level converter, under the same operating conditions. About 50 % lower dv/dt means significant improvement in the device reliability. In total, there are 14 H-bridge inverter cells in each phase leg. In order to drive such a converter, 14 phase-shifted carriers are required and adjacent carriers are shifted by (180/14) degrees. The phase voltage and line voltage waveforms of the 29-level 33-kV converter are depicted in Figs. 7.59 and 7.60.

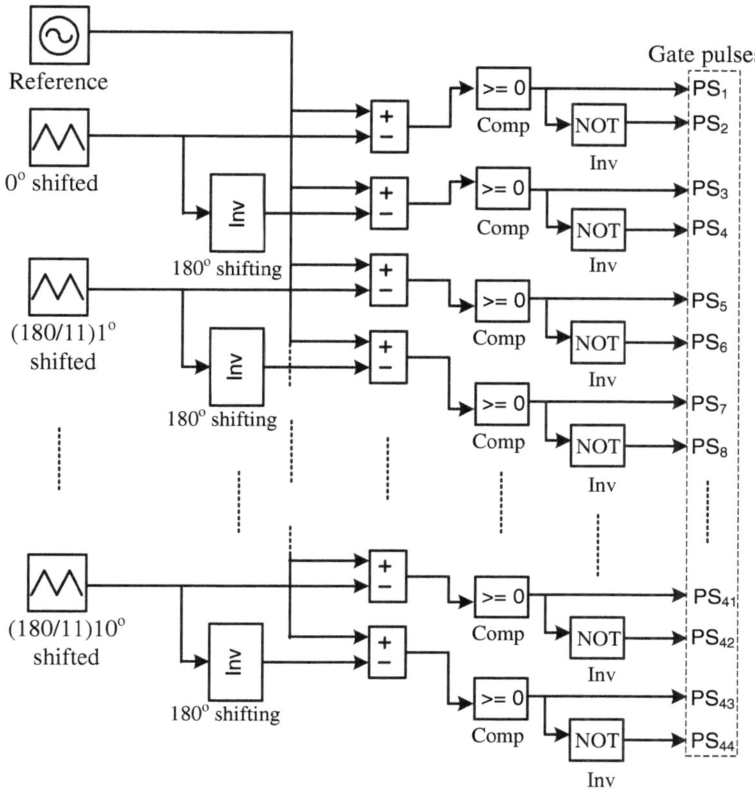

Fig. 7.55 Phase-shifted carrier-based switching scheme for 23-level MMC converter

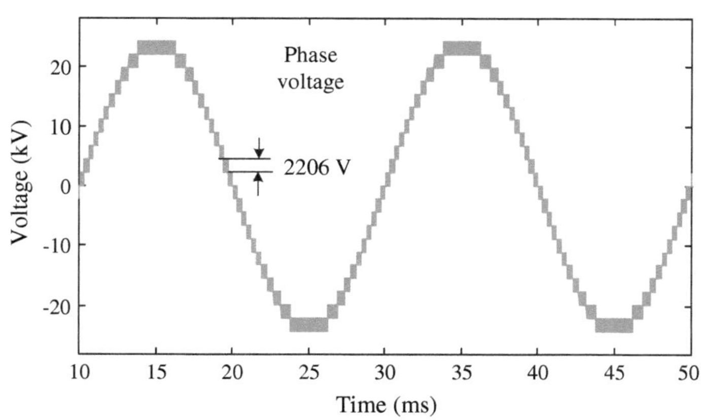

Fig. 7.56 Phase voltage of 23-level MMC converter

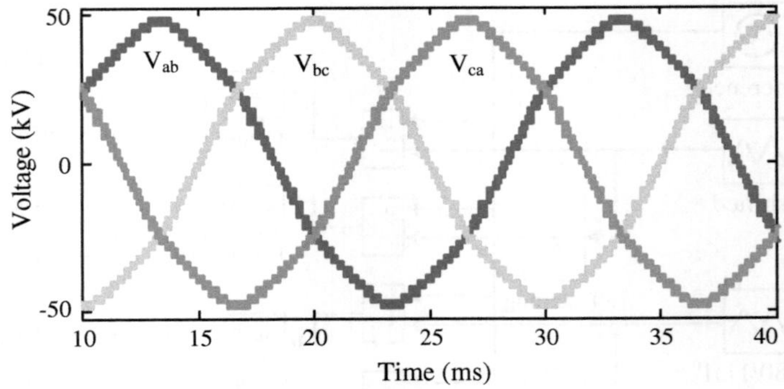

Fig. 7.57 Line voltages of 23-level MMC converter

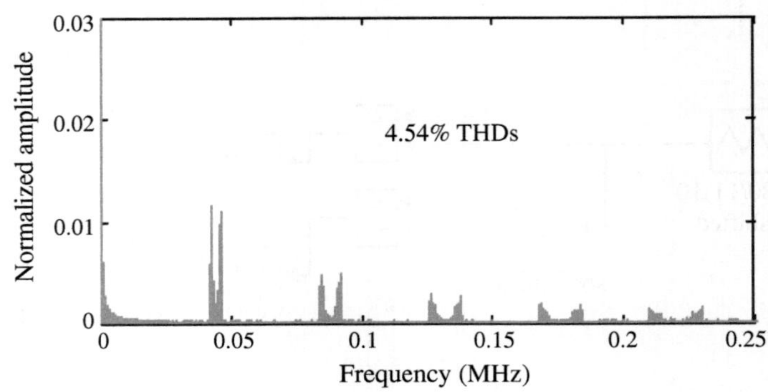

Fig. 7.58 Line voltage frequency spectrum of 23-level MMC converter

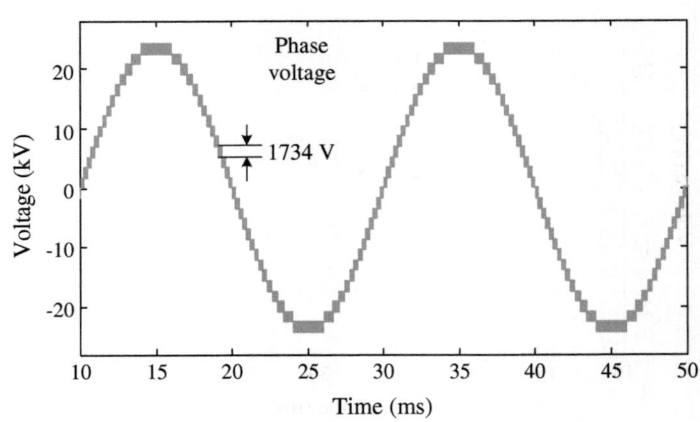

Fig. 7.59 Phase voltage of 29-level MMC converter

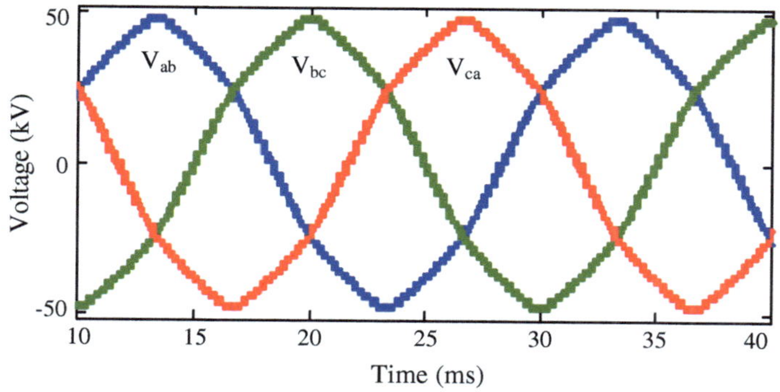

Fig. 7.60 Line voltages of 29-level MMC converter

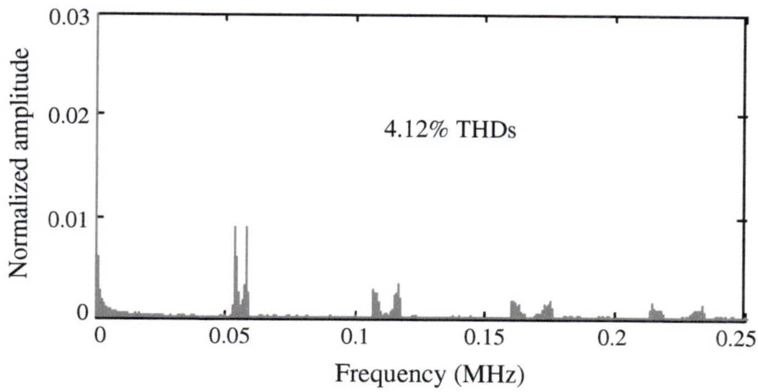

Fig. 7.61 Line voltage frequency spectrum of 29-level MMC converter

The line peak-to-peak voltage consists of 56 voltage levels, and each level contributes 1,734 V to the peak-to-peak line voltage. The line voltage contains 4.12 % THDs, which satisfies the grid requirements and may enable line filter-less grid connection. Figure 7.61 plots the frequency spectrum of the 29-level 33-kV converter. Compared with the 23-level converter, about 9 % additional harmonic reduction can be achieved with the 29-level converter. The switching scheme involves about 154 ALOs, which is 27 % more complex than that of a 23-level converter control scheme. The cost of 168 IGBTs is about AUD 229,992.00, which is about 12 % lower than that of a similar 15-level converter using IGBTs of one level higher voltage rating.

7.3.4 Design and Analysis of 43-Level 33-kV Converter

The H-bridge inverter DC-link voltage for a 33-kV 43-level converter is 1,156 V. The available cheap and matured 2.5-kV IGBT can be used for constructing the 33-kV 43-level converter, because this IGBT is recommended for application of

1,200 V in maximum. About 96 % DVUF can be obtained with the 2.5-kV IGBTs. In total, there are 21 H-bridge inverter cells in each phase leg and 252 active switching devices are required for the 3-phase 43-level converter. Figure 7.62 shows the circuit diagram of 43-level converter. In order to drive 252 IGBTs, 21 phase-shifted carriers are required and adjacent carriers are shifted by (180/21) degrees. Figure 7.63 shows the block diagram of the switching control scheme for the 3-phase 43-level converter.

A total of 231 ALOs are involved with the switching scheme. Although three times more switching devices are used in the 43-level converter, the semiconductor cost is about 41 % lower than that of the 15-level converter using IGBTs of one level higher voltage rating, due to the cheap cost of low-voltage-rated devices. The line peak-to-peak voltage consists of 84 voltage levels, and each level contributes

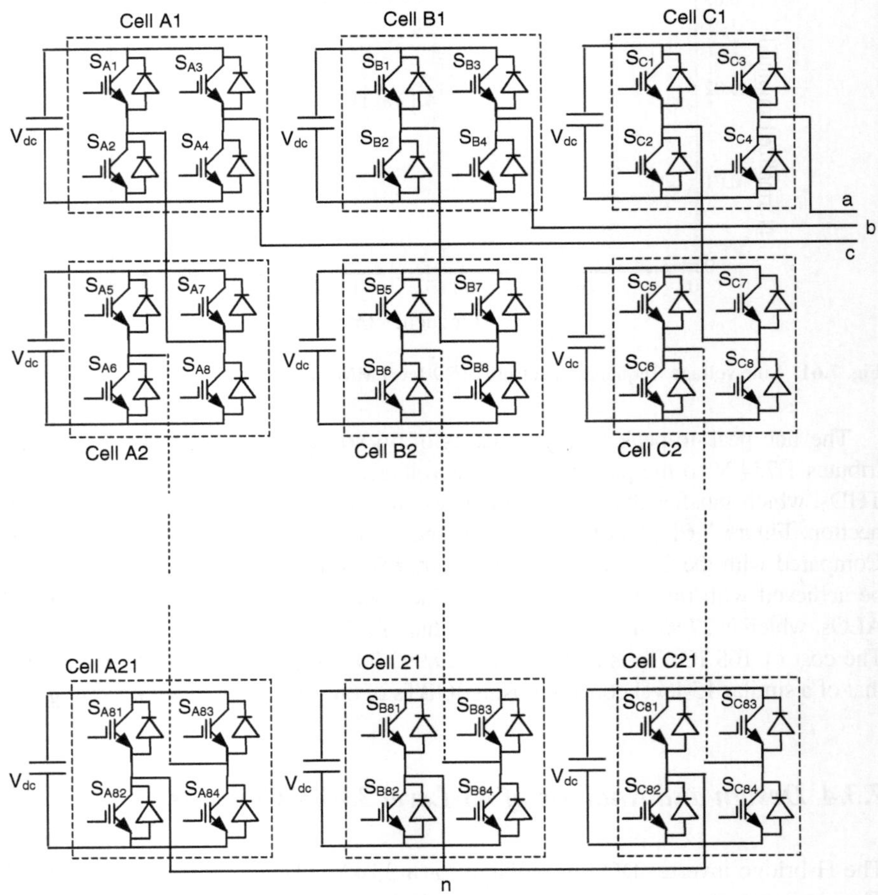

Fig. 7.62 Circuit diagram of a 43-level MMC converter

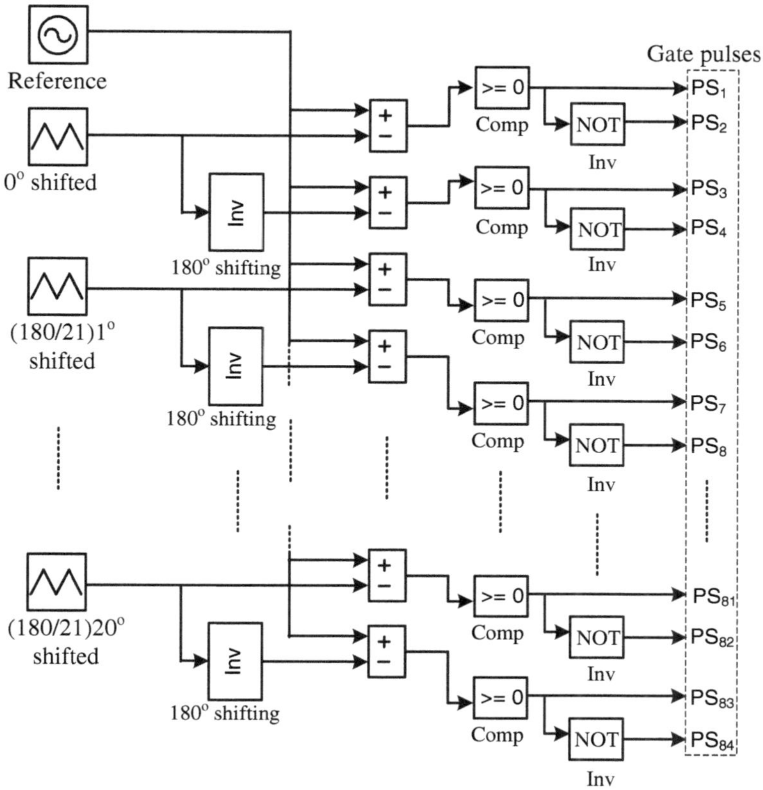

Fig. 7.63 Switching control scheme of 43-level MMC converter

1,156 V to the peak-to-peak line voltage. Figure 7.64 shows voltage waveform across an IGBT in a 43-level 33-kV converter. Each H-bridge inverter cell produces two voltage levels as in a 2-level converter. Figure 7.65 shows the output voltage waveform of an H-bridge inverter cell in the 43-level 33-kV converter. Figure 7.66 shows the phase voltage of 43-level 33-kV converter. Due to the small step size, the line voltage waveforms are found to be very consistent with the reference sine waveforms. Figure 7.67 plots the line voltage waveform. The output power quality of a 43-level converter is good enough to feed directly into the grid. The line voltage of THDs is about 3.61 %, which strongly satisfies the 5 % limit by IEEE1547 and IEC61727 standards. The frequency spectrum of line voltage is depicted in Fig. 7.68. Compared with the 15-level converter, the 43-level converter provides 44 % better quality output power. As the number of active switching devices has increased three times, the complexity of the control scheme of a 43-level converter has also increased three times more than that of a 15-level converter. Current through an IGBT is shown in Fig. 7.69.

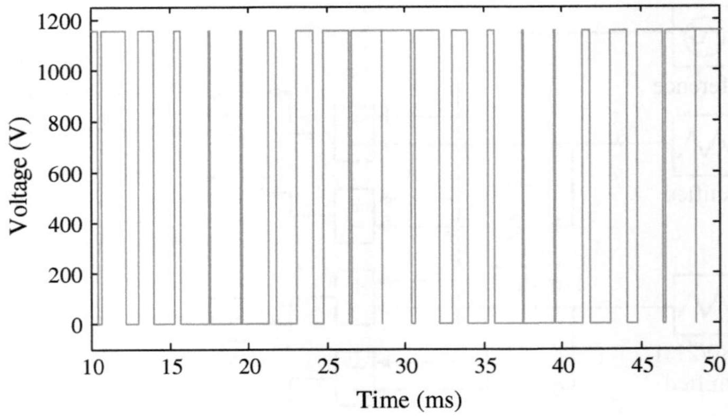

Fig. 7.64 Voltage across an IGBT in 43-level 33-kV MMC converter

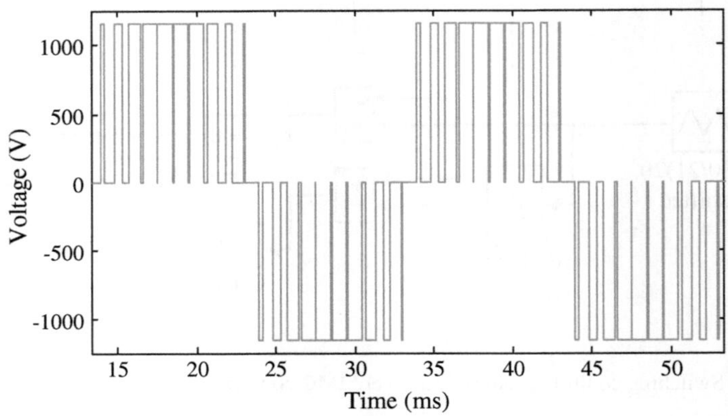

Fig. 7.65 Output voltage of an H-bridge inverter cell in 43-level 33-kV MMC converter

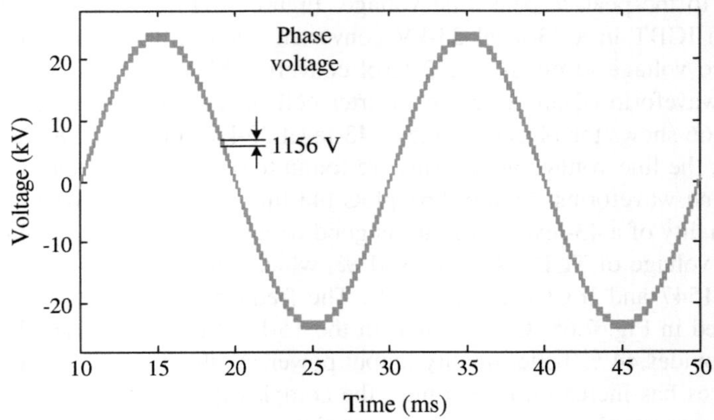

Fig. 7.66 Phase voltage of 43-level MMC converter

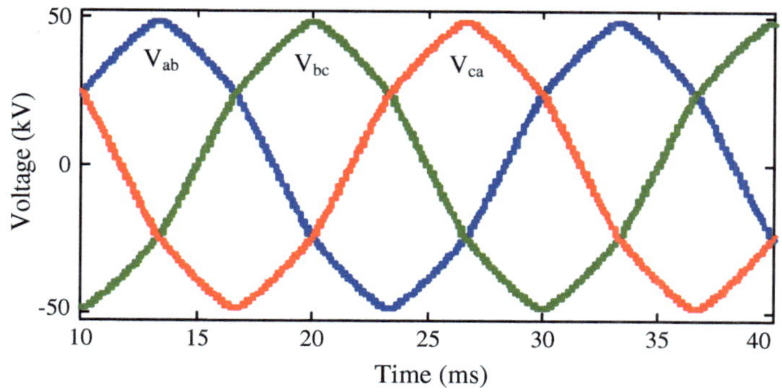

Fig. 7.67 Line voltages of 43-level MMC converter

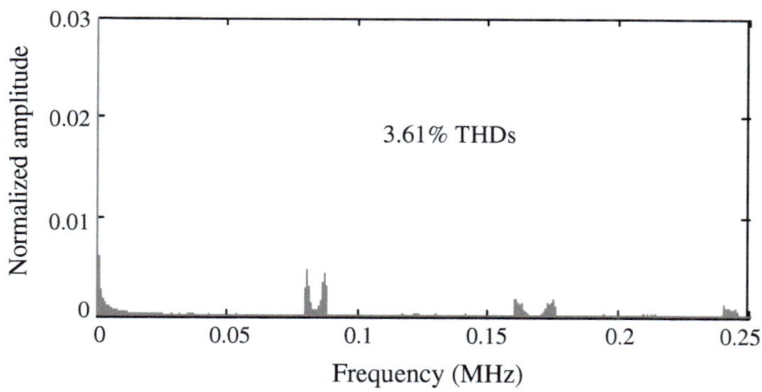

Fig. 7.68 Line voltage frequency spectrum of 43-level MMC converter

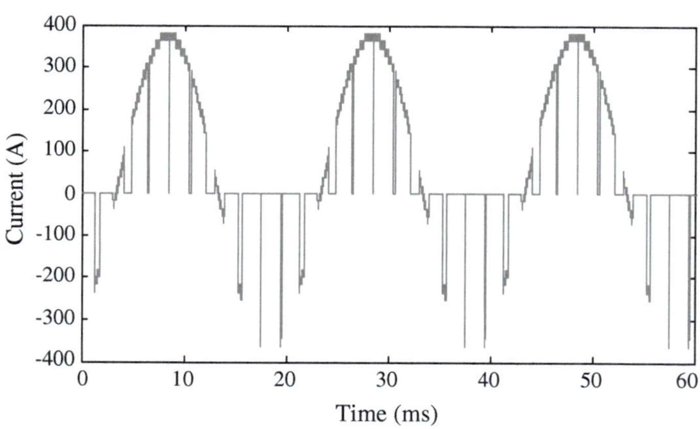

Fig. 7.69 IGBT current of 43-level MMC converter

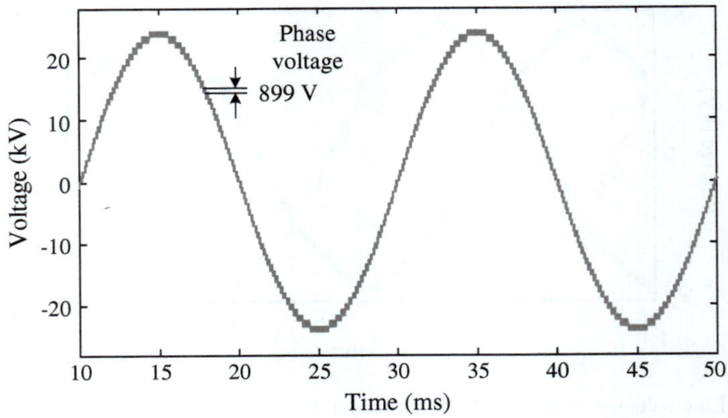

Fig. 7.70 Phase voltage of 55-level MMC converter

7.3.5 Design and Analysis of 55-Level 33-kV Converter

The H-bridge inverter DC-link voltage for a 33-kV 55-level converter is 899 V. The available cheap and mature 1.7-kV IGBT can be used to design the 33-kV 55-level converter, because this IGBT is recommended for application of 900 V in maximum. An almost 100 % DVUF can be obtained with the 1.7-kV IGBTs. In total, there are 27 H-bridge inverter cells in each phase leg and 324 active switching devices are required for the 3-phase 55-level converter. The 1.7-kV IGBT is cheap, and the total semiconductor cost of the 55-level converter is about AUD 110,030.00, which is 22 % lower than that of the 43-level converter using IGBTs of one level higher voltage rating. The IGBTs price of 2.5 kV or lower is in linear

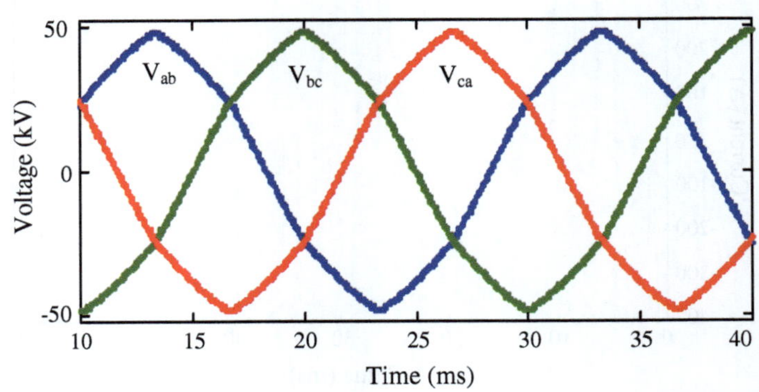

Fig. 7.71 Line voltages of 55-level MMC converter

range. Therefore, the reduction to the total price is small. The line peak-to-peak voltage consists of 108 voltage levels, and each level contributes 899 V to the peak-to-peak line voltage. Due to the small step size, the line voltage waveforms are found to be very consistent with the reference sine waveforms. Figures 7.70 and 7.71 plot the phase and line voltage waveforms, respectively. The line voltage frequency spectrum is shown in Fig. 7.72. The 55-level 33-kV voltage waveform contains about 3.47 % THDs, which is only 3.8 % lower than that of the 43-level converter. However, the switching scheme of the 55-level converter consists of 297 ALOs. Compared with the 43-level converter, the control scheme of the 55-level converter requires 66 more ALOs; that is, the control scheme is 28 % more complex.

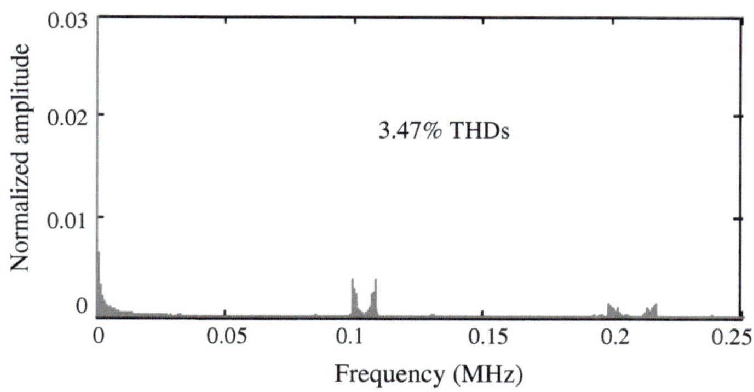

Fig. 7.72 Line voltage frequency spectrum of 55-level MMC converter

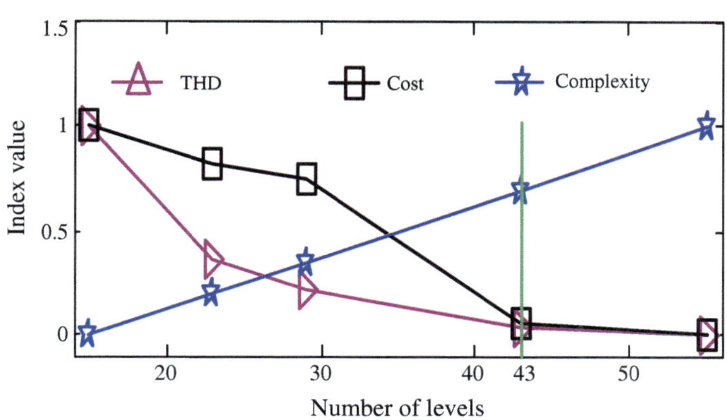

Fig. 7.73 Graphical representation converter parameters [10, 21]

7.3.6 Selection of Optimal Number of Levels for 33-kV Converter

The H-bridge inverter cell communication voltage of a 13-level converter is about 4,044 V. The highest voltage rating of a commercially available IGBT is 6.5 kV, which is recommended for application of a maximum voltage of 3,600 V. Therefore, a 13-level or lower-level converter cannot be used to construct the 33-kV converter. Each H-bridge inverter cell communication voltage of a 15-level topology-based 33-kV converter is 3,467 V which may be supported by the 6.5-kV IGBT. Owing to this, at least 15-level topology is required to design a 33-kV converter. The output power quality of a 55-level inverter is good enough to directly feed into the 33-kV AC grid. The cheap 1.7-kV IGBT can be used to design the 55-level inverter. There is no significant performance improvement or cost reduction with converters of more than 55 levels. Moreover, the control complexity increases with the number of converter levels. For this reason, 15-level to 55-level MMC converter topologies are considered for a 33-kV inverter system.

The ALOs for switching section, THDs of output power, and cost of semiconductors are calculated as tabulated in Table 7.11. The number of ALOs is used to compare the complexity of the converters. The THDs are calculated in the MATLAB/Simulink environment. Using (3-29), the normalized index values are calculated. Table 7.12 summarizes the normalized index values of Table 7.11. Based on Table 7.12, the performance graphs are plotted in Fig. 7.73. For the 33-kV converter, the total index value is the lowest at the 43 level, because there is no significant output power quality improvement and semiconductor cost reduction for converters with more than 43 levels. In addition, the component number and control complexity increase linearly with the increase in the number of levels. Therefore, 43-level topology is considered as the optimal for 33-kV converter systems. Figure 7.74 shows total indexes of different 33-kV MMC converters.

Table 7.11 Converter comparison for a 33-kV system [10, 21]

No. of levels	15	23	29	43	55
IGBTs	84	132	168	252	324
THD (%)	6.40	4.54	4.12	3.61	3.47
Cost (AU$)	258,552	237,600	229,992	141,200	110,030
ALOs	77	121	154	231	297

Table 7.12 Total indexes of different 33-kV MMC converters [10, 21]

No. of levels	15	23	29	43	55
Performance	1.00	0.36	0.22	0.04	0.00
IGBTs cost	1.00	0.85	0.80	0.20	0.00
Complexity	0.00	0.20	0.35	0.70	1.00
Total index	2.00	1.41	1.37	0.94	1.00

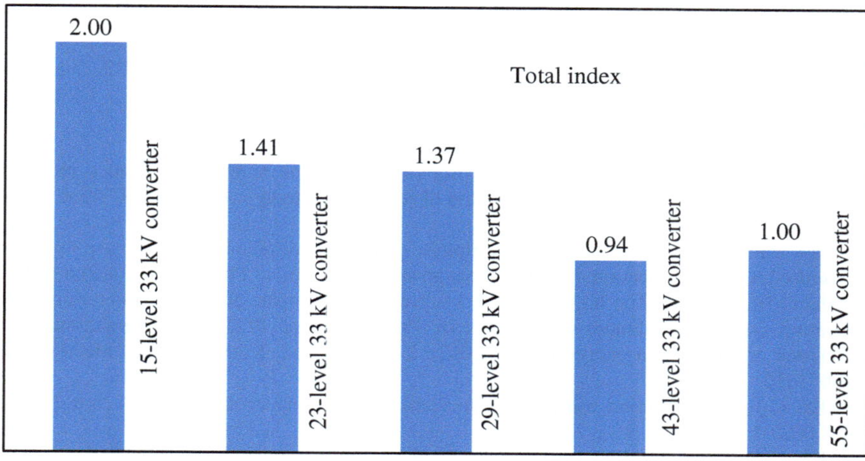

Fig. 7.74 Total indexes of different 33-kV converters

7.4 Summary

In this chapter, the designs of 11-kV and 33-kV converter systems have been analyzed with a focus on selecting the optimal number of levels for a medium-voltage converter. All possible MMC converter topologies have been considered for each converter system. In order to ensure a cost-effective design, the DVUFs were calculated and only the selected converters with high DVUFs were considered. During the design process, the availability of semiconductor devices was considered in the first instance. After checking the availability of devices, the converter systems were designed taking into account three main factors: the specified converter output power quality, complexity of the switching controller, and cost of the semiconductors. The investigation has shown that the 19-level MMC converter is the optimal choice for an 11-kV converter system and the 43-level topology is the optimal choice for a 33-kV inverter system.

References

1. Islam MR, Guo YG, Zhu JG (2014) A review of offshore wind turbine nacelle: technical challenges, and research and developmental trends. Renew Sustain Energy Rev 33:161–176
2. Islam MR, Guo YG, Zhu JG (2013) Power converters for wind turbines: current and future development. In: Mendez-Vilas A (eds) Materials and processes for energy: communicating current research and technological developments. Energy Book Series-2013 edn. Spain, pp 559–571
3. Solutions-SEMISTACK Renewable Energy (2011) Three phase inverter. Catalogue 2011/2012, Semikron International, May 2011, pp 94–95
4. Solutions-SEMIKUBE (2011) Three phase rectifier and inverter. Catalogue 2011/2012, Semikron International, May 2011, pp 89–90

5. Mitsubishi Electric, Mitsubishi IGBT Modules CM100MXA-24S, [Online]. Available at: http://www.ineltron.de/english/mitsubishi-data/IGBT/CM100MXA-24S.pdf
6. Solutions-SKiiPRACK (2011) Four-quadrant converter. Catalogue 2011/2012, Semikron International, May 2011, pp 88–89
7. See http://www.abb.com/ for information about ABB central inverter
8. See http://www.siemens.com/ for information about SINVERT PVS
9. Islam MR, Guo YG, Zhu JG (2014) A high-frequency link multilevel cascaded medium-voltage converter for direct grid integration of renewable energy systems. IEEE Trans Power Electron 29(8):4167–4182
10. Islam MR, Guo YG, Zhu JG (2014) A multilevel medium-voltage inverter for step-up-transformer-less grid connection of photovoltaic power plants. IEEE J Photovoltaics 4(3):881–889
11. Islam MR, Guo YG, Zhu JG (2012) A Transformer-less compact and light wind turbine generating system for offshore wind farms. In: Proceedings of 2012 IEEE international conference on power and energy (PECon2012), 2–5 Dec 2012, Kota Kinabalu, Malaysia, pp 605–610
12. ABB (2013) Distribution transformer. Available online: http://www.coronabd.com, [access 18.03.2013]
13. ABB (2013) Liquid-filled transformer. Available online: http://www05.abb.com, [access 18.03.2013]
14. Pauwels SLIM Transformer (2013) SLIM transformer inside the world's highest wind turbine. Available online: http://www.pauwels.com, [access 18.02.2013]
15. Gerdes G, Tiedemann A, Zeelenberg S. Case study: a survey for the analysis of the experiences and lessons learned by developers of offshore wind farms. Available online: http://www.offshore-power.net, [access 10.03.2013]
16. Islam MR, Guo YG, Zhu JG (2013) A medium-frequency transformer with multiple secondary windings for medium-voltage converter based wind turbine generating systems. J Appl Phys 113(17):17A324-1–17A324-3
17. Islam MR, Guo YG, Zhu JG (2011) Performance and cost comparison of NPC, FC and SCHB multilevel converter topologies for high-voltage applications. In: 2011 international conference on electrical machines and systems, Beijing, China, 20–23 Aug 2011, pp 1–6
18. Islam MR, Guo YG, Zhu JG, Dorrell D (2011) Design and comparison of 11 kV multilevel voltage source converters for local grid based renewable energy systems. In Proceedings of 37th annual conference of the IEEE Industrial Electronics Society (IECON 2011), 7–10 Nov 2011, Melbourne, Australia, pp 3596–3601
19. Islam MR, Guo YG, Jafari M, Malekjamshidi, Z, Zhu JG, (2011) A 43-level 33 kV 3-phase modular multilevel cascaded converter for direct grid integration of renewable generation systems. In IEEE innovative smart grid technologies conference (ISGT) Asia, 20–23 May 2014, Kuala Lumpur, Malaysia
20. Mitsubishi Electric, Powerex Power Semiconductor Solutions, "IGBT/HVIGBT," [Online] Available at: http://www.pwrx.com
21. Islam MR, Guo YG, Zhu JG (2014) Multilevel converters for step-up transformer-less direct grid integration of renewable generation units with medium voltage smart microgrids. In: Large scale renewable power generation: advances in technologies for generation, transmission and storage. Springer, Berlin, pp 127–149

Chapter 8
Conclusions and Future Works

Abstract This chapter concludes the book. In summary, this book has focused on the optimal design of medium-voltage power converters for step-up-transformer-less direct grid integration of renewable power generation systems and development of a scaled down 1 kV laboratory prototype test platform to describe the implementation process of medium-voltage converter topologies. The design and prototype development concept can be used to model the 11–33 kV system taking into account some additional safety and reliability issues. Further, a few directions have been recommended for future research and development.

8.1 Introduction

Distribution of wind and solar photovoltaic (PV) energy throughout the world should be first priority in solving our energy and environment crisis. Scientists all around the world are now seeking energy solution from different renewable resources and till today only wind and solar PV energy sources are found to be suitable for future large-scale generation. By generating electric power from our abundance renewable, such as wind and solar energy sources, we can solve a big portion of energy deficiency. Moreover, the energy sector of the world is growing curiously. It is facing an accelerating compound crisis of the globally established fossil fuels. Immediate different breakthroughs for wind and solar PV power generation technologies are necessity to mitigate our fast growing energy demand. It is expected that more than 80 % future wind and solar PV power generation systems will be connected with the grids by 2030. Therefore, it is essential for scientists and researchers to find the most effective converter technologies for the grid integration of wind and solar PV power generation systems. Conventional grid integration technique utilizes the power-frequency step-up-transformer, filter, and booster. These heavy and large size power-frequency step-up-transformers significantly increase the weight and volume of the renewable power generation systems. Because of the heavy weight and large size of the power-frequency transformer, the wind turbine generator and PV inverter system can be expensive and complex

M.R. Islam et al., *Power Converters for Medium Voltage Networks*,
Green Energy and Technology, DOI 10.1007/978-3-662-44529-7_8

in terms of installation and maintenance especially in offshore and remote area applications.

8.2 Conclusions

In order to select the most suitable power converter topology, an extensive literature review was carried out on the existing renewable power generation technologies covering generation, conversion, transmission, and distribution issues. The recent commercial medium and large-scale renewable power generation systems were mainly considered. As these require large areas of land, they are usually installed in offshore or remote areas, far from cities and industrial area where is electricity is consumed. The available power converter topologies developed in the recent decades were investigated for their offshore and remote area applications. The recently introduced smart micro-grid concept for efficient power transmission and distribution was also considered in the literature review.

Although several types of multilevel converter topologies have been developed in the last few decades, most of them are not suitable for medium-voltage high-power offshore and remote applications. In order to find suitable multilevel converter topology, an extensive analysis was carried out on different multilevel converter topologies taking into account the specified converter performance, control complexity, cost, and market availability of the power semiconductors.

Because of some special features (e.g., number of components scale linearly with the number of levels, and individual modules are identical and completely modular in construction hence enabling a high-level number attainability), the MMC converter topology can be viewed as the best possible candidate for medium voltage applications. However, the MMC converter requires multiple isolated and balanced DC sources. A high-frequency magnetic link operated at a few kHz to MHz was therefore developed to generate multiple isolated and balanced DC sources for all the H-bridge inverter cells of the MMC converter from a single source and a comprehensive electromagnetic analysis was conducted to verify the feasibility of this new technology. Compared with the power-frequency transformers, the high-frequency magnetic link has much smaller and lighter magnetic cores and windings, thus much lower costs.

The multilevel converter requires a number of switching and control PWM signals, which cannot be generated by the available DSP because the available DSP can at present only provide about six pairs of PWM channels. Unlike the DSP which runs a sequential program in its microprocessor, an FPGA may run all the operations in parallel with the clock signal. In this instance, the FPGA is the natural choice for medium voltage multilevel converters. Various design techniques and software environments are available for the modeling of switching control schemes with FPGA technology. Most of the techniques require special software such as HDL coder, System Generator, PSIM, and ModelSim, which increase the development time and cost. The most common software, such as the

MATLAB/Simulink and Xilinx ISE-based design technique, was used to reduce the developmental time and cost of the switching controller.

A high-frequency magnetic-link MMC medium-voltage converter could be the possible option for step-up-transformer-less direct grid integration of renewable sources. To verify the feasibility of the proposed system, a scaled down 1.73 kVA laboratory prototype system with modular 5-level cascaded converter was developed and reported, which converts 210 V DC (rectified generator voltage) into 3-phase 1 kV rms 50 Hz AC. A Metglas amorphous alloy 2605SA1-based 10 kHz magnetic link was used to generate the isolated and balanced 6 DC supplies for the 3-phase 5-level MMC converter.

The high number of levels of the converter means that medium-voltage attainability is possible so as to connect the renewable generation units to the medium-voltage grid directly. The proposed MMC converter can also provide better output power quality. Since the component number and control complexity increase linearly with the number of levels, the optimal selection of the number of converter levels is important for the best performance/cost ratio of the medium-voltage converter systems. An 11 kV and a 33 kV system were designed and analyzed for optimal selection of the converter number of levels taking into account the specified system performance, control complexity, cost, and market availability of the power semiconductors.

8.3 Future Works

Although the reduced scaled prototype design, development, and experimentation were described in this book in details, further work is required for the commercial success of the high-frequency magnetic-link MMC medium-voltage converter. The authors will be continuing their research for successful implementation of a real-scale 11–33 kV converter for step-up-transformer-less direct grid connection of wind and solar PV power generation systems. The future research findings will be published and incorporated into a new edition of this book, with the following issues taken into account:

- In order to improve the efficiency of the high-frequency magnetic-link section, advanced soft switching techniques can be implemented with controlled rectifiers rather than diode-based rectifiers;
- It would be an interesting topic to have a detailed study and analysis of the electromagnetic design of the multi-input–multi-output high-frequency magnetic link to improve the efficiency of the system;
- While only a few traditional modulation schemes have been considered in this research book, it would also be interesting to study the effects of different advanced modulation schemes for the MMC converter;
- Since, the grid synchronization has not been considered in this book, a suitable grid synchronization system could be described in future;

- Although the high-frequency magnetic link can minimize the voltage imbalance and common mode issues, it would also be an interesting article to develop a special modulation scheme with voltage imbalance and common mode minimization capabilities; and
- In order to verify the feasibility and ensure the safety, reliability, and stability of the real-scale high-frequency magnetic-link MMC medium-voltage converter, a full voltage (e.g., 11–33 kV) test platform should be developed.

MIX
Papier aus verantwortungsvollen Quellen
Paper from responsible sources
FSC® C105338

If you have any concerns about our products,
you can contact us on
ProductSafety@springernature.com

In case Publisher is established outside the EU,
the EU authorized representative is:
Springer Nature Customer Service Center GmbH
Europaplatz 3, 69115 Heidelberg, Germany

Printed by Libri Plureos GmbH
in Hamburg, Germany